D1478599

The Biodiesel Handbook

Editors

Gerhard Knothe
National Center for Agricultural Utilization Research
Agricultural Research Service
U.S. Department of Agriculture
Peoria, Illinois, U.S.A.

Jon Van Gerpen
Department of Biological and Agricultural Engineering
University of Idaho
Moscow, Idaho, U.S.A.

Jürgen Krahl
University of Applied Sciences
Coburg, Germany

Champaign, Illinois

AOCS Mission Statement

To be the global forum for professionals interested in lipids and related materials through the exchange of ideas, information science, and technology.

The paper used in this book is acid-free and falls within the guidelines established to ensure permanence and durability.

Library of Congress Cataloging-in-Publication Data

The biodiesel handbook / editors, Gerhard Knothe, Jon Van Gerpen, Jurgen Krahl.
p. cm.
Includes bibliographical references and index.
ISBN 1-893997-79-0 (acid-free paper)
1. Biodiesel fuels. I. Knothe, Gerhard. II. Van Gerpen, Jon Harlan. III. Krahl, Jurgen, 1962-

TP359.B46B56 2005
662'.669--dc22

2004027236

Printed in the United States of America.
08 07 06 05 04 5 4 3 2 1

Preface

The technical concept of using vegetable oils or animal fats or even used oils as a renewable diesel fuel is a fascinating one. Biodiesel is now the form in which these oils and fats are being used as neat diesel fuel or in blends with petroleum-based diesel fuels.

The concept itself may appear simple, but that appearance is deceiving since the use of biodiesel is fraught with numerous technical issues. Accordingly, many researchers around the world have dealt with these issues and in many cases devised unique solutions. This book is an attempt to summarize these issues, to explain how they have been dealt with, and to present data and technical information. Countless legislative and regulatory efforts around the world have helped pave the way toward the widespread application of the concept. This book addresses these issues also. To complete the picture, chapters on the history of vegetable oil-based diesel fuels, the basic concept of the diesel engine, and glycerol, a valuable byproduct of biodiesel production, are included.

We hope that the reader may find the information in this book useful and stimulating and that most of the significant issues regarding biodiesel are adequately addressed. If a reader notices an error or inconsistency or has a suggestion to improve a possible future edition of this book, he or she is encouraged to contact us.

This book has been compiled from the contributions of many authors, who graciously agreed to do so. We express our deepest appreciation to all of them. We also sincerely thank the staff of AOCS Press for their professionalism and cooperation in bringing the book to print.

Gerhard Knothe
Jon Van Gerpen
Jürgen Krahl

November 4, 2004

Contributing Authors

Gerhard Knothe, USDA, ARS, NCAUR, Peoria, IL 61604

Jon Van Gerpen, Department of Biological and Agricultural Engineering, University of Idaho, Moscow, Idaho 83844

Michael J. Haas, USDA, ARS, ERRC, Wyndmoor, PA 19038

Thomas A. Foglia, USDA, ARS, ERRC, Wyndmoor, PA 19038

Robert O. Dunn, USDA, ARS, NCAUR, Peoria, IL 61604

Heinrich Prankl, BLT–Federal Institute of Agricultural Engineering, A 3250 Wieselburg, Austria

Leon Schumacher, Department of Biological Engineering, University of Missouri-Columbia, Columbia, MO 65211

C.L. Peterson, Department of Biological and Agricultural Engineering (Emeritus), University of Idaho, Moscow, ID 83844

Gregory Möller, Department of Food Science and Technology, University of Idaho, Moscow, ID 83844

Neil A. Bringe, Monsanto Corporation, St. Louis, MO 63167

Robert L. McCormick, National Renewable Energy Laboratory, Golden, CO 80401

Teresa L. Alleman, National Renewable Energy Laboratory, Golden, CO 80401

Jürgen Krahl, University of Applied Sciences, Coburg, Germany

Axel Munack, Institute of Technology and Biosystems Engineering, Federal Agricultural Research Center, Braunschweig, Germany

Olaf Schröder, Institute of Technology and Biosystems Engineering, Federal Agricultural Research Center, Braunschweig, Germany

Hendrik Stein, Institute of Technology and Biosystems Engineering, Federal Agricultural Research Center, Braunschweig, Germany

Jürgen Bünger, Center of Occupational and Social Medicine, University of Göttingen, Göttingen, Germany

Steve Howell, MARC-IV Consulting Incorporated, Kearney, MO 64060

Joe Jobe, National Biodiesel Board, Jefferson City, MO 65101

Dieter Bockey, Union for Promoting Oilseed and Protein Plants, 10117 Berlin, Germany

Jürgen Fischer, ADM/Ölmühle Hamburg, Hamburg, Germany

Werner Körbitz, Austrian Biofuels Institute, Vienna, Austria

Sven O. Gärtner, IFEU-Institute for Energy and Environmental Research, Heidelberg, Germany

Guido A. Reinhardt, IFEU-Institute for Energy and Environmental Research, Heidelberg, Germany

Donald B. Appleby, Procter & Gamble Chemicals, Cincinnati, OH 45241

Contents

1

Introduction

Gerhard Knothe

Introduction: What Is Biodiesel?

The major components of vegetable oils and animal fats are triacylglycerols (TAG; often also called triglycerides). Chemically, TAG are esters of fatty acids (FA) with glycerol (1,2,3-propanetriol; glycerol is often also called glycerine; see Chapter 11). The TAG of vegetable oils and animal fats typically contain several different FA. Thus, different FA can be attached to one glycerol backbone. The different FA that are contained in the TAG comprise the FA profile (or FA composition) of the vegetable oil or animal fat. Because different FA have different physical and chemical properties, the FA profile is probably the most important parameter influencing the corresponding properties of a vegetable oil or animal fat.

To obtain biodiesel, the vegetable oil or animal fat is subjected to a chemical reaction termed *transesterification*. In that reaction, the vegetable oil or animal fat is reacted in the presence of a catalyst (usually a base) with an alcohol (usually methanol) to give the corresponding alkyl esters (or for methanol, the methyl esters) of the FA mixture that is found in the parent vegetable oil or animal fat. Figure 1 depicts the transesterification reaction.

Biodiesel can be produced from a great variety of feedstocks. These feedstocks include most common vegetable oils (e.g., soybean, cottonseed, palm, peanut, rapeseed/canola, sunflower, safflower, coconut) and animal fats (usually tallow) as well as waste oils (e.g., used frying oils). The choice of feedstock depends largely on geography. Depending on the origin and quality of the feedstock, changes to the production process may be necessary.

Biodiesel is miscible with petrodiesel in all ratios. In many countries, this has led to the use of blends of biodiesel with petrodiesel instead of neat biodiesel. It is important to note that these blends with petrodiesel are *not* biodiesel. Often blends with petrodiesel are denoted by acronyms such as B20, which indicates a blend of 20% biodiesel with petrodiesel. Of course, the untransesterified vegetable oils and animal fats should also not be called “biodiesel.”

Methanol is used as the alcohol for producing biodiesel because it is the least expensive alcohol, although other alcohols such as ethanol or *iso*-propanol may yield a biodiesel fuel with better fuel properties. Often the resulting products are also called fatty acid methyl esters (FAME) instead of biodiesel. Although other alcohols can by definition yield biodiesel, many now existing standards are designed in such a fashion that only methyl esters can be used as biodiesel if the standards are observed correctly.

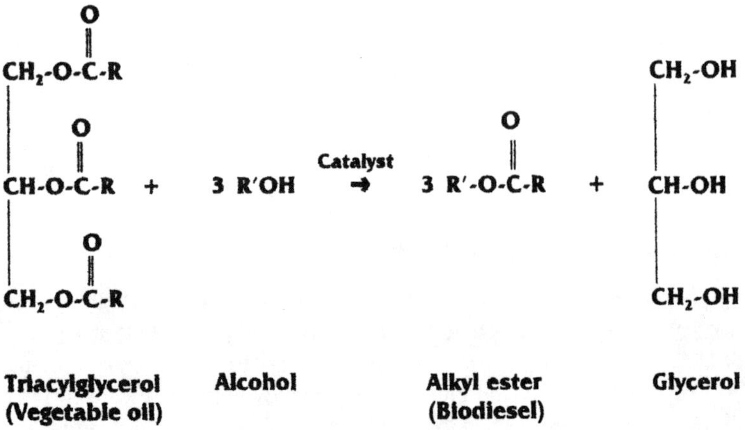

$$
\begin{array}{l} CH_2\text{-O-}\overset{O}{\overset{\|}{C}}\text{-R} \\ | \\ CH\text{-O-}\overset{O}{\overset{\|}{C}}\text{-R} \\ | \\ CH_2\text{-O-}\overset{O}{\overset{\|}{C}}\text{-R} \end{array} + 3\ R'OH \xrightarrow{\text{Catalyst}} 3\ R'\text{-O-}\overset{O}{\overset{\|}{C}}\text{-R} + \begin{array}{l} CH_2\text{-OH} \\ | \\ CH\text{-OH} \\ | \\ CH_2\text{-OH} \end{array}
$$

Triacylglycerol (Vegetable oil) **Alcohol** **Alkyl ester (Biodiesel)** **Glycerol**

Fig. 1. The transesterification reaction. R is a mixture of various fatty acid chains. The alcohol used for producing biodiesel is usually methanol ($R' = CH_3$).

Biodiesel has several distinct advantages compared with petrodiesel in addition to being fully competitive with petrodiesel in most technical aspects:

- Derivation from a renewable domestic resource, thus reducing dependence on and preserving petroleum.
- Biodegradability.
- Reduction of most exhaust emissions (with the exception of nitrogen oxides, NO_x).
- Higher flash point, leading to safer handling and storage.
- Excellent lubricity, a fact that is steadily gaining importance with the advent of low-sulfur petrodiesel fuels, which have greatly reduced lubricity. Adding biodiesel at low levels (1–2%) restores the lubricity.

Some problems associated with biodiesel are its inherent higher price, which in many countries is offset by legislative and regulatory incentives or subsidies in the form of reduced excise taxes, slightly increased NO_x exhaust emissions (as mentioned above), stability when exposed to air (oxidative stability), and cold flow properties that are especially relevant in North America. The higher price can also be (partially) offset by the use of less expensive feedstocks, which has sparked interest in materials such as waste oils (e.g., used frying oils).

Why Are Vegetable Oils and Animal Fats Transesterified to Alkyl Esters (Biodiesel)?

The major reason that vegetable oils and animal fats are transesterified to alkyl esters (biodiesel) is that the kinematic viscosity of the biodiesel is much closer to that of petrodiesel. The high viscosity of untransesterified oils and fats leads to operational problems in the diesel engine such as deposits on various engine parts. Although there are engines and burners that can use untransesterified oils, the vast majority of engines require the lower-viscosity fuel.

Why Can Vegetable Oils and Animal Fats and Their Derivatives Be Used as (Alternative) Diesel Fuel?

The fact that vegetable oils, animal fats, and their derivatives such as alkyl esters are suitable as diesel fuel demonstrates that there must be some similarity to petrodiesel fuel or at least to some of its components. The fuel property that best shows this suitability is called the *cetane number* (see Chapter 6.1).

In addition to ignition quality as expressed by the cetane scale, several other properties are important for determining the suitability of biodiesel as a fuel. Heat of combustion, pour point, cloud point, (kinematic) viscosity, oxidative stability, and lubricity are among the most important of these properties.

2

The History of Vegetable Oil-Based Diesel Fuels

Gerhard Knothe

Rudolf Diesel

It is generally known that vegetable oils and animal fats were investigated as diesel fuels well before the energy crises of the 1970s and early 1980s sparked renewed interest in alternative fuels. It is also known that Rudolf Diesel (1858–1913), the inventor of the engine that bears his name, had some interest in these fuels. However, the early history of vegetable oil-based diesel fuels is often presented inconsistently, and "facts" that are not compatible with Diesel's own statements are encountered frequently.

Therefore, it is appropriate to begin this history with the words of Diesel himself in his book *Die Entstehung des Dieselmotors* (1) [*The Development* (or *Creation* or *Rise* or *Coming*) *of the Diesel Engine*] in which he describes when the first seed of developing what was to become the diesel engine was planted in his mind. In the first chapter of the book entitled "The Idea," Diesel states: "When my highly respected teacher, Professor Linde, explained to his listeners during the lecture on thermodynamics in 1878 at the *Polytechnikum* in Munich (note: now the Technical University of Munich) that the steam engine only converts 6–10% of the available heat content of the fuel into work, when he explained Carnot's theorem and elaborated that during the isothermal change of state of a gas all transferred heat is converted into work, I wrote in the margin of my notebook: 'Study, if it isn't possible to practically realize the isotherm!' At that time I challenged myself! That was not yet an invention, not even the idea for it. From then on, the desire to realize the ideal Carnot process determined my existence. I left the school, joined the practical side, had to achieve my standing in life. The thought constantly pursued me."

This statement by Diesel clearly shows that he approached the development of the diesel engine from a thermodynamic point of view. The objective was to develop an efficient engine. The relatively common assertion made today that Diesel developed "his" engine specifically to use vegetable oils as fuel is therefore incorrect.

In a later chapter of his book entitled "Liquid Fuels," Diesel addresses the use of vegetable oils as a fuel: "For [the] sake of completeness it needs to be mentioned that already in the year 1900 plant oils were used successfully in a diesel engine. During the Paris Exposition in 1900, a small diesel engine was operated on arachide (peanut) oil by the French Otto Company. It worked so well that only a few insiders knew about this inconspicuous circumstance. The engine was built for petroleum and was used for the plant oil without any change. In this case also, the consumption experiments resulted in heat utilization identical to petroleum." A total of five diesel engines were shown at the Paris Exposition, according to a biography (2) of Diesel by his son, Eugen Diesel, and one of them was apparently operating on peanut oil.

The statements in Diesel's book can be compared to a relatively frequently cited source on the initial use of vegetable oils, a biography entitled *Rudolf Diesel, Pioneer of the Age of Power* (3). In this biography, the statement is made that "as the nineteenth century ended, it was obvious that the fate and scope of the internal-combustion engine were dependent on its fuel or fuels. At the Paris Exposition of 1900, a Diesel engine, built by the French Otto Company, ran wholly on peanut oil. Apparently none of the onlookers was aware of this. The engine, built especially for that type of fuel, operated exactly like those powered by other oils."

Unfortunately, the bibliography for the corresponding chapter in the biography by Nitske and Wilson (3) does not clarify where the authors obtained this information nor does it list references to the writings by Diesel discussed here. Thus, according to Nitske and Wilson, the peanut oil-powered diesel engine at the 1900 World's Fair in Paris was built specifically to use that fuel, which is not consistent with the statements in Diesel's book (1) and the literature cited below. Furthermore, the above texts from the biography (3) and Diesel's book (1) imply that it was not Diesel who conducted the demonstration and that he was not the source of the idea of using vegetable oils as fuel. According to Diesel, the idea for using peanut oil appears to have originated instead within the French government (see text below). However, Diesel conducted related tests in later years and appeared supportive of the concept.

A *Chemical Abstracts* search yielded references to other papers by Diesel in which he reflected in greater detail on that event in 1900. Two references (4,5) relate to a presentation Diesel made to the Institution of Mechanical Engineers (of Great Britain) in March 1912. (Apparently in the last few years of his life, Diesel spent considerable time traveling to give presentations, according to the biography by Nitske and Wilson.) Diesel states in these papers (4,5) that "at the Paris Exhibition in 1900 there was shown by the Otto Company a small Diesel engine, which, at the request of the French Government, ran on Arachide (earth-nut or pea-nut) oil, and worked so smoothly that only very few people were aware of it. The engine was constructed for using mineral oil, and was then worked on vegetable oil without any alterations being made. The French Government at the time thought of testing the applicability to power production of the Arachide, or earth-nut, which grows in considerable quantities in their African colonies, and which can be easily cultivated there, because in this way the colonies could be supplied with power and industry from their own resources, without being compelled to buy and import coal or liquid fuel. This question has not been further developed in France owing to changes in the Ministry, but the author resumed the trials a few months ago. It has been proved that Diesel engines can be worked on earth-nut oil without any difficulty, and the author is in a position to publish, on this occasion for the first time, reliable figures obtained by tests: Consumption of earth-nut oil, 240 grammes (0.53 lb) per brake horsepower-hour; calorific power of the oil, 8600 calories (34,124 British thermal units) per kg, thus fully equal to tar oils; hydrogen 11.8 percent. This oil is almost as effective as the natural mineral oils, and as it can also be used for lubricating oil, the whole work can be carried out with a single kind of oil produced directly on the spot. Thus this engine becomes a really independent engine for the tropics."

Diesel continued that (note the prescient concluding statement), "similar successful experiments have also been made in St. Petersburg with castor oil; and animal oils, such as train-oil, have been used with excellent results. The fact that fat oils from vegetable sources can be used may seem insignificant today, but such oils may perhaps become in course of time of the same importance as some natural mineral oils and the tar products are now. Twelve years ago, the latter were not more developed than the fat oils are today, and yet how important they have since become. One cannot predict what part these oils will play in the Colonies in the future. In any case, they make it certain that motor-power can still be produced from the heat of the sun, which is always available for agricultural purposes, even when all our natural stores of solid and liquid fuels are exhausted."

The following discussion is based on numerous references available mainly from searching *Chemical Abstracts* or from a publication summarizing literature before 1949 on fuels from agricultural sources (6). Because many of the older references are not readily available, the summaries in *Chemical Abstracts* were used as information source in these cases.

Background and Fuel Sources

The aforementioned background in the papers by Diesel (4,5) on using vegetable oils to provide European tropical colonies, especially those in Africa, with a certain degree of energy self-sufficiency can be found in the related literature until the 1940s. Palm oil was often considered as a source of diesel fuel in the "historic" studies, although the diversity of oils and fats as sources of diesel fuel, an important aspect again today, and striving for energy independence were reflected in other "historic" investigations. Most major European countries with African colonies, i.e., Belgium, France, Italy, and the UK, with Portugal apparently making an exception, had varying interest in vegetable oil fuels at the time, although several German papers, primarily from academic sources (Technische Hochschule Breslau), were also published. Reports from other countries also reflect a theme of energy independence.

Vegetable oils were also used as emergency fuels and for other purposes during World War II. For example, Brazil prohibited the export of cottonseed oil so that it could be substituted for imported diesel fuel (7). Reduced imports of liquid fuel were also reported in Argentina, necessitating the commercial exploitation of vegetable oils (8). China produced diesel fuel, lubricating oils, "gasoline," and "kerosene," the latter two by a cracking process, from tung and other vegetable oils (9,10). However, the exigencies of the war caused hasty installation of cracking plants based on fragmentary data (9). Researchers in India, prompted by the events of World War II, extended their investigations on 10 vegetable oils for development as domestic fuels

(11). Work on vegetable oils as diesel fuel ceased in India when petroleum-based diesel fuel again became easily available at low cost (12). The Japanese battleship *Yamato* reportedly used edible refined soybean oil as bunker fuel (13).

Concerns about the rising use of petroleum fuels and the possibility of resultant fuel shortages in the United States in the years after World War II played a role in inspiring a "dual fuel" project at The Ohio State University (Columbus, OH), during which cottonseed oil (14), corn oil (15), and blends thereof with conventional diesel fuel were investigated. In a program at the Georgia School of Technology (now Georgia Institute of Technology, Atlanta, GA), neat vegetable oils were investigated as diesel fuel (16). Once again, energy security perspectives have become a significant driving force for the use of vegetable oil-based diesel fuels, although environmental aspects (mainly reduction of exhaust emissions) play a role at least as important as that of energy security. For example, in the United States, the Clean Air Act Amendments of 1990 and the Energy Policy Act of 1992 mandate the use of alternative or "clean" fuels in regulated truck and bus fleets. Amendments to the Energy Policy Act enacted into law in 1998, which provide credits for biodiesel use (also in blends with conventional diesel fuel), are a major reason for the significant increase in the use of biodiesel in the United States.

In modern times, biodiesel is derived, or has been reported to be producible from many different sources, including vegetable oils, animal fats, used frying oils, and even soapstock. Generally, factors such as geography, climate, and economics determine which vegetable oil is of greatest interest for potential use in biodiesel fuels. Thus, in the United States, soybean oil is considered to be a prime feedstock; in Europe, it is rapeseed (canola) oil, and in tropical countries, it is palm oil. As noted above, different feedstocks were investigated in the "historic" times. These included palm oil, soybean oil, cottonseed oil, castor oil, and a few less common oils, such as babassu (17) and crude raisinseed oil (18); nonvegetable sources such as industrial tallow (19) and even fish oils (20–25) were also investigated. In numerous reports, especially from France and Belgium, dating from the early 1920s, palm oil was probably the feedstock that received the most attention, although cottonseed and some other oils were tested (26–38). The availability of palm oil in tropical locations again formed the background as mentioned above. Eleven vegetable oils from India (peanut, karanj, punnal, polang, castor, kapok, mahua, cottonseed, rapeseed, coconut, and sesame) were investigated as fuels (11). A Brazilian study reported on 14 vegetable oils that were investigated (17). Walton (39) summarized results on 20 vegetable oils (castor, grapeseed, maize, camelina, pumpkinseed, beechnut, rapeseed, lupin, pea, poppyseed, peanut, hemp, linseed, chestnut, sunflower seed, palm, olive, soybean, cottonseed, and shea butter). He also pointed out (39) that "at the moment the source of supply of fuels is in a few hands, the operator has little or no control over prices or qualities, and it seems unfortunate that at this date, as with the petrol engine, the engine has to be designed to suit the fuel whereas, strictly speaking, the reverse should obtain—the fuel should be refined to meet the design of an ideal engine."

Although environmental aspects played virtually no role in promoting the use of vegetable oils as fuel in "historic" times and no emissions studies were conducted, it is still worthwhile to note some allusions to this subject from that time. (i) "In case further development of vegetable oils as fuel proves practicable, it will simplify the fuel problems of many tropical localities remote from mineral fuel, and where the use of wood entails much extra labor and other difficulties connected with the various heating capacities of the wood's use, to say nothing of the risk of indiscriminate deforestation" (27). (ii) "It might be advisable to mention, at this juncture, that, owing to the altered combustion characteristics, the exhaust with all these oils is invariably quite clean and the characteristic diesel knock is virtually eliminated" (39). (iii) Observations by other authors included: "invisible" or "slightly smoky" exhausts when running an engine on palm oil (29); clearer exhaust gases (34); in the case of use of fish oils as diesel fuels, the exhaust was described as colorless and practically odorless (23). The visual observations of yesterday have been confirmed in "modern" times for biodiesel fuel. Numerous recent studies showed that biodiesel fuel reduces most exhaust emissions.

Technical Aspects

Many "historic" publications discussed the satisfactory performance of vegetable oils as fuels or fuel sources, although it was often noted that their higher costs relative to petroleum-derived fuel would prevent widespread use.

The kinematic viscosity of vegetable oils is about an order of magnitude greater than that of conventional, petroleum-derived diesel fuel. High viscosity causes poor atomization of the fuel in the engine's combustion chambers and ultimately results in operational problems, such as engine deposits. Since the renewal of interest in vegetable oil-derived fuels during the late 1970s, four possible solutions to the problem of high viscosity were investigated: transesterification, pyrolysis, dilution with conventional petroleum-derived diesel fuel, and microemulsification (40). Transesterification is the most common method and leads to monoalkyl esters of vegetable oils and fats, now called biodiesel when used for fuel purposes. As mentioned in Chapter 1 of this book, methanol is usually used for transesterification because in many countries, it is the least expensive alcohol.

The high viscosity of vegetable oils as a major cause of poor fuel atomization resulting in operational problems such as engine deposits was recognized early (29,41–45). Although engine modifications such as higher injection pressure were considered (41,46), reduction of the high viscosity of vegetable oils usually was achieved by heating the vegetable oil fuel (29,41–44,47). Often the engine was started on petrodiesel; after a few minutes of operation, it was then switched to the vegetable oil fuel, although a successful cold-start using high-acidity peanut oil was reported (48). Advanced injection timing was a technique also employed (49). Seddon (47) gives an interesting practical account of a truck that operated successfully on different vegetable oils using preheated fuel. The preheating technique was also applied in a study on the feasibility of using vegetable oils in the transportation facilities needed for developing the tin mines of Nigeria (47,50).

It was also recognized that the performance of the vegetable oil-based fuels generally was satisfactory; however, power output was slightly lower than with petroleum-based diesel fuel and fuel consumption was slightly higher (16,23, 25,28,30,32,35,39,41,43,44,50–52), although engine load-dependent or opposite effects were reported (8,14,15,53). Ignition lag was reportedly reduced with engines using soybean oil (52). In many of these publications, it was noted that the diesel engines used operated more smoothly on vegetable oils than on petroleum-based diesel fuel. Due to their combustion characteristics, vegetable oils with a high oxygen content were suggested, thus making it practical to use gas turbines as prime movers (54).

Fuel quality issues were also addressed. It was suggested that when "the acid content of the vegetable oil fuels is maintained at a minimum no adverse results are experienced either on the injection equipment or on the engine" (50; see also 47). Relatedly, other authors discussed that the effect of free fatty acids, moisture, and other contaminants on fuel properties is an important issue (11). The effects of different kinds of vegetable oils on the corrosion of neat metals and lubrication oil dilution and contamination, for example, were studied (44).

Pyrolysis, cracking, or other methods of decomposition of vegetable oils to yield fuels of varying nature is an approach that accounts for a significant amount of the literature in "historic" times. Artificial "gasoline," "kerosene," and "diesel" were obtained in China from tung oil (9) and other oils (10). Other oils used in such an approach included fish oils (20–22), as well as linseed oil (55), castor oil (56), palm oil (57), cottonseed oil (58), and olive oil (59). Numerous reports from several countries including China, France, and Japan were concerned with obtaining fuels by the cracking of vegetable oils or related processes (60–93). The other approaches, i.e., dilution with petrodiesel and, especially, microemulsification, appear to have received little or no attention during the "historic" times. However, some experiments on blending of conventional diesel fuel with cottonseed oil (14,94), corn oil (15), and turnip, sunflower, linseed, peanut, and cottonseed oil (8) were described. Blends of aqueous ethanol with "vegetable gasoline" were reported (95). Ethanol was also used to improve the atomization and combustion of highly viscous castor oil (96).

In addition to powering vehicles, the use of vegetable oils for other related purposes received some attention. The possibility of deriving fuels as well as lubricating oils and greases from vegetable oils in the French African colonies was discussed (97). The application of vegetable oils as fuels for heating and power purposes was examined (98). At least one critique of the use of vegetable oils, particularly olive oil, for fuel and lubricant use was published (99). Along with the technical literature in journals and reports, several patents from the "historic" times dealt with vegetable oils or their derivatives as fuels, obtained mainly through cracking or pyrolysis (100–106).

The First "Biodiesel"

Walton (39) recommended that "to get the utmost value from vegetable oils as fuel it is academically necessary to split off the triglycerides and to run on the residual fatty acid. Practical experiments have not yet been carried out with this; the problems are likely to be much more difficult when using free fatty acids than when using the oils straight from the crushing mill. It is obvious that the glycerides have no fuel value and in addition are likely, if anything, to cause an excess of carbon in comparison with gas oil."

Walton's statement points in the direction of what is now termed "biodiesel" by recommending the elimination of glycerol from the fuel, although esters are not mentioned. In this connection, some remarkable work performed in Belgium and its former colony, the Belgian Congo (known after its independence for a long time as Zaire), deserves more recognition than it has received. It appears that Belgian patent 422,877, granted on Aug. 31, 1937 to G. Chavanne (University of Brussels, Belgium) (107), constitutes the first report on what is today known as biodiesel. It describes the use of ethyl esters of palm oil (although other oils and methyl esters are mentioned) as diesel fuel. These esters were obtained by acid-catalyzed transesterification of the oil (base catalysis is now more common). This work was described later in more detail (108).

Of particular interest is a related extensive report published in 1942 on the production and use of palm oil ethyl ester as fuel (109). That work described what was probably the first test of an urban bus operating on biodiesel. A bus fueled with palm oil ethyl ester served the commercial passenger line between Brussels and Louvain (Leuven) in the summer of 1938. The performance of the bus operating on that fuel reportedly was satisfactory. It was noted that the viscosity difference between the esters and conventional diesel fuel was considerably less than that between the parent oil and conventional diesel fuel. Also, the article pointed out that the esters are miscible with other fuels. That work also discussed what is probably the first cetane number (CN) testing of a biodiesel fuel. In the report, the CN of palm oil ethyl ester was reported as ~83 (relative to a high-quality standard with CN 70.5, a low-quality standard of CN 18, and diesel fuels with CN of 50 and 57.5). Thus, those results agree with "modern" work reporting relatively high CN for such biodiesel fuels. A later paper by another author reported the autoignition temperature of various alkyl esters of palm oil fatty acids (110). In more recent times, the use of methyl esters of sunflower oil to reduce the viscosity of vegetable oil was reported at several technical conferences in 1980 and 1981 (39–41) and marks the beginning of the rediscovery and eventual commercialization of biodiesel.

A final thought should be given to the term "biodiesel" itself. A *Chemical Abstracts* search (using the "SciFinder" search engine with "biodiesel" as the key word) yielded the first use of the term "biodiesel" in the technical literature in a Chinese paper published in 1988 (111). The next paper using this term appeared in 1991 (112); from then on, the use of the word "biodiesel" in the literature expanded exponentially.

References

1. Diesel, R., *Die Entstehung des Dieselmotors*, Verlag von Julius Springer, Berlin, 1913.
2. Diesel, E., *Diesel: Der Mensch, Das Werk, Das Schicksal*, Hanseatische Verlagsgesellschaft, Hamburg, 1937.
3. Nitske, W.R., and C.M. Wilson, *Rudolf Diesel, Pioneer of the Age of Power*, University of Oklahoma Press, Norman, Oklahoma, 1965.
4. Diesel, R., The Diesel Oil-Engine, *Engineering 93:* 395–406 (1912); *Chem. Abstr. 6:* 1984 (1912).
5. Diesel, R., The Diesel Oil-Engine and Its Industrial Importance Particularly for Great Britain, *Proc. Inst. Mech. Eng.* 179–280 (1912); *Chem. Abstr. 7:* 1605 (1913).
6. Wiebe, R., and J. Nowakowska, The Technical Literature of Agricultural Motor Fuels, *USDA Bibliographic Bulletin No. 10*, Washington, 1949, pp. 183–195.
7. Anonymous, Brazil Uses Vegetable Oil for Diesel Fuel, *Chem. Metall. Eng. 50:* 225 (1943).
8. Martinez de Vedia, R., Vegetable Oils as Diesel Fuels, *Diesel Power Diesel Transp. 22:* 1298–1301, 1304 (1944).
9. Chang, C.-C., and S.-W. Wan, China's Motor Fuels from Tung Oil, *Ind. Eng. Chem. 39:* 1543–1548 (1947); *Chem. Abstr. 42:* 1037 h (1948).
10. Cheng, F.-W., China Produces Fuels from Vegetable Oils, *Chem. Metall. Eng. 52:* 99 (1945).
11. Chowhury, D.H., S.N. Mukerji, J.S. Aggarwal, and L.C. Verman, Indian Vegetable Fuel Oils for Diesel Engines, *Gas Oil Power 37:* 80–85 (1942); *Chem. Abstr. 36:* 5330^9 (1942).
12. Amrute, P.V., Ground-Nut Oil for Diesel Engines, *Australasian Eng.* 60–61 (1947). *Chem. Abstr. 41:* 6690 d (1947).
13. Reference 1250 (p. 195) in present Reference 6.
14. Huguenard, C.M., Dual Fuel for Diesel Engines Using Cottonseed Oil, M.S. Thesis, The Ohio State University, Columbus, OH, 1951.
15. Lem, R.F.-A., Dual Fuel for Diesel Engines Using Corn Oil with Variable Injection Timing, M.S. Thesis, The Ohio State University, Columbus, OH, 1952.
16. Baker, A.W., and R.L. Sweigert, A Comparison of Various Vegetable Oils as Fuels for Compression-Ignition Engines, *Proc. Oil & Gas Power Meeting of the ASME* 40–48 (1947).
17. Pacheco Borges, G., Use of Brazilian Vegetable Oils as Fuel, *Anais Assoc. Quím. Brasil 3:* 206–209 (1944); *Chem. Abstr. 39:* 5067^8 (1945).
18. Manzella, A., L'Olio di Vinaccioli quale Combustibile Succedaneo della NAFTA (Raisin Seed Oil as a Petroleum Substitute), *Energia Termical 4:* 92–94 (1936); *Chem. Abstr. 31:* 7274^9 (1937).
19. Lugaro, M.E., and F. de Medina, The Possibility of the Use of Animal Oils and Greases in Diesel Motors, *Inst. Sudamericano Petróleo, Seccion Uruguaya, Mem. Primera Conf. Nacl. Aprovisionamiento y Empleo Combustibles 2:* 159–175 (1944); *Chem. Abstr. 39:* 5431^8 (1945).
20. Kobayashi, K., Formation of Petroleum from Fish Oils, Origin of Japanese Petroleum, *J. Chem. Ind. (Japan) 24:* 1–26 (1921); *Chem. Abstr. 15:* 2542 (1921).
21. Kobayashi, K., and E. Yamaguchi, Artificial Petroleum from Fish Oils, *J. Chem. Ind. (Japan) 24:* 1399–1420 (1921); *Chem. Abstr. 16:* 2983 (1922).

22. Faragher, W.F., G. Egloff, and J.C. Morrell, The Cracking of Fish Oil, *Ind. Eng. Chem. 24:* 440–441 (1932); *Chem. Abstr. 26:* 2882 (1932).
23. Lumet, G., and H. Marcelet, Utilization of Marine Animal and Fish Oils (as Fuels) in Motors, *Compt. Rend. 185:* 418–420 (1927); *Chem. Abstr. 21:* 3727 (1927).
24. Marcelet, H., Heat of Combustion of Some Oils from Marine Animals, *Compt. Rend. 184:* 604-605 (1927); *Chem. Abstr. 21:* 1890 (1927).
25. Okamura, K., Substitute Fuels for High-Speed Diesel Engines, *J. Fuel Soc. Japan 19:* 691–705 (1940); *Chem. Abstr. 35:* 1964[7] (1941).
26. Mayné, R., Palm Oil Motors, *Ann. Gembloux 26:* 509–515 (1920); *Chem. Abstr. 16:* 3192.
27. Ford, G.H., Vegetable Oils as Engine Fuel, *Cotton Oil Press 5:* 38 (1921); *Chem. Abstr. 15:* 3383 (1921).
28. Lazennec, I., Palm Oil as Motor Fuel, *Ind. Chim. 8:* 262 (1921); *Chem. Abstr. 15:* 3383 (1921).
29. Mathot, R.E., Vegetable Oils for Internal Combustion Engines, *Engineer 132:* 138–139 (1921); *Chem. Abstr. 15:* 3735 (1921).
30. Anonymous, Palm Oil as a Motor Fuel, *Bull. Imp. Inst. 19:* 515 (1921); *Chem. Abstr. 16:* 2769 (1922).
31. Anonymous, Tests on the Utilization of Vegetable Oils as a Source of Mechanical Energy, *Bull. Mat. Grasses Inst. Colon. Marseille* 4–14 (1921); *Chem. Abstr. 16:* 3192 (1922).
32. Mathot, R.E., Utilization of Vegetable Oils as Motor Fuels, *Bull. Mat. Grasses Inst. Colon. Marseille* 116–128 (1921); *Chem. Abstr. 17:* 197 (1923).
33. Goffin, Tests of an Internal Combustion Motor Using Palm Oil as Fuel, *Bull. Mat. Grasses Inst. Colon. Marseille* 19–24 (1921); *Chem. Abstr. 16:* 3192 (1922).
34. Leplae, E., Substitution of Vegetable Oil for Paraffin as Fuel for Motors and Tractors in the Colonies, *La Nature 2436:* 374–378 (1920); *Chem. Abstr. 16:* 4048 (1922).
35. Anonymous, The Utilization of Palm Oil as a Motor Fuel in the Gold Coast, *Bull. Imp. Inst. 20:* 499–501 (1922); *Chem. Abstr. 17:* 1878 (1923).
36. Mathot, R.E., Mechanical Traction in the (French) Colonies, *Chim. Ind.* (Special Number)*:* 759–763 (1923); *Chem. Abstr. 17:* 3243 (1923).
37. Delahousse, P., Tests with Vegetable Oils in Diesel and Semi-Diesel Engines, *Chim. Ind.* (Special Number)*:* 764–766 (1923); *Chem. Abstr. 17:* 3243 (1923).
38. Lumet, G., Utilization of Vegetable Oils, *Chal. Ind.* (Special Number)*:* 190–195 (1924); *Chem. Abstr. 19:* 1189 (1925).
39. Walton, J., The Fuel Possibilities of Vegetable Oils, *Gas Oil Power 33:* 167–168 (1938); *Chem. Abstr. 33:* 833[6] (1939).
40. Schwab, A.W., M.O. Bagby, and B. Freedman, Preparation and Properties of Diesel Fuels from Vegetable Oils, *Fuel 66:* 1372–1378 (1987).
41. Schmidt, A.W., Pflanzenöle als Dieselkraftstoffe, *Tropenpflanzer 35:* 386–389 (1932); *Chem. Abstr. 27:* 1735 (1933).
42. Schmidt, A.W., Engine Studies with Diesel Fuel (Motorische Untersuchungen mit Dieselkraftstoffen), *Automobiltechn. Z. 36:* 212–214 (1933); *Chem. Abstr. 27:* 4055 (1933).
43. Schmidt, A.W., and K. Gaupp, Pflanzenöle als Dieselkraftstoffe, *Tropenpflanzer 37:* 51–59 (1934); *Chem. Abstr. 28:* 6974[7] (1934).
44. Gaupp, K., Pflanzenöle als Dieselkraftstoffe (*Chem. Abstr.* translation: Plant Oils as Diesel Fuel), *Automobiltech. Z. 40:* 203–207 (1937); *Chem. Abstr. 31:* 8876[5] (1937).

45. Boiscorjon d'Ollivier, A., French Production of Soybean Oil. (La Production métropolitaine des Oléagineux: 'Le Soja'), *Rev. Combust. Liq. 17:* 225–235 (1939); *Chem. Abstr. 34:* 3937^7 (1940).
46. Tatti, E., and A. Sirtori, Use of Peanut Oil in Injection, High-Compression, High-Speed Automobile Motors, *Energia Termica 5:* 59–64 (1937); *Chem. Abstr. 32:* 2318^8 (1938).
47. Seddon, R.H., Vegetable Oils in Commercial Vehicles, *Gas Oil Power 37:* 136–141, 146 (1942); *Chem. Abstr. 36:* 6775^7 (1942).
48. Gautier, M., Use of Vegetable Oils in Diesel Engines, *Rev. Combust. Liq. 11:* 19–24 (1933); *Chem. Abstr. 27:* 4372 (1933).
49. Gautier, M., Vegetable Oils and the Diesel Engines, *Rev. Combust. Liq. 11:* 129–136 (1935); *Chem. Abstr. 29:* 4611^9 (1935).
50. Smith, D.H., Fuel by the Handful, *Bus and Coach 14:* 158–159 (1942).
51. Gauthier, M., Utilization of Vegetable Oil as Fuel in Diesel Engines, *Tech. Moderne 23:* 251–256 (1931); *Chem. Abstr. 26:* 278 (1932).
52. Hamabe, G., and H. Nagao, Performance of Diesel Engines Using Soybean Oil as Fuel, *Trans. Soc. Mech. Eng. (Japan) 5:* 5–9 (1939); *Chem. Abstr. 35:* 4178^9 (1941).
53. Manzella, G., Peanut Oil as Diesel Engine Fuel, *Energia Term. 3:* 153–160 (1935); *Chem. Abstr. 30:* 2347^7 (1936).
54. Gonzaga, L., The Role of Combined Oxygen in the Efficiency of Vegetable Oils as Motor Fuel, *Univ. Philipp. Nat. Appl. Sci. Bull. 2:* 119–124 (1932); *Chem. Abstr. 27:* 833 (1933).
55. Mailhe, A., Preparation of a Petroleum from a Vegetable Oil, *Compt. Rend. 173:* 358–359 (1921); *Chem. Abstr. 15:* 3739 (1921).
56. Melis, B., Experiments on the Transformation of Vegetable Oils and Animal Fats to Light Fuels, *Atti Congr. Naz. Chim. Ind.* 238–240 (1924); *Chem. Abstr. 19:* 1340 (1925).
57. Morrell, J.C., G. Egloff, and W.F. Faragher, Cracking of Palm Oil, *J. Chem. Soc. Chem. Ind. 51:* 133–4T (1932); *Chem. Abstr. 26:* 3650 (1932).
58. Egloff, G., and J.C. Morrell, The Cracking of Cottonseed Oil, *Ind. Eng. Chem. 24:* 1426–1427 (1932); *Chem. Abstr. 27:* 618 (1933).
59. Gomez Aranda, V., A Spanish Contribution to the Artificial Production of Hydrocarbons, *Ion 2:* 197–205 (1942); *Chem. Abstr. 37:* 1241^3 (1943).
60. Kobayashi, K., Artificial Petroleum from Soybean, Coconut, and Chrysalis Oils and Stearin, *J. Chem. Ind. (Japan) 24:* 1421–1424 (1921); *Chem. Abstr. 16:* 2983 (1922).
61. Mailhe, A., Preparation of Motor Fuel from Vegetable Oils, *J. Usines Gaz. 46:* 289–292 (1922); *Chem. Abstr. 17:* 197 (1923).
62. Sato, M., Preparation of a Liquid Fuel Resembling Petroleum by the Distillation of the Calcium Salts of Soybean Oil Fatty Acids, *J. Chem. Ind. (Japan) 25:* 13–24 (1922); *Chem. Abstr. 16:* 2984 (1922).
63. Sato, M., Preparation of Liquid Fuel Resembling Petroleum by Distilling the Calcium Soap of Soybean Oil, *J. Soc. Chem. Ind. (Japan) 26:* 297–304 (1923); *Chem. Abstr. 18:* 1375 (1924).
64. Sato, M., and K.F. Tseng, The Preparation of Fuel Oil by the Distillation of the Lime Soap of Soybean Oil. III. Experiments Using Oxides and Carbonates of Alkali Metals as Saponifying Agent, *J. Soc. Chem. Ind. (Japan) 29:* 109–115 (1926); *Chem. Abstr. 20:* 2759 (1926).

65. Sato, M., Preparation of Fuel Oil by the Dry Distillation of Calcium Soap of Soybean Oil. IV. Comparison with Magnesium Soap, *J. Soc. Chem. Ind. (Japan) 30:* 242–245 (1927); *Chem. Abstr. 21:* 2371 (1927).
66. Sato, M., Preparation of Fuel Oil by the Dry Distillation of Calcium Soap of Soybean Oil. V. Hydrogenation of the Distilled Oil, *J. Soc. Chem. Ind. (Japan) 30:* 242–245 (1927); *Chem. Abstr. 21:* 2371 (1927).
67. Sato, M., The Preparation of Fuel Oil by the Dry Distillation of Calcium Soap of Soybean Oil. VI. The Reaction Mechanism of Thermal Decomposition of Calcium and Magnesium Salts of Some Higher Fatty Acids, *J. Soc. Chem. Ind. (Japan) 30:* 252–260 (1927); *Chem. Abstr. 21:* 2372 (1927).
68. Sato, M., and C. Ito, The Preparation of Fuel Oil by the Dry Distillation of Calcium Soap of Soybean Oil. VI. The Reaction Mechanism of Thermal Decomposition of Calcium and Magnesium Salts of Some Higher Fatty Acids, *J. Soc. Chem. Ind. (Japan) 30:* 261–267 (1927); *Chem. Abstr. 21:* 2372 (1927).
69. de Sermoise, C., The Use of Certain Fuels in Diesel Motors, *Rev. Combust. Liq. 12:* 100–104 (1934); *Chem. Abstr. 28:* 4861^{6} (1934).
70. Koo, E.C., and S.-M. Cheng, The Manufacture of Liquid Fuel from Vegetable Oils, *Chin. Ind. 1:* 2021–2039 (1935); *Chem. Abstr. 30:* 837^{8} (1936).
71. Koo, E.C., and S.-M. Cheng, First Report on the Manufacture of Gasoline from Rapeseed Oil, *Ind. Res. (China) 4:* 64–69 (1935); *Chem. Abstr. 30:* 2725^{4} (1936).
72. Koo, E.C., and S.-M. Cheng, Intermittent Cracking of Rapeseed Oil (article in Chinese), *J. Chem. Eng. (China) 3:* 348–353 (1936); *Chem. Abstr. 31:* 2846^{2} (1937).
73. Ping, K., Catalytic Conversion of Peanut Oil into Light Spirits, *J. Chin. Chem. Soc. 3:* 95–102 (1935). *Chem. Abstr. 29:* 4612^{5} (1935).
74. Ping, K., Further Studies on the Liquid-Phase Cracking of Vegetable Oils, *J. Chin. Chem. Soc. 3:* 281–287 (1935); *Chem. Abstr. 29:* 7683^{1} (1935).
75. Ping, K., Cracking of Peanut Oil, *J. Chem. Eng. (China) 3:* 201–210 (1936); *Chem. Abstr. 31:* 238^{5} (1937).
76. Ping, K., Light Oils from Catalytic Pyrolysis of Vegetable Seeds. I. Castor Beans, *J. Chem. Eng. (China) 5:* 23–34 (1938); *Chem. Abstr. 33:* 7136^{7} (1939).
77. Tu, C.-M., and C. Wang, Vapor-Phase Cracking of Crude Cottonseed Oil, *J. Chem. Eng. (China) 3:* 222–230 (1936); *Chem. Abstr. 31:* 238^{3} (1937).
78. Tu, C.-M., and F.-Y. Pan, The Distillation of Cottonseed Oil Foot, *J. Chem. Eng. (China) 3:* 231–239 (1936); *Chem. Abstr. 31:* 238^{4} (1937).
79. Chao, Y.-S., Studies on Cottonseeds. III. Production of Gasoline from Cottonseed-Oil Foot, *J. Chem. Eng. (China) 4:* 169–172 (1937); *Chem. Abstr. 31:* 6911^{8} (1937).
80. Banzon, J., Coconut Oil, I. Pyrolysis, *Philipp. Agric. 25:* 817–832 (1937); *Chem. Abstr. 31:* 4518^{3} (1937).
81. Michot-Dupont, F., Fuels Obtained by the Destructive Distillation of Crude Oils Seeds, *Bull. Assoc. Chim. 54:* 438–448 (1937); *Chem. Abstr. 31:* 4787^{6} (1937).
82. Cerchez, V. Th., Conversion of Vegetable Oils into Fuels, *Mon. Pétrole Roumain 39:* 699–702 (1938); *Chem. Abstr. 32:* 8741^{5} (1938).
83. Friedwald, M., New Method for the Conversion of Vegetable Oils to Motor Fuel, *Rev. Pétrolifère* (No. 734): 597–599 (1937); *Chem. Abstr. 31:* 5607^{7} (1937).
84. Dalal, N.M., and T.N. Mehta, Cracking of Vegetable Oils, *J. Indian Chem. Soc. 2:* 213–245 (1939); *Chem. Abstr. 34:* 6837^{5} (1940).

85. Chang, C.H., C.D. Shiah, and C.W. Chan, Effect of the Addition of Lime on the Cracking of Vegetable Oils, *J. Chin. Chem. Soc. 8:* 100–107 (1941); *Chem. Abstr. 37:* 6108^4 (1943).
86. Suen, T.-J., and K.C. Wang, Clay Treatment of Vegetable Gasoline, *J. Chin. Chem. Soc. 8* :93–99 (1941); *Chem. Abstr. 37:* 6108^6 (1943).
87. Sun, Y.C., Pressure Cracking of Distillation Bottoms from the Pyrolysis of Mustard Seed, *J. Chin. Chem. Soc. 8:* 108–111 (1941); *Chem. Abstr. 37:* 6108^5 (1943).
88. Lo, T.-S., Some Experiments on the Cracking of Cottonseed Oil, *Science (China) 24:* 127–138 (1940); *Chem. Abstr. 34:* 6040^4 (1940).
89. Lo, T.-S., and L.-S. Tsai, Chemical Refining of Cracked Gasoline from Cottonseed Oil, *J. Chin. Chem. Soc. 9:* 164–172 (1942); *Chem. Abstr. 37:* 6919^2 (1943).
90. Lo, T.-S., and L.-S. Tsai, Further Study of the Pressure Distillate from the Cracking of Cottonseed Oil, *J. Chem. Eng. (China) 9:* 22–27 (1942); *Chem. Abstr. 40:* 2655^8 (1946).
91. Bonnefoi, J., Nature of the Solid, Liquid, and Gaseous Fuels Which Can Be Obtained from the Oil-Palm Fruit, *Bull. Mat. Grasses Inst. Coloniale Marseille 27:* 127–134 (1943); *Chem. Abstr. 39:* 3141^1 (1945).
92. François, R., Manufacture of Motor Fuels by Pyrolysis of Oleaginous Seeds, *Techpl. Appl. Pétrole 2:* 325–327 (1947); *Chem. Abstr. 41:* 6037 d (1947).
93. Otto, R.B., Gasoline Derived from Vegetable Oils, *Bol. Divulgação Inst. Óleos 3:* 91–99 (1945); *Chem. Abstr. 41:* 6690 f (1947).
94. Tu, C.-M., and T.-T. Ku, Cottonseed Oil as a Diesel Oil, *J. Chem. Eng. (China) 3:* 211–221 (1936); *Chem. Abstr. 31:* 237^9 (1937).
95. Suen, T.-J., and L.-H. Li, Miscibility of Ethyl Alcohol and Vegetable Gasoline, *J. Chin. Chem. Soc. 8:* 76–80 (1941); *Chem. Abstr. 37:* 249^3 (1943).
96. Ilieff, B., Die Pflanzenöle als Dieselmotorbrennstoffe, *Österr. Chem.-Ztg. 42:* 353–356 (1939); *Chem. Abstr. 34:* 607^4 (1940).
97. Jalbert, J., Colonial Motor Fuels and Lubricants from Plants, *Carburants Nat. 3:* 49–56 (1942); *Chem. Abstr. 37:* 6107^1 (1943).
98. Charles, Application of Vegetable Oils as Fuels for Heating and Power Purposes, *Chim. Ind. (Special Number):* 769–774 (1923); *Chem. Abstr. 17:* 3242 (1923).
99. Fachini, S., The Problem of Olive Oils as Fuels and Lubricants, *Chim. Ind. (Special Number):* 1078–1079 (1933); *Chem. Abstr. 28:* 283^8 (1934).
100. Physical Chemistry Research Co., Distilling Oleaginous Vegetable Materials. French Patent 756,544, December 11, 1933; *Chem. Abstr. 28:* 2507^2 (1934).
101. Physical Chemistry Research Co., Motor Fuel. French Patent 767,362, July 17, 1934; *Chem. Abstr. 29:* 2695^2 (1935).
102. Legé, E.G.M.R., Fuel Oils. French Patent 812,006, April 28, 1937; *Chem. Abstr. 32:* 1086 (4); see also addition 47,961, August 28, 1937; *Chem. Abstr. 32:* 4773^2 (1938).
103. Jean, J.W., Motor Fuels, U.S. Patent 2,117,609 (May 17, 1938); *Chem. Abstr. 32:* 5189^2 (1938).
104. Standard Oil Development Co., Motor Fuels, British Patent 508,913 (July 7, 1939); *Chem. Abstr. 34:* 3054^2 (1940).
105. Bouffort, M.M.J., Converting Fatty Compounds into Petroleum Oils, French Patent 844,105 (1939); *Chem. Abstr. 34:* 7598^7 (1942).
106. Archer, H.R.W., and A. Gilbert Tomlinson, Coconut Products, Australian Patent 113,672 (August 13, 1941); *Chem. Abstr. 36:* 3348^1 (1942).

107. Chavanne, G., Procédé de Transformation d'Huiles Végétales en Vue de Leur Utilisation comme Carburants (Procedure for the Transformation of Vegetable Oils for Their Uses as Fuels), Belgian Patent 422,877 (August 31, 1937); *Chem. Abstr. 32:* 4313^2 (1938).
108. Chavanne, G., Sur un Mode d'Utilization Possible de l'Huile de Palme à la Fabrication d'un Carburant Lourd (A Method of Possible Utilization of Palm Oil for the Manufacture of a Heavy Fuel), *Bull. Soc. Chim. 10:* 52–58 (1943); *Chem. Abstr. 38:* 2183^9 (1944).
109. van den Abeele, M., L'Huile de Palme: Matière Première pour la Préparation d'un Carburant Lourd Utilisable dans les Moteurs à Combustion Interne (Palm Oil as Raw Material for the Production of a Heavy Motor Fuel), *Bull. Agric. Congo Belge 33:* 3–90 (1942); *Chem. Abstr. 38:* 2805^1 (1944).
110. Duport, R., Auto-Ignition Temperatures of Diesel Motor Fuels (Étude sur la Température d'Auto-Inflammation des Combustibles pour Moteurs Diesel), *Oléagineux 1:* 149–153 (1946); *Chem. Abstr. 43:* 2402 d (1949).
111. Wang, R., Development of Biodiesel Fuel, *Taiyangneng Xuebao 9:* 434–436 (1988); *Chem. Abstr. 111:* 26233 (1989).
112. Bailer, J., and K. de Hueber, Determination of Saponifiable Glycerol in "Bio-Diesel," *Fresenius J. Anal. Chem. 340:* 186 (1991); *Chem. Abstr. 115:* 73906 (1991).

3

The Basics of Diesel Engines and Diesel Fuels

Jon Van Gerpen

Introduction

The diesel engine has been the engine of choice for heavy-duty applications in agriculture, construction, industrial, and on-highway transport for >50 yr. Its early popularity could be attributed to its ability to use the portion of the petroleum crude oil that had previously been considered a waste product from the refining of gasoline. Later, the diesel's durability, high torque capacity, and fuel efficiency ensured its role in the most demanding applications. Although diesels have not been widely used in passenger cars in the United States (<1%), they have achieved widespread acceptance in Europe with >33% of the total market (1).

In the United States, on-highway diesel engines now consume >30 billion gallons of diesel fuel per year, and virtually all of this is in trucks (2). At the present time, only a minute fraction of this fuel is biodiesel. However, as petroleum becomes more expensive to locate and extract, and environmental concerns about diesel exhaust emissions and global warming increase, "biodiesel" is likely to emerge as one of several potential alternative diesel fuels.

To understand the requirements of a diesel fuel and how biodiesel can be considered a desirable substitute, it is important to understand the basic operating principles of the diesel engine. This chapter describes these principles, particularly in light of the fuel used and the ways in which biodiesel provides advantages over conventional petroleum-based fuels.

Diesel Combustion

The operating principles of diesel engines are significantly different from those of the spark-ignited engines that dominate the U.S. passenger car market. In a spark-ignited engine, fuel and air that are close to the chemically correct, or *stoichiometric*, mixture are inducted into the engine cylinder, compressed, and then ignited by a spark. The power of the engine is controlled by limiting the quantity of fuel-air mixture that enters the cylinder using a flow-restricting valve called a *throttle*. In a diesel engine, also known as a *compression-ignited* engine, only air enters the cylinder through the intake system. This air is compressed to a high temperature and pressure, and then finely atomized fuel is sprayed into the air at high velocity. When it contacts the high-temperature air, the fuel vaporizes quickly, mixes with the air, and undergoes a series of spontaneous chemical reactions that result in a self-ignition or *autoignition*. No spark plug is required, although some diesel engines are equipped with electrically heated glow plugs to assist with starting the engine under cold conditions. The power of the engine is controlled by varying the volume of fuel injected into the cylinder; thus, there is no need for a throttle. Figure 1 shows a cross section of the diesel combustion chamber with the fuel injector positioned between the intake and exhaust valves.

The timing of the combustion process must be precisely controlled to provide low emissions with optimum fuel efficiency. This timing is determined by the fuel injection timing plus the short time period between the start of fuel injection and the autoignition, called the *ignition delay*. When autoignition occurs, the portion of the fuel that had been prepared for combustion burns very rapidly during a period known as *premixed combustion*. When the fuel that had been prepared during the ignition delay is exhausted, the remaining fuel burns at a rate determined by the mixing of the fuel and air. This period is known as *mixing-controlled* combustion.

The heterogeneous fuel-air mixture in the cylinder during the diesel combustion process contributes to the formation of soot particles, one of the most difficult challenges for diesel engine designers. These particles are formed in high-temperature regions of the combustion chamber in which the air-fuel ratio is fuel-rich and consists mostly of carbon with small amounts of hydrogen and inorganic compounds. Although the mechanism is still not understood, biodiesel reduces the amount of soot produced and this appears to be associated with the bound oxygen in the fuel (3). The particulate level in the engine exhaust is composed of these soot particles along with high-molecular-weight hydrocarbons that adsorb to the particles as the gas temperature decreases during the expansion process and in the exhaust pipe. This hydrocarbon material, called the *soluble organic fraction*, usually increases when biodiesel is used, offsetting some of the decrease in soot (4). Biodiesel's low volatility apparently causes a small portion of the fuel to survive

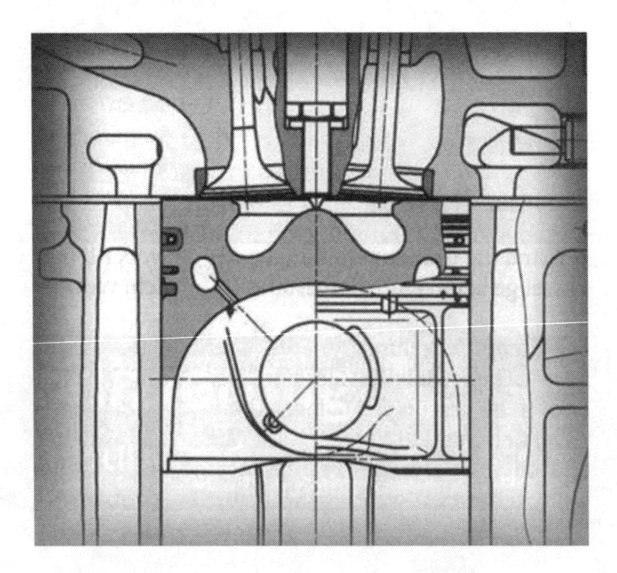

Fig. 1. Cross section of a diesel engine combustion chamber.

the combustion process, probably by coating the cylinder walls, where it is then released during the exhaust process.

A second difficult challenge for diesel engine designers is the emission of oxides of nitrogen (NO_x). NO_x emissions are associated with high gas temperatures and lean fuel conditions; in contrast to most other pollutants, they usually increase when biodiesel is used (4). NO_x contribute to smog formation and are difficult to control in diesel engines because reductions in NO_x tend to be accompanied by increases in particulate emissions and fuel consumption. Although the bound oxygen on the biodiesel molecule may play a role in creating a leaner air-fuel ratio in NO_x formation regions, the dominant mechanism seems to be the effect of changes in the physical properties of biodiesel, such as the speed of sound and bulk modulus, on the fuel injection timing (5).

One of the most important properties of a diesel fuel is its readiness to autoignite at the temperatures and pressures present in the cylinder when the fuel is injected. The laboratory test that is used to measure this tendency is the *cetane number* (CN) test (ASTM D 613). The test compares the tendency to autoignite of the test fuel with a blend of two reference fuels, cetane (hexadecane) and heptamethylnonane. Fuels with a high CN will have short ignition delays and a small amount of premixed combustion because little time is available to prepare the fuel for combustion. Most biodiesel fuels have higher CN than petroleum-based diesel fuels. Biodiesel fuels from more saturated feedstocks have higher CN than those from less saturated feedstocks (6). Biodiesel from soybean oil is usually reported to have a CN of 48–52, whereas biodiesel from yellow grease, containing more saturated esters, is normally between 60 and 65 (7). For more details, see Chapter 4.1 and the tables in Appendix A.

Energy Content (Heat of Combustion). The energy content of the fuel is not controlled during manufacturing. The actual benefit of the lower heating value for diesel fuel will vary depending on the refinery in which it was produced, the time of year, and the source of the petroleum feedstock because all of these variables affect the composition of the fuel. Diesel fuels with high percentages of aromatics tend to have high energy contents per liter even though the aromatics have low heating values per kilogram. Their high density more than compensates for their lower energy content on a weight basis. This is of special importance for diesel engines because fuel is metered to the engine volumetrically. A fuel with a lower energy content per liter will cause the engine to produce less peak power. At part load conditions, the engine operator will still be able to meet the demand for power but a greater volume of fuel will have to be injected. The fuel injection system may advance the fuel injection timing when the fuel flow rate increases, and this can cause an increase in the NO_x emissions. In addition to the compressibility effects mentioned earlier, this effect is another reason for the higher NO_x emissions observed with biodiesel (8).

Biodiesel fuels do not contain aromatics, but they contain methyl esters with different levels of saturation. Unsaturated esters have a lower energy content on a weight basis, but due to their higher density, they have more energy per unit volume. For example, methyl stearate has a higher heating value of 40.10 MJ/kg, which is 0.41% higher than that of methyl oleate (39.93 MJ/kg). However, on a volume basis (at 40°C), methyl stearate has an energy content of 34.07 MJ/L, which is 0.7% less than that of methyl oleate (34.32 MJ/L) (9,10). These differences are small enough that feedstock differences are difficult to detect in actual use.

Biodiesel has a lower energy content (lower heating value of 37.2 MJ/kg for soy biodiesel) than No. 2 diesel fuel (42.6 MJ/kg). On a weight basis, the energy level is 12.5% less. Because biodiesel is more dense than diesel fuel, the energy content is only 8% less on a per gallon basis (32.9 vs. 36.0 MJ/L). Because diesel engines will inject equal volumes of fuel, diesel engine operators may see a power loss of ~8.4%. In some cases, the power loss may be even less than this because biodiesel's higher viscosity can decrease the amount of fuel that leaks past the plungers in the diesel fuel injection pump, leaving more fuel to be injected.

Tests showed that the actual efficiency at which the energy in the fuel is converted to power is the same for biodiesel and petroleum-based diesel fuel (11). Therefore, the brake specific fuel consumption (BSFC), which is the fuel flow rate divided by the engine's output power and is the parameter most often used by engine manufacturers to characterize fuel economy, will be at least 12.5% higher for biodiesel. The values for heat of combustion of various fatty materials taken from the literature are given in the tables in Appendix A.

Emissions. Under ideal circumstances, all of the carbon in the diesel fuel will burn to carbon dioxide, and all of the hydrogen will burn to water vapor. In most cases, virtually all of the fuel follows this path. However, if sulfur is present in the fuel, it will be oxidized to sulfur dioxide and sulfur trioxide. These oxides of sulfur can react with water vapor to form sulfuric acid and other sulfate compounds. The sulfates can form particles in the exhaust and elevate the exhaust particulate level. In 1993, the U.S. Environmental Protection Agency (EPA) mandated that diesel fuel should contain no more than 500 ppm of sulfur. This was a factor of 10 reduction in sulfur level and greatly reduced sulfur as a source of exhaust particulate. In 2006, the EPA has mandated a new reduction in sulfur to 15 ppm. This will eliminate sulfur as a component of exhaust particulate and allow the introduction of catalytic after-treatment for diesel engines. Sulfur is a powerful catalyst poison and limits the options available for controlling emissions on future engines. Biodiesel from soybean oil is very low in sulfur. However, biodiesel from some animal fat feedstocks has sulfur levels that exceed the 2006 mandate and will require further treatment.

Aromatics are a class of hydrocarbon compounds that are characterized by stable chemical structures. They are usually present in diesel fuel at levels between 25 and 35%. They are considered desirable by diesel engine operators because they provide greater energy per liter of fuel; however, they may contribute to higher emissions of particulate and NO_x, and have lower CN. In the early 1990s, the California Air Resources Board implemented standards that limited the aromatic content of diesel fuels sold in California to 10%. The board later allowed the aromatic content to be higher if fuel producers could show that their fuels produced equivalent or lower emissions than the low-aromatic fuel. Biodiesel contains no aromatic compounds.

Low-Temperature Operation. Diesel fuel contains small amounts of long-chain hydrocarbons, called waxes, that crystallize at temperatures within the normal diesel engine operating range. If temperatures are low enough, these wax crystals will agglomerate, plug fuel filters, and prevent engine operation. At a low enough temperature, the fuel will actually solidify. This phenomenon also occurs with biodiesel. The saturated fatty acids produce methyl esters that will start to crystallize at ~0ºC for soybean oil and as high as 13–15°C for animal fats and frying oils (12,13). The most common measure of this tendency to crystallize is the cloud point (CP). This is the temperature at which the onset of crystallization is observed visually as a cloudiness in the fuel. A more extreme test is the pour point (PP), which is the lowest temperature at which the fuel can still be poured from a vessel. ASTM D 2500 and D 97 are used to determine the CP and PP of the fuels, respectively. Other tests are used to measure the tendency of the fuel to plug fuel filters.

Additives, known as PP depressants, can be used to inhibit the agglomeration of the wax crystals, which then lowers the point at which fuel filter plugging occurs. It is also common to add No. 1 diesel fuel to No. 2 diesel fuel to lower its operating point. No. 1 diesel fuel has a very low level of waxes and dilutes the waxes in No. 2 diesel fuel, which lowers the temperature at which they cause the fuel to solidify. Both No. 1 and No. 2 diesel fuels can be blended with biodiesel to lower the operating temperature of the fuel. Biodiesel used at the 1–2% level as a lubricity additive does not seem to have any measurable effect on the CP. The allowable operating temperature for B20 blends is higher than that for the original diesel fuel, but many B20 users have been able to operate in cold climates without problems.

Viscosity. Fuel viscosity is specified in the standard for diesel fuel within a fairly narrow range. Hydrocarbon fuels in the diesel boiling range easily meet this viscosity specification. Most diesel fuel injection systems compress the fuel for injection using a simple piston and cylinder pump called the plunger and barrel. To develop the high pressures required in modern injection systems, the clearances between the plunger and barrel are ~0.0001″ (0.0025 cm). Despite this small clearance, a substantial fraction of the fuel leaks past the plunger during compression. If fuel viscosity is low, the leakage will correspond to a power loss for the engine. If fuel viscosity is high, the injection pump will be unable to supply sufficient fuel to fill the pumping chamber. Again, the effect will be a loss in power. The viscosity range for typical biodiesel fuels overlaps the diesel fuel range, with some biodiesel fuels having viscosities above the limit (14). If fuel viscosity is extremely excessive, as is the case with vegetable oils, there will be a degradation of the spray in the cylinder causing poor atomization, contamination of the lubricating oil, and the production of black smoke. More details on viscosity are given in Chapter 6.2, and data appear in the tables of Appendix A.

Corrosion. Many of the parts in the diesel fuel injection system are made of high-carbon steels; thus, they are prone to corrosion when in contact with water. Water damage is a leading cause of premature failure of fuel injection systems. Diesel fuel containing excessive water that enters the injection system can cause irreversible damage in a very short time. Many diesel engines are equipped with water separators that cause small water droplets to coalesce until they are large enough to drop out of the fuel flow where they can be removed. There are some reports that these water separators are not effective when used with biodiesel.

Water can be present in fuels as dissolved water and free water. Petroleum-based diesel fuel can absorb only ~50 ppm of dissolved water, whereas biodiesel can absorb as much as 1500 ppm (15). Although this dissolved water can affect the stability of the fuel, free water is more strongly associated with corrosion concerns. ASTM D 2709 is used to measure the total amount of free water and sediment in a diesel fuel sample. The method uses a centrifuge to collect the water and the specifications on both diesel fuel and biodiesel limit the amount of water and sediment to 0.05%.

Some compounds in diesel fuel, especially sulfur compounds, can be corrosive. Because copper compounds are particularly susceptible to this type of corrosion, copper is used as an indicator of the tendency of the fuel to cause corrosion. In ASTM D 130, polished copper strips are soaked in the fuel to characterize the tendency to corrode metals. Although some tarnish is typically allowed, corrosion causes the fuel to fail the test.

Sediment. Diesel fuel filters are designed to capture particles that are >10 μm in size. Some newer engines are even equipped with filters that capture particles as small as 2 μm. These filters should stop foreign materials from entering the fuel injection system. However, when fuels are exposed to high temperatures and the oxygen in air, they can undergo chemical changes that form compounds that are insoluble in the fuel. These compounds form varnish deposits and sediments that can plug orifices and coat moving parts, causing them to stick. Several test procedures were developed that attempt to measure the tendency of diesel fuels to produce these sediments, such as ASTM D 2274, but none have gained the acceptance required to be included in the diesel fuel specification (ASTM D 975). Because of its high concentration of unsaturated compounds, biodiesel is expected to be more susceptible to oxidative degradation than petroleum-based diesel fuel.

Inorganic materials present in the fuel may produce ash that can be abrasive and contribute to wear between the piston and cylinder. ASTM D 482 is used to characterize ash from diesel fuels. The ASTM specification for biodiesel, D 6751, requires that ASTM D 874 be used. This method measures sulfated ash, which is specified because it is more sensitive to ash from sodium and potassium. These metals originate from the catalyst used in the biodiesel production process and are likely to be the main sources for ash in biodiesel.

When fuel is exposed to high temperatures in the absence of oxygen, it can pyrolyze to a carbon-rich residue. Although this should not occur in the cylinder of a properly operating engine, some injection systems have the potential to create a region within the injection nozzle in which this residue can collect and limit the range of motion of moving parts. Various test procedures such as ASTM D 189, D 524, and D 4530 were developed as an attempt to predict the tendency of a fuel to form in-cylinder carbon deposits. Unfortunately, it is difficult to reproduce in-cylinder conditions in a test; thus, the correlation of these procedures with actual engine deposits is limited.

Diesel fuel injection systems have closely fitting parts that are subjected to high loads. These parts require lubrication to prevent rapid wear. All diesel injection systems rely on the fuel itself to provide this lubrication. Although the mechanism remains a topic for debate, it is known that as refiners reduce the sulfur content of diesel fuel, the ability of the fuel to provide the necessary lubrication decreases. The property that characterizes the ability of the fuel to lubricate is the *lubricity*. There are two methods that are commonly used to measure diesel fuel lubricity, the scuffing load ball on cylinder lubricity evaluator (SLBOCLE: ASTM D 6078-99) and the high frequency reciprocating rig (HFRR: ASTM D 6079-99) but both procedures have been widely criticized. This is primarily due to the lack of correlation between the test procedures and the large amount of test-to-test variability. Biodiesel has excellent lubricity, and as little as 1–2% biodiesel can raise the lubricity of a poor lubricity fuel to an acceptable level (16).

Flashpoint. Diesel engine operators are accustomed to treating diesel fuel as if it were nonflammable. The volatilities of both No. 1 and No. 2 diesel fuel are low enough that the air-vapor mixture above the fuel is below the flammability limit. The property that characterizes this behavior is the *flashpoint* (FlP). The FlP is the temperature at which the fuel will give off enough vapor to produce a flammable mixture: 52–66°C for diesel fuel and below –40°C for gasoline. An important advantage of biodiesel is that its very high flashpoint, >150ºC, indicates that it presents a very low fire hazard.

New Technologies

Requirements for lower emissions and continued demands for improved fuel economy have driven the engine industry to technical advances that incorporate state-of-the-art electronics and manufacturing technology. Electronically controlled cam-actuated unit injection has pushed the limits for fuel injection pressures to

>2000 bar. The rapid mixing provided by the high spray velocity resulting from this extreme injection pressure provides low particulate formation and virtually complete soot oxidation while allowing retarded injection timing settings for reduced NO_x.

The introduction of common rail injection systems for light- and medium-duty engines has allowed new flexibility in programming the injection event. These systems allow multiple injections within a single engine cycle. A common strategy is to start the combustion with two brief injections, called the pilot- and preinjections. These injections produce an environment in the cylinder so that when the main injection occurs, the ignition delay will be shorter, the amount of premixed combustion will be less, and the NO_x production will be reduced. These small injections that precede the main injection also reduce engine noise and vibration. Immediately after the main injection, a small amount of fuel may be injected to assist in oxidizing the soot particles. Then, later in the expansion process, a postinjection provides the elevated exhaust hydrocarbon level required by the after-treatment equipment. The high degree of control offered by common rail injection systems would have been useless without the use of an electronic control unit. The application of powerful on-board computers to diesel engines initially lagged behind their use on spark-ignition engines, but current engines have corrected this deficiency.

With the exception of some oxidation catalysts, diesel engines have traditionally not used exhaust after-treatment for emission control. The three-way catalyst technology that is widely used for spark-ignited vehicles is not suitable for use on diesel engines because it requires a near stoichiometric fuel-air mixture to obtain simultaneous reductions in carbon monoxide, unburned hydrocarbons, and oxides of nitrogen. Diesels always operate with excess oxygen; thus, the reducing catalyst required to eliminate NO_x cannot operate. The oxidation catalysts provided on some diesel engines are able to reduce particulate levels by oxidizing some of the adsorbed hydrocarbons from the soot particles, but they are not effective at reducing the solid portion of the particulate, and they do nothing to reduce NO_x.

Recent innovations include catalyzed diesel particulate filters or traps. These devices force the exhaust to pass through a porous ceramic material that captures the exhaust particles. The surface of the ceramic is coated with a catalyst that oxidizes the particles as they are collected. NO_x traps and absorbers are also being developed. These devices catalytically convert the NO_x to stable compounds that are collected within the catalyst and then periodically removed during regeneration cycles. The catalysts used in both the particulate traps and the NO_x absorbers are very sensitive to fuel sulfur. As mentioned earlier, to allow this technology to develop, the U.S. EPA mandated a reduction in fuel sulfur from 500 to 15 ppm by 2006.

To improve the engine's air supply, variable geometry turbochargers have been developed to extend the engine operating range over which adequate air is provided to keep particulate emissions low. Air-to-air after-coolers are also used to lower intake air temperatures to reduce both NO_x and particulate emissions.

Little is known about biodiesel use in advanced technology engines. Although the addition of exhaust after-treatment systems to control particulate and NO_x emissions may reduce one of the driving forces for biodiesel use, there is no indication that biodiesel will not be fully compatible with the new engine systems.

References

1. Broge, J.L., Revving Up For Diesel, *Automotive Eng. Int. 110:* 40–49 (2002).
2. Energy Information Administration, Official Energy Statistics from the U.S. Government, www.eia.doe.gov.
3. McCormick, R.L., J.D. Ross, and M.S. Graboski, Effect of Several Oxygenates on Regulated Emissions from Heavy-Duty Diesel Engines, *Environ. Sci. Technol. 31:* 1144–1150 (1997).
4. Sharp, C.A., S.A. Howell, and J. Jobe, The Effect of Biodiesel Fuels on Transient Emissions from Modern Diesel Engines, Part I. Regulated Emissions and Performance, *SAE Paper No. 2000-01-1967*, 2000.
5. Tat, M.E., J.H. Van Gerpen, S. Soylu, M. Canakci, A. Monyem, and S. Wormley, The Speed of Sound and Isentropic Bulk Modulus of Biodiesel at 21°C from Atmospheric Pressure to 35 MPa, *J. Am. Oil Chem. Soc. 77:* 285–289 (2000).
6. Knothe, G., M.O. Bagby, and T.W. Ryan, III, Cetane Numbers of Fatty Compounds: Influence of Compound Structure and of Various Potential Cetane Improvers, *SAE Paper 971681, (SP-1274)*, 1997.
7. Van Gerpen, J., Cetane Number Testing of Biodiesel, Liquid Fuels and Industrial Products from Renewable Resources, in *Proceedings of the Third Liquid Fuels Conference*, Nashville, Sept. 15–17, 1996.
8. Tat, M.E, and J.H. Van Gerpen, Fuel Property Effects on Biodiesel, Presented at the American Society of Agricultural Engineers 2003 Annual Meeting, *ASAE Paper 036034*, Las Vegas, July 27–30, 2003.
9. Freedman, B., and M.O. Bagby, Heats of Combustion of Fatty Esters and Triglycerides, *J. Am. Oil Chem. Soc. 66:* 1601–1605 (1989).
10. Weast, R.C., ed., *Handbook of Chemistry and Physics*, 51st edn., Chemical Rubber Company, Cleveland, 1970–1971.
11. Monyem, A., and J.H. Van Gerpen, The Effect of Biodiesel Oxidation on Engine Performance and Emissions, *Biomass Bioenergy 4:* 317–325 (2001).
12. Lee, I., L.A. Johnson, and E.G. Hammond, Use of Branched-Chain Esters to Reduce the Crystallization Temperature of Biodiesel, *J. Am. Oil Chem. Soc. 72:* 1155–1160 (1995).
13. Dunn, R.O., and M.O. Bagby, Low-Temperature Properties of Triglyceride-Based Diesel Fuels: Transesterified Methyl Esters and Petroleum Middle Distillate/Ester Blends, *J. Am. Oil Chem. Soc. 72:* 895–904 (1995).
14. Tat, M.E., and J.H. Van Gerpen, The Kinematic Viscosity of Biodiesel and Its Blends with Diesel Fuel, *J. Am. Oil Chem. Soc. 76:* 1511–1513 (1999).
15. Van Gerpen, J.H., E.G. Hammond, L. Yu, and A. Monyem, Determining the Influence of Contaminants on Biodiesel Properties, *Society of Automotive Engineers Technical Paper Series No. 971685*, Warrendale, PA, 1997.
16. Schumacher, L.G., and B.T. Adams, Using Biodiesel as a Lubricity Additive for Petroleum Diesel Fuel, *ASAE Paper 026085*, July 2002.

4

Biodiesel Production

4.1

Basics of the Transesterification Reaction

Jon Van Gerpen and Gerhard Knothe

Introduction

Four methods to reduce the high viscosity of vegetable oils to enable their use in common diesel engines without operational problems such as engine deposits have been investigated: blending with petrodiesel, pyrolysis, microemulsification (cosolvent blending), and transesterification (1). Transesterification is by far the most common method and will be dealt with in this chapter. Only the transesterification reaction leads to the products commonly known as biodiesel, i.e., alkyl esters of oils and fats. The other three methods are discussed in Chapter 10.

The most commonly prepared esters are methyl esters, largely because methanol is the least expensive alcohol, although there are exceptions in some countries. In Brazil, for example, where ethanol is less expensive, ethyl esters are used as fuel. In addition to methanol and ethanol, esters of vegetable oils and animal fats with other low molecular weight alcohols were investigated for potential production and their biodiesel properties. Properties of various esters are listed in the tables in Appendix A. Table 1 of this chapter contains a list of C_1–C_4 alcohols and their relevant properties. Information on vegetable oils and animal fats used as starting materials in the transesterification reaction as well as on resulting individual esters and esters of oils and fats appears in Appendix A.

In addition to vegetable oils and animal fats, other materials such as used frying oils can also be suitable for biodiesel production; however, changes in the reaction procedure frequently have to be made due to the presence of water or free fatty acids (FFA) in the materials. The present section discusses the transesterification reaction as it is most commonly applied to (refined) vegetable oils and related work. Alternative feedstocks and processes, briefly indicated here, will be discussed later. The general scheme of the transesterification reaction was presented in the introduction and is given here again in Figure 1.

Di- and monoacylglycerols are formed as intermediates in the transesterification reaction. Figure 2 qualitatively depicts conversion vs. reaction time for a transesterification reaction taking into account the intermediary di- and monoacylglycerols. Actual details in this figure, such as the final order of concentration of

TABLE 1
Properties of C_1–C_4 Alcohols[a]

	Formula	Molecular weight	Boiling point (°C)	Melting point (°C)	Density (g.mL)
Methanol	CH_3OH	32.042	65	–93.9	$0.7914^{20/4}$
Ethanol	C_2H_5OH	46.069	78.5	–117.3	$0.7893^{20/4}$
1-Propanol	$CH_2OH\text{-}CH_2\text{-}CH_3$	60.096	97.4	–126.5	$0.8035^{20/4}$
2-Propanol (*iso*-Propanol)	$CH_3\text{-}CHOH\text{-}CH_3$	60.096	82.4	–89.5	$0.7855^{20/4}$
1-Butanol (*n*-Butanol)	$CH_3\text{-}CH_2\text{-}CH_2\text{-}CH_2OH$	74.123	117.2	–89.5	$0.8098^{20/4}$
2-Butanol	$CH_3\text{-}CHOH\text{-}CH_2\text{-}CH_3$	74.123	99.5	—	$0.8080^{20/4}$
2-Methyl-1-propanol (*iso*-butanol)	$CH_2OH\text{-}CH(CH_3)\text{-}CH_2\text{-}CH_3$	74.123	108	—	$0.8018^{20/4}$
2-Methyl-2-propanol (*tert*-butanol)	$CH_3\text{-}C(CH_3)OH\text{-}CH_3$	74.123	82.3	25.5	$0.7887^{20/4}$

[a]*Source:* Reference 60.

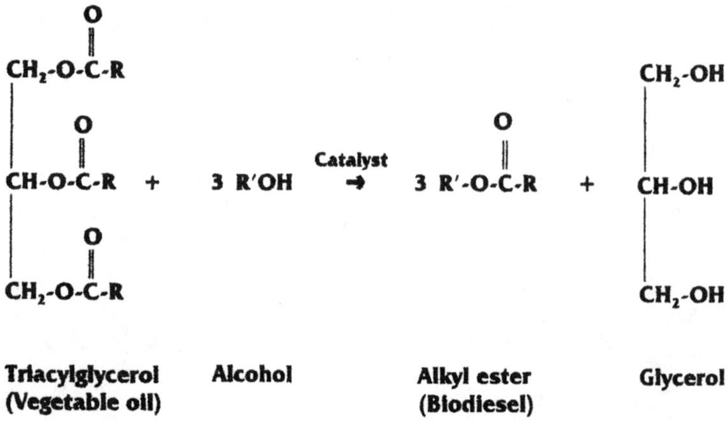

$$\begin{array}{l} CH_2\text{-}O\text{-}C(=O)\text{-}R \\ | \\ CH\text{-}O\text{-}C(=O)\text{-}R \\ | \\ CH_2\text{-}O\text{-}C(=O)\text{-}R \end{array} \quad + \quad 3\ R'OH \quad \xrightarrow{\text{Catalyst}} \quad 3\ R'\text{-}O\text{-}C(=O)\text{-}R \quad + \quad \begin{array}{l} CH_2\text{-}OH \\ | \\ CH\text{-}OH \\ | \\ CH_2\text{-}OH \end{array}$$

Triacylglycerol (Vegetable oil) **Alcohol** **Alkyl ester (Biodiesel)** **Glycerol**

Fig. 1. The transesterification reaction. R is a mixture of various fatty acid chains. The alcohol used for producing biodiesel is usually methanol ($R' = CH_3$).

the various glycerides at the end of the reaction and concentration maximums for di- and monoacylglycerols, may vary from reaction to reaction depending on conditions. The scale of the figure can also vary if concentration (in mol/L) is plotted vs. time instead of conversion.

Several reviews dealing with the production of biodiesel by transesterification have been published (2–10). Accordingly, the production of biodiesel by transesterification has been the subject of numerous research papers. Generally, transesterification can proceed by base or acid catalysis (for other transesterification processes, see the next section). However, in homogeneous catalysis, alkali catalysis (sodium or potassium hydroxide; or the corresponding alkoxides) is a much more rapid process than acid catalysis (11–13).

In addition to the type of catalyst (alkaline vs. acidic), reaction parameters of base-catalyzed transesterification that were studied include the molar ratio of alcohol to vegetable oil, temperature, reaction time, degree of refinement of the vegetable oil, and effect of the presence of moisture and FFA (12). For the transesterification to give maximum yield, the alcohol should be free of moisture and the FFA content of the oil should be <0.5% (12). The absence of moisture in the transesterification reaction is important because according to the equation (shown for methyl esters),

$$R\text{-}COOCH_3 + H_2O \rightarrow R\text{-}COOH + CH_3OH \qquad (R = \text{alkyl})$$

hydrolysis of the formed alkyl esters to FFA can occur. Similarly, because triacylglycerols are also esters, the reaction of the triacylglycerols with water can form FFA. At 32°C, transesterification was 99% complete in 4 h when using an alkaline catalyst (NaOH or NaOMe) (12). At ≥60°C, using an alcohol:oil molar ratio of at

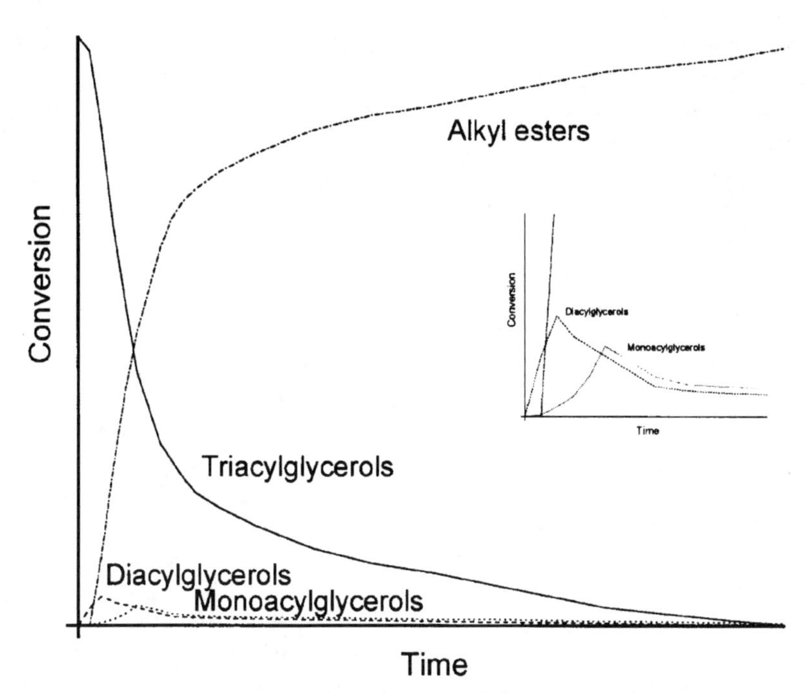

Fig. 2. Qualitative plot of conversion in a progressing transesterification reaction indicating relative concentrations of vegetable oil (triacylglycerols), intermediary di- and monoacylglycerols, as well as methyl ester product. Actual details can vary from reaction to reaction as mentioned in the text.

least 6:1 and fully refined oils, the reaction was complete in 1 h, yielding methyl, ethyl, or butyl esters (12). Although the crude oils could be transesterified, ester yields were reduced because of gums and extraneous material present in the crude oils. These parameters (60°C reaction temperature and 6:1 methanol:oil molar ratio) have become a standard for methanol-based transesterification. Similar molar ratios and temperatures were reported in earlier literature (14–17). Other alcohols (ethanol and butanol) require higher temperatures (75 and 114°C, respectively) for optimum conversion (12). Alkoxides in solution with the corresponding alcohol [made either by reacting the metal directly with alcohol or by electrolysis of salts and subsequent reaction with alcohol (18)] have the advantage over hydroxides that the water-forming reaction according to the equation

$$R'OH + XOH \rightarrow R'OX + H_2O \qquad (R' = \text{alkyl};\ X = \text{Na or K})$$

cannot occur in the reaction system, thus ensuring that the transesterification reaction system remains as water free as possible. This reaction, however, is the one forming the transesterification-causing alkoxide when using NaOH or KOH as catalysts. The catalysts are hygroscopic; precautions, such as blanketing with nitrogen, must be taken to prevent contact with moisture. The use of alkoxides reportedly also results in glycerol of higher purity after the reaction.

Effects similar to those discussed above were observed in studies on the transesterification of beef tallow (19,20). FFA and, even more importantly, water should be kept as low as possible (19). NaOH reportedly was more effective than the alkoxide (19); however, this may have been a result of the reaction conditions. Mixing was important due to the immiscibility of NaOH/MeOH with beef tallow, with smaller NaOH/MeOH droplets resulting in faster transesterification (20). Ethanol is more soluble in beef tallow which increased yield (21), an observation that should hold for other feedstocks as well.

Other work reported the use of both NaOH and KOH in the transesterification of rapeseed oil (22). Recent work on producing biodiesel from waste frying oils employed KOH. With the reaction conducted at ambient pressure and temperature, conversion rates of 80–90% were achieved within 5 min, even when stoichiometric amounts of methanol were employed (23). In two transesterifications (with more MeOH/KOH steps added to the methyl esters after the first step), the ester yields were 99%. It was concluded that an FFA content up to 3% in the feedstock did not affect the process negatively, and phosphatides up to 300 ppm phosphorus were acceptable. The resulting methyl ester met the quality requirements for Austrian and European biodiesel without further treatment. In a study similar to previous work on the transesterification of soybean oil (11,12), it was concluded that KOH is preferable to NaOH in the transesterification of safflower oil of Turkish origin (24). The optimal conditions were given as 1 wt% KOH at 69 ± 1°C with a 7:1 alcohol:vegetable oil molar ratio to give 97.7% methyl ester yield in 18 min. Depending on the vegetable oil and its component fatty acids influencing FFA content, adjustments to the alcohol:oil molar ratio and the amount of catalyst may be required as was reported for the alkaline transesterification of *Brassica carinata* oil (25).

In principle, transesterification is a reversible reaction, although in the production of vegetable oil alkyl esters, i.e., biodiesel, the back reaction does not occur or is negligible largely because the glycerol formed is not miscible with the product, leading to a two-phase system. The transesterification of soybean oil with methanol or 1-butanol was reported to proceed (26) with pseudo-first-order or second-order kinetics, depending on the molar ratio of alcohol to soybean oil (30:1 pseudo first order, 6:1 second order; NaOBu catalyst), whereas the reverse reaction was second order (26). However, the originally reported kinetics (26) were reinvestigated (27–30) and differences were found. The methanolysis of sunflower oil at a molar ratio of methanol:sunflower oil of 3:1 was reported to begin with second-order kinetics but then the rate decreased due to the formation of glycerol (27). A shunt reaction (a reaction in which all three positions of the triacylglycerol react virtually simultaneously to give three alkyl ester molecules and glycerol) originally proposed (26) as part of the forward reaction was shown to be unlikely, that second-order kinetics are not followed, and that miscibility phenomena (27–30) play a significant role. The reason is that the vegetable oil starting material and methanol are not well miscible. The miscibility phenomenon results in a lag time in the formation of methyl esters as indicated qualitatively in Figure 2. The formation of glycerol from triacylglycerols proceeds stepwise *via* the di- and monoacylglycerols, with a fatty acid alkyl ester molecule being formed in each step. From the observation that diacylglycerols reach their maximum concentration before the monoacylglycerols, it was concluded that the last step, formation of glycerol from monoacylglycerols, proceeds more rapidly than the formation of monoacylglycerols from diacylglycerols (31).

The addition of cosolvents such as tetrahydrofuran (THF) or methyl *tert*-butyl ether (MTBE) to the methanolysis reaction was reported to significantly accelerate the methanolysis of vegetable oils as a result of solubilizing methanol in the oil to a rate comparable to that of the faster butanolysis (29–34). This is to overcome the limited miscibility of alcohol and oil at the early reaction stage, creating a single phase. The technique is applicable for use with other alcohols and for acid-catalyzed pretreatment of high FFA feedstocks. However, molar ratios of alcohol:oil and other parameters are affected by the addition of the cosolvents. There is also some additional complexity due to recovering and recycling the cosolvent, although this can be simplified by choosing a cosolvent with a boiling point near that of the alcohol being used. In addition, there may be some hazards associated with its most common cosolvents, THF and MTBE.

Other possibilities for accelerating the transesterification are microwave (35) or ultrasonic (36,37) irradiation. Factorial experiment design and surface response methodology were applied to different production systems (38) and are also discussed in the next section. A continuous pilot plant-scale process for producing methyl esters with conversion rates >98% was reported (39,40) as well as a discontinuous two-stage process with a total methanol:acyl (from triacylglycerols) ratio of 4:3 (41). Other basic materials, such as alkylguanidines, which were anchored to or entrapped in various supporting materials such as polystyrene and zeolite (42), also catalyze transesterification. Such systems may provide for easier catalyst recovery and reuse.

Industrial Production

The chemistry described above forms the basis of the industrial production of biodiesel. Also, biodiesel processing and quality are closely related. The processes used to refine the feedstock and convert it to biodiesel determine whether the fuel will meet the applicable specifications. This section briefly describes the processing and production of biodiesel and how these determine fuel quality. The emphasis is on processing as it is conducted in the United States, where most biodiesel is produced by reacting soybean oil or used cooking oils with methanol and the standard for fuel quality is ASTM D 6751-02.

For alkali-catalyzed transesterification, Figure 3 shows a schematic diagram of the processes involved in biodiesel production from feedstocks containing low lev-

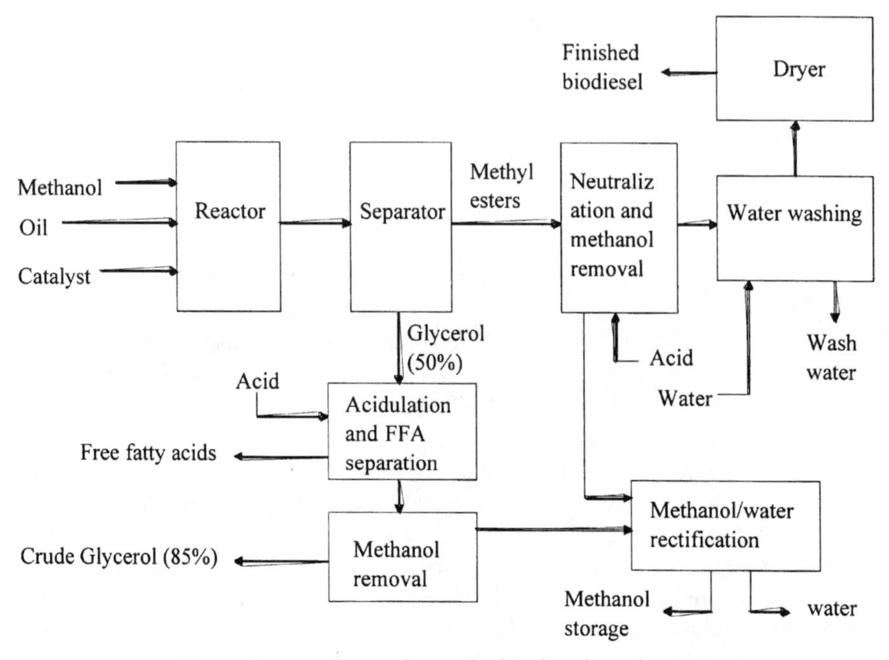

Fig. 3. Process flow scheme for biodiesel production.

els of FFA. These include soybean oil, canola (rapeseed) oil, and the higher grades of waste restaurant oils. Alcohol, catalyst, and oil are combined in a reactor and agitated for ~1 h at 60°C. Smaller plants often use batch reactors (43) but most larger plants (>4 million L/yr) use continuous flow processes involving continuous stirred-tank reactors (CSTR) or plug flow reactors (44). The reaction is sometimes done in two steps in which ~80% of the alcohol and catalyst is added to the oil in a first-stage CSTR. Then, the product stream from this reactor goes through a glycerol removal step before entering a second CSTR. The remaining 20% of the alcohol and catalyst is added in this second reactor. This system provides a very complete reaction with the potential of using less alcohol than single-step systems.

After the reaction, glycerol is removed from the methyl esters. Due to the low solubility of glycerol in the esters, this separation generally occurs quickly and can be accomplished with either a settling tank or a centrifuge. The excess methanol tends to act as a solubilizer and can slow the separation. However, this excess methanol is usually not removed from the reaction stream until after the glycerol and methyl esters are separated due to concern about reversing the transesterification reaction. Water may be added to the reaction mixture after the transesterification is complete to improve the separation of glycerol (43,45).

Some authors (46–51) state that it is possible to react the oil and methanol without a catalyst, which eliminates the need for the water washing step. However, high temperatures and large excesses of methanol are required. The difficulty of reproducing the reaction kinetics results of other researchers was noted (49) and was attributed to catalytic effects at the surfaces of the reaction vessels; it was also noted that these effects would be exacerbated at higher temperatures. Not including the effect of surface reactions could cause difficulties when scaling up reactors due to the decrease in the ratio of reactor surface area to volume. Kreutzer (52) described how higher pressures and temperatures (90 bar, 240°C) can transesterify the fats without prior removal or conversion of the FFA. However, most biodiesel plants use lower temperatures, near atmospheric pressure, and longer reaction times to reduce equipment costs.

Returning to Figure 3, after separation from the glycerol, the methyl esters enter a neutralization step and then pass through a methanol stripper, usually a vacuum flash process or a falling film evaporator, before water washing. Acid is added to the biodiesel product to neutralize any residual catalyst and to split any soap that may have formed during the reaction. Soaps will react with the acid to form water-soluble salts and FFA according to the following equation:

R-COONa	+	HAc	→	R-COOH	+	NaAc
Sodium soap		Acid		Fatty acid		Salt

The salts will be removed during the water washing step and the FFA will stay in the biodiesel. The water washing step is intended to remove any remaining catalyst, soap, salts, methanol, or free glycerol from the biodiesel. Neutralization before washing reduces the amount of water required and minimizes the potential for emulsions to form when the wash water is added to the biodiesel. After the wash process, any remaining water is removed from the biodiesel by a vacuum flash process.

The glycerol stream leaving the separator is only ~50% glycerol. It contains some of the excess methanol and most of the catalyst and soap. In this form, the glycerol has little value and disposal may be difficult. The methanol content requires the glycerol to be treated as hazardous waste. The first step in refining the glycerol is usually to add acid to split the soaps into FFA and salts. The FFA are not soluble in the glycerol and will rise to the top where they can be removed and recycled. Mittelbach and Koncar (53) described a process for esterifying these FFA and then returning them to the transesterification reaction stream. The salts remain with the glycerol, although depending on the chemical compounds present, some may precipitate out. One frequently touted option is to use potassium hydroxide as the reaction catalyst and phosphoric acid for neutralization so that the salt formed is potassium phosphate, which can be used for fertilizer. After acidulation and separation of the FFA, the methanol in the glycerol is removed by a vacuum flash process, or another type of evaporator. At this point, the glycerol should have a purity of ~85% and is typically sold to a glycerol refiner. The glycerol refining process takes the purity up to 99.5–99.7% using vacuum distillation or ion exchange processes.

Methanol that is removed from the methyl ester and glycerol streams will tend to collect any water that may have entered the process. This water should be removed in a distillation column before the methanol is returned to the process. This step is more difficult if an alcohol such as ethanol or isopropanol is used that forms an azeotrope with water. Then, a molecular sieve is used to remove the water.

Acid-Catalyzed Pretreatment

Special processes are required if the oil or fat contains significant amounts of FFA. Used cooking oils typically contain 2–7% FFA, and animal fats contain 5–30% FFA. Some very low-quality feedstocks, such as trap grease, can approach 100% FFA. When an alkali catalyst is added to these feedstocks, the FFA react with the catalyst to form soap and water as shown in the reaction below:

R-COOH	+	KOH	→	R-COOK	+	H_2O
Fatty acid		Potassium hydroxide		Potassium soap		Water

Up to ~5% FFA, the reaction can still be catalyzed with an alkali catalyst, but additional catalyst must be added to compensate for that lost to soap. The soap created during the reaction is either removed with the glycerol or washed out during the water wash. When the FFA level is >5%, the soap inhibits separation of the glycerol from the methyl esters and contributes to emulsion formation during the water wash. For these cases, an acid catalyst such as sulfuric acid can be used to esterify the FFA to methyl esters as shown in the following reaction:

R-COOH	+	CH_3OH	→	R-$COOCH_3$	+	H_2O
Fatty acid		Methanol		Methyl ester		Water

This process can be used as a pretreatment to convert the FFA to methyl esters, thereby reducing the FFA level (Fig. 4). Then, the low-FFA pretreated oil can be transesterified with an alkali catalyst to convert the triglycerides to methyl esters (54). As shown in the reaction, water is formed and, if it accumulates, it can stop the reaction well before completion. It was proposed (55) to allow the alcohol to separate from the pretreated oil or fat after the reaction. Removal of this alcohol also removes the water formed by the esterification reaction and allows for a second step of esterification; alternatively, one may proceed directly to alkali-catalyzed transesterification. Note that the methanol-water mixture will also contain some dissolved oil and FFA that should be recovered and reprocessed. Pretreatment with an acidic ion-exchange resin has also been described (56). It was shown (57,58) that acid-catalyzed esterification can be used to produce biodiesel from low-grade by-products of the oil refining industry such as soapstock. Soapstock, a mixture of water, soaps, and oil, is dried, saponified, and then esterified with methanol or some other simple alcohol using an inorganic acid as a catalyst. The procedure relies on a large excess of alcohol, and the

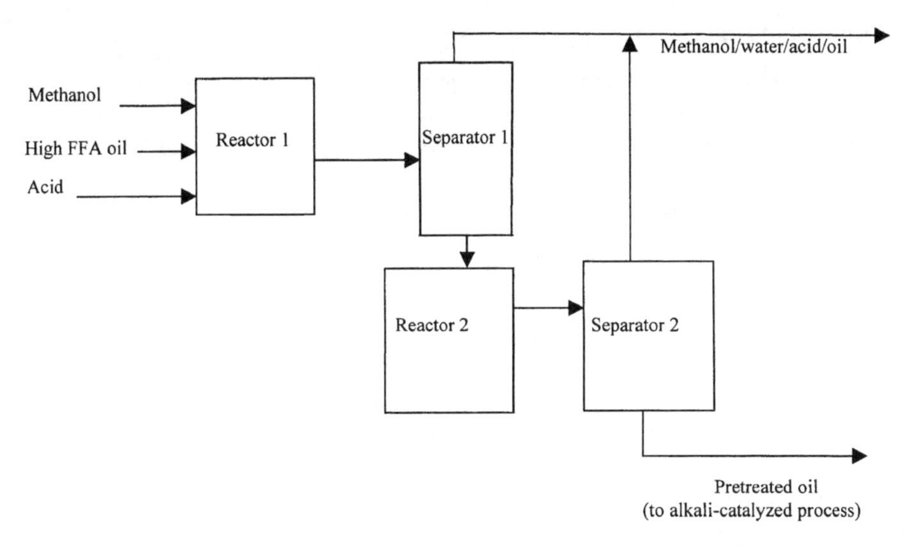

Fig. 4. Pretreatment process for feedstocks high in free fatty acids (FFA).

cost of recovering this alcohol determines the feasibility of the process. More information is given in the next section.

Fuel Quality

The primary criterion for biodiesel quality is adherence to the appropriate standard. Standards are listed in Appendix B. Generally, the fuel quality of biodiesel can be influenced by several factors, including the quality of the feedstock, the fatty acid composition of the parent vegetable oil or animal fat, the production process, the other materials used in this process, and postproduction parameters.

When specifications are met, the biodiesel can be used in most modern engines without modifications while maintaining the engine's durability and reliability. Even when used in low-level blends with petrodiesel fuel, biodiesel is expected to meet the standard before being blended. Although some properties in the standards, such as cetane number and density, reflect the properties of the chemical compounds that make up biodiesel, other properties provide indications of the quality of the production process. Generally, the parameters given in ASTM D6751 are defined by other ASTM standards and those in EN 14214 by other European or international (ISO) standards. However, other test methods, such as those developed by professional oleochemical organizations, such as the American Oil Chemists' Society (AOCS), may also be suitable (or even more appropriate because they were developed for fats and oils and not for petroleum-derived materials addressed in the ASTM standards). This discussion will focus on the most important issues for ensuring product quality for biodiesel as it relates to production as well as some postproduction parameters.

Production Process Factors

The most important issue during biodiesel production is the completeness of the transesterification reaction. The basic chemical process that occurs during the reaction is indicated in Figure 2 with the reaction proceeding stepwise from triacylglycerols to glycerol and alkyl esters with each step producing a fatty acid alkyl ester.

Even after a fully "complete" transesterification reaction, small amounts of tri-, di-, and monoacylglycerols will remain in the biodiesel product. The glycerol portion of the acylglycerols is summarily referred to as *bound glycerol*. When the bound glycerol is added to the free glycerol remaining in the product, the sum is known as the *total glycerol*. Limits for bound and total glycerol are usually included in biodiesel standards. For example, ASTM D6751 requires <0.24% total glycerol in the final biodiesel product as measured using a gas chromatographic (GC) method described in ASTM D 6584. Because the glycerol portion of the original oil is usually ~10.5%, this level of total glycerol corresponds to 97.7% reaction completion. Other methods can be used to measure total glycerol such as high-performance liquid chromatography (HPLC) (e.g., AOCS Recommended Practice Ca 14b-96: Quantification of Free Glycerine in Selected Glycerides and Fatty Acid Methyl Esters by HPLC with Laser Light-Scattering Detection) or a chemical procedure such as that described in AOCS Official Method Ca 14–56 (Total, Free and Combined Glycerol Iodometric Method). However, only the GC procedures are acceptable for demonstrating compliance with standards.

Free Glycerol. Glycerol is essentially insoluble in biodiesel so that almost all glycerol is easily removed by settling or centrifugation. Free glycerol may remain either as suspended droplets or as the very small amount that does dissolve in the biodiesel. Alcohols can act as cosolvents to increase the solubility of glycerol in the biodiesel. Most glycerol should be removed from the biodiesel product during the water washing process. Water-washed fuel is generally very low in free glycerol, especially if hot water is used for washing. Distilled biodiesel tends to have a greater problem with free glycerol due to glycerol carry-over during distillation. Fuel with excessive free glycerol will usually have a problem with glycerol settling out in storage tanks, creating a very viscous mixture that can plug fuel filters and cause combustion problems in the engine.

Residual Alcohol and Residual Catalyst. Because alcohols such as methanol and ethanol as well as the alkaline catalysts are more soluble in the polar glycerol phase, most will be removed when the glycerol is separated from the biodiesel. However, the biodiesel typically contains 2–4% methanol after the separation, which may constitute as much as 40% of the excess methanol from the reaction. Most processors will recover this methanol using a vacuum stripping process. Any methanol remaining after this stripping process should be removed by the water washing process. Therefore, the residual alcohol level in the biodiesel should be very low. A specific value for the allowable methanol level is specified in European biodiesel standards (0.2% in EN 14214), but is not included in the ASTM standard; however, the flash point specification in both standards limits the alcohol level. Tests showed that as little as 1% methanol in the biodiesel can lower its flashpoint from 170°C to <40°C. Therefore, by including a flashpoint specification of 130°C, the ASTM standard limits the amount of alcohol to a very low level (<0.1%). Residual alcohol left in biodiesel will generally be too small to have a negative effect on fuel performance. However, lowering the flashpoint presents a potential safety hazard because the fuel may have to be treated more like gasoline (which has a low flashpoint) than diesel fuel.

Most of the residual catalyst is removed with the glycerol. Like the alcohol, remaining catalyst in the biodiesel should be removed during the water washing. Although a value for residual catalyst is not included in the ASTM standard, it will be limited by the specification on sulfated ash. Excessive ash in the fuel can lead to engine deposits and high abrasive wear levels. The European standard EN 14214 places limits on calcium and magnesium as well as the alkali metals sodium and potassium.

Postproduction Factors

Water and Sediment. These two items are largely housekeeping issues for biodiesel. Water can be present in two forms, either as dissolved water or as suspended water droplets. Although biodiesel is generally insoluble in water, it actually takes up considerably more water than petrodiesel fuel. Biodiesel can contain as much as 1500 ppm of dissolved water, whereas diesel fuel usually takes up only ~50 ppm (59). The standards for diesel fuel (ASTM D 975) and biodiesel (ASTM D 6751) both limit the amount of water to 500 ppm. For petrodiesel fuel, this actually allows a small amount of suspended water. However, biodiesel must be kept dry. This is a challenge because many diesel storage tanks have water on the bottom due to condensation. Suspended water is a problem in fuel injection equipment because it contributes to the corrosion of the closely fitting parts in the fuel injection system. Water can also contribute to microbial growth in the fuel. This problem can occur in both biodiesel and petrodiesel fuel and can result in acidic fuel and sludge that will plug fuel filters.

Sediment may consist of suspended rust and dirt particles or it may originate from the fuel as insoluble compounds formed during fuel oxidation. Some biodiesel users have noted that switching from petrodiesel to biodiesel causes an increase in sediment originating from deposits on the walls of fuel tanks that had previously contained petrodiesel fuel. Because its solvent properties are different from those of petrodiesel fuel, biodiesel may loosen these sediments and cause fuel filter plugging during the transition period.

Storage Stability. Storage stability refers to the ability of the fuel to resist chemical changes during long-term storage; it is a major issue with biodiesel and is discussed at length in Chapter 6.4. Contact with air (oxidative stability) and water (hydrolytic stability) are the major factors affecting storage stability. Oxidation is usually accompanied by an increase in the acid value and viscosity of the fuel. Often these changes are accompanied by a darkening of the biodiesel color from yellow to brown and the development of a "paint" smell. In the presence of water, the esters can hydrolyze to long-chain FFA, which also cause the acid value to increase. The methods generally applied to petrodiesel fuels for assessing this issue, such as ASTM D 2274, were shown to be incompatible with biodiesel, and this remains an issue for research. Chapter 6.4 discusses some methods for assessing the oxidative stability of biodiesel that were or are being evaluated.

Antioxidant additives such as butylated hydroxytoluene and *t*-butylhydroquinone were found to enhance the storage stability of biodiesel. Biodiesel produced from soybean oil naturally contains some antioxidants (tocopherols, e.g., vitamin E), providing some protection against oxidation (some tocopherol is lost during refining of the oil before biodiesel production). Any fuel that is going to be stored for an extended period of time, whether it is petrodiesel or biodiesel, should be treated with an antioxidant additive.

Quality Control. All biodiesel production facilities should be equipped with a laboratory so that the quality of the final biodiesel product can be monitored. To monitor the completeness of the reaction according to the total glycerol level specified requires GC analysis as called for in biodiesel standards. Analytical methods, including GC and other procedures, are discussed in more detail in Chapter 5.

It is also important to monitor the quality of the feedstocks, which can often be limited to acid value and water contents, tests that are not too expensive. Another strategy used by many producers is to draw a sample of the oil (or alcohol) from each delivery and use that sample to produce biodiesel in the laboratory. This test can be fairly rapid (1–2 h) and can indicate whether serious problems may occur in the plant.

References

1. Schwab, A.W., M.O. Bagby, and B. Freedman, Preparation and Properties of Diesel Fuels from Vegetable Oils, *Fuel 66:* 1372–1378 (1987).
2. Bondioli, P., The Preparation of Fatty Acid Esters by Means of Catalytic Reactions, *Topics Catalysis 27:* 77–82 (2004).
3. Hoydonckx, H.E., D.E. De Vos, S.A. Chavan, and P.A. Jacobs, Esterification and Transesterification of Renewable Chemicals, *Topics Catalysis 27:* 83–96 (2004).
4. Demirbas, A., Biodiesel Fuels from Vegetable Oils via Catalytic and Non-Catalytic Supercritical Alcohol Transesterifications and Other Methods: A Survey, *Energy Convers. Manag. 44:* 2093–2109 (2003).
5. Shah, S., S. Sharma, and M.N. Gupta, Enzymatic Transesterification for Biodiesel Production, *Indian J. Biochem. Biophys. 40:* 392–399 (2003).
6. Haas, M.J., G.J. Piazza, and T.A. Foglia, Enzymatic Approaches to the Production of Biodiesel Fuels, in *Lipid Biotechnology*, edited by T.M. Kuo and H.W. Gardner, Marcel Dekker, New York, 2002, pp. 587–598.

7. Fukuda, H., A. Kondo, and H. Noda, Biodiesel Fuel Production by Transesterification of Oils, *J. Biosci. Bioeng. 92:* 405–416 (2001).
8. Ma, F., and M.A. Hanna, Biodiesel Production: A Review, *Bioresour. Technol. 70:* 1–15 (1999).
9. Schuchardt, U., R. Sercheli, and R.M. Vargas, Transesterification of Vegetable Oils: A Review, *J. Braz. Chem. Soc. 9:* 199–210 (1998).
10. Gutsche, B., Technologie der Methylesterherstellung—Anwendung für die Biodiesel-produktion (Technology of Methyl Ester Production and Its Application to Biofuels), *Fett/Lipid 99:* 418–427 (1997).
11. Freedman, B., and E.H. Pryde, Fatty Esters from Vegetable Oils for Use as a Diesel Fuel, in *Vegetable Oil Fuels, Proceedings of the International Conference on Plant and Vegetable Oils as Fuels*, Fargo, ND, 1982, ASAE Publication 4-82, pp. 117–122.
12. Freedman, B., E.H. Pryde, and T.L. Mounts, Variables Affecting the Yields of Fatty Esters from Transesterified Vegetable Oils, *J. Am. Oil Chem. Soc. 61:* 1638–1643 (1984).
13. Canakci, M., and J. Van Gerpen, Biodiesel Production via Acid Catalysis, *Trans. ASAE 42:* 1203–1210 (1999).
14. Feuge, R.O., and A.T. Gros, Modification of Vegetable Oils. VII. Alkali Catalyzed Interesterification of Peanut Oil with Ethanol, *J. Am. Oil Chem. Soc. 26:* 97–102 (1949).
15. Gauglitz, E.J., Jr., and L.W. Lehman, The Preparation of Alkyl Esters from Highly Unsaturated Triglycerides, *J. Am. Oil Chem. Soc. 40:* 197–198 (1963).
16. Lehman, L.W., and E.J. Gauglitz, Jr., The Preparation of Alkyl Esters from Highly Unsaturated Triglycerides. II, *J. Am. Oil Chem. Soc. 43:* 383–384 (1966).
17. Kurz, H., Zur katalytischen Umesterung fetter Oele durch alkoholische Kalilauge (The Catalytic Alcoholysis of Fatty Oils with Alcoholic Potash), *Fette Seifen 44:* 144–145 (1937).
18. Markolwitz, M., Consider Europe's Most Popular Catalyst, *Biodiesel Magazine 1:* 20–22 (2004).
19. Ma, F., L.D. Clements, and M.A. Hanna, The Effects of Catalyst, Free Fatty Acids, and Water on Transesterification of Beef Tallow, *Trans. ASAE 41:* 1261–1264 (1998).
20. Ma, F., L.D. Clements, and M.A. Hanna, The Effect of Mixing on Transesterification of Beef Tallow, *Bioresour. Technol. 69:* 289–293 (1999).
21. Ma, F., L.D. Clements, and M.A. Hanna, Biodiesel Fuel from Animal Fat. Ancillary Studies on Transesterification from Beef Tallow, *Ind. Eng. Chem. Res. 37:* 3768–3771 (1998).
22. Mittelbach, M., M. Wörgetter, J. Pernkopf, and H. Junek, Diesel Fuel Derived from Vegetable Oils: Preparation and Use of Rape Oil Methyl Ester, *Energy Agric. 2:* 369–384 (1983).
23. Ahn, E., M. Koncar, M. Mittelbach, and R. Marr, A Low-Waste Process for the Production of Biodiesel, *Sep. Sci. Technol. 30:* 2021–2033 (1995).
24. Isigigur, A., F. Karaosmanoôlu, and H.A. Aksoy, Methyl Ester from Safflower Seed Oil of Turkish Origin as a Biofuel for Diesel Engines, A*ppl. Biochem. Biotechnol. 45–46:* 103–122 (1994).
25. Dorado, M.P., E. Ballisteros, F.J. Lopez, and M. Mittelbach, Optimization of Alkali-Catalyzed Transesterification of *Brassica carinata* Oil for Biodiesel Production, *Energy Fuels 18:* 77–83 (2004).
26. Freedman, B., R.O. Butterfield, and E.H. Pryde, Transesterification Kinetics of Soybean Oil, *J. Am. Oil Chem. Soc. 63:* 1375–1380 (1986).

27. Mittelbach, M., and B. Trathnigg, Kinetics of Alkaline Catalyzed Methanolysis of Sunflower Oil, *J. Am. Oil Chem. Soc. 92:* 145–148 (1990).
28. Noureddini, H., and D. Zhu, Kinetics of Transesterification of Soybean Oil, *J. Am. Oil Chem. Soc. 74:* 1457–1463 (1997).
29. Boocock, D.G.B., S.K. Konar, V. Mao, and H. Sidi, Fast One-Phase Oil-Rich Processes for the Preparation of Vegetable Oil Methyl Esters, *Biomass Bioenergy 11:* 43–50 (1996).
30. Boocock, D.G.B., S.K. Konar, V. Mao, C. Lee, and S. Buligan, Fast Formation of High-Purity Methyl Esters from Vegetable Oils, *J. Am. Oil Chem. Soc. 75:* 1167–1172 (1998).
31. Komers, K., R. Stloukal, J. Machek, and F. Skopal, Biodiesel from Rapeseed Oil, Methanol and KOH 3. Analysis of Composition of Actual Reaction Mixture, *Eur. J. Lipid. Sci. Technol. 103:* 359–362 (2001).
32. Boocock, D.G.B., S.K. Konar, and H. Sidi, Phase Diagrams for Oil/Methanol/Ether Mixtures, *J. Am. Oil Chem. Soc. 73:* 247–1251 (1996).
33. Zhou, W., S.K. Konar, and D.G.B. Boocock, Ethyl Esters from the Single-Phase Base-Catalyzed Ethanolysis of Vegetable Oils, *J. Am. Oil Chem. Soc. 80:* 367–371 (2003).
34. Boocock, D.G.B., Single-Phase Process for Production of Fatty Acid Methyl Esters from Mixtures of Triglycerides and Fatty Acids, Canadian Patent 2,381,394 (2001).
35. Breccia, A., B. Esposito, G. Breccia Fratadocchi, and A. Fini, Reaction Between Methanol and Commercial Seed Oils Under Microwave Irradiation, *J. Microwave Power Electromagn. Energy 34:* 3–8 (1999).
36. Stavarache, C., M. Vinatoru, R. Nishimura, and Y. Maeda, Conversion of Vegetable Oil to Biodiesel Using Ultrasonic Irradiation, *Chem. Lett. 32:* 716–717 (2003).
37. Lifka, J., and B. Ondruschka, Einfluss des Stofftransportes auf die Herstellung von Biodiesel (Influence of Mass Transfer on the Production of Biodiesel), *Chem. Ing. Techn. 76:* 168–171 (2004).
38. Vicente, G., A. Coteron, M. Martinez, and J. Aracil, Application of the Factorial Design of Experiments and Response Methodology to Optimize Biodiesel Production, *Ind. Crops Prod. 8:* 29–35 (1998).
39. Noureddini, H., D. Harkey, and V. Medikonduru, A Continuous Process for the Conversion of Vegetable Oils into Methyl Esters of Fatty Acids, *J. Am. Oil Chem. Soc. 75:* 1775–1783 (1998).
40. Peterson, C.L., J.I. Cook, J.C. Thompson, and J.S. Taberski, Continuous Flow Biodiesel Production, *Appl. Eng. Agric. 18:* 5–11 (2002).
41. Cvengroš, J., and F. Povazanec, Production and Treatment of Rapeseed Oil Methyl Esters as Alternative Fuels for Diesel Engines, *Bioresour. Technol. 55:* 145–152 (1996).
42. Sercheli, R., R.M. Vargas, and U. Schuchardt, Alkylguanidine-Catalyzed Heterogeneous Transesterification of Soybean Oil, *J. Am. Oil Chem. Soc. 76:* 1207–1210 (1999).
43. Stidham, W.D., D.W. Seaman, and M.F. Danzer, Method for Preparing a Lower Alkyl Ester Product from Vegetable Oil, U.S. Patent 6,127,560 (2000).
44. Assman, G., G. Blasey, B. Gutsche, L. Jeromin, J. Rigal, R. Armengand, and B. Cormary, Continuous Progress for the Production of Lower Alkyl Esters, U.S. Patent 5,514,820 (1996).
45. Wimmer, T., Process for the Production of Fatty Acid Esters of Lower Alcohols, U.S. Patent 5,399,731 (1995).

46. Saka, S., and K. Dadan, Transesterification of Rapeseed Oils in Supercritical Methanol to Biodiesel Fuels, in *Proceedings of the 4th Biomass Conference of the Americas*, edited by R.P. Overend and E. Chornet, Oakland, CA, 1999.
47. Saka, S., and D. Kusdiana, Biodiesel Fuel from Rapeseed Oil as Prepared in Supercritical Methanol, *Fuel 80:* 225–231 (2001).
48. Kusdiana, D., and S. Saka, Kinetics of Transesterification in Rapeseed Oil to Biodiesel Fuel as Treated in Supercritical Methanol, *Fuel 80:* 693–698 (2001).
49. Dasari, M.A., M.J. Goff, and G.J. Suppes, Non-Catalytic Alcoholysis Kinetics of Soybean Oil, *J. Am. Oil Chem. Soc. 80:* 189–192 (2003).
50. Warabi, Y., D. Kusdiana, and S. Saka, Reactivity of Triglycerides and Fatty Acids of Rapeseed Oil in Supercritical Alcohols, *Bioresour. Technol. 91:* 283–287 (2004).
51. Diasakou, M., A. Louloudi, and N. Papayannakos, Kinetics of the Non-Catalytic Transesterification of Soybean Oil, *Fuel 77:* 1297–1302 (1998).
52. Kreutzer, U.R., Manufacture of Fatty Alcohols Based on Natural Fats and Oils, *J. Am. Oil Chem. Soc. 61:* 343–348 (1984).
53. Mittelbach, M., and M. Koncar, Method for the Preparation of Fatty Acid Alkyl Esters, U.S. Patent 5,849,939 (1998).
54. Keim, G.I., Treating Fats and Fatty Oils, U.S. Patent 2,383,601 (1945).
55. Kawahara, Y., and T. Ono, Process for Producing Lower Alcohol Esters of Fatty Acids, U.S. Patent 4,164,506 (1979).
56. Jeromin, L., E. Peukert, and G. Wollman, Process for the Pre-Esterification of Free Fatty Acids in Fats and Oils, U.S. Patent 4,698,186 (1987).
57. Haas, M.J., P.J. Michalski, S. Runyon, A. Nunez, and K.M. Scott, Production of FAME from Acid Oil, a By-Product of Vegetable Oil Refining, *J. Am. Oil Chem. Soc. 80:* 97–102 (2003).
58. Haas, M.J., S. Bloomer, and K. Scott, Process for the Production of Fatty Acid Alkyl Esters, U.S. Patent 6,399,800 (2002).
59. Van Gerpen, J.H., E.H. Hammond, L. Yu, and A. Monyem, Determining the Influence of Contaminants on Biodiesel Properties, SAE Technical Paper Series 971685, SAE, Warrendale, PA, 1997.
60. Weast, R.C., M.J. Astle, and W.H. Beyer, eds., *Handbook of Chemistry and Physics*, 66th edn., CRC Press, Boca Raton, FL, 1985–1986.

4.2

Alternate Feedstocks and Technologies for Biodiesel Production

Michael J. Haas and Thomas A. Foglia

Introduction

The purpose of this chapter is to examine alternate feedstocks and technologies for the production of biodiesel. To undertake such an analysis, we begin with a consideration of the status quo of biodiesel production. This will facilitate subsequent examination of the forces that drive the choice of alternate feedstocks and conversion technologies.

Biodiesel Production: The Status Quo

Lipid Reactant

Throughout the world, the typical lipid feedstocks for biodiesel production are refined vegetable oils. Within this group, the oil of choice varies with location according to availability; the most abundant lipid is generally the most common feedstock. The reasons for this are not only the desire to have an ample supply of product fuel, but also because of the inverse relation between supply and cost. Refined oils can be relatively expensive under the best of conditions, compared with petroleum products, and the choice of oil for biodiesel production depends on local availability and corresponding affordability. Thus, rapeseed and sunflower oils are used in the European Union (1), palm oil predominates in biodiesel production in tropical countries (2,3), and soybean oil (4) and animal fats is the major feedstocks in the United States. Fatty acid (FA) ester production was also demonstrated from a variety of other feedstocks, including the oils of coconut (5), rice bran (6,7), safflower (8), palm kernel (9), *Jatropha curcas* (10), Ethiopian mustard (*Brassica carinata*) (11), and the animal fats, tallow (12–14), and lard (15). Indeed, any animal or plant lipid should be a ready substrate for the production of biodiesel. Such factors as supply, cost, storage properties, and engine performance will determine whether a particular potential feedstock is actually adopted for commercial fuel production.

Governmental decisions can affect this choice of feedstock, in that a governmental subsidy program favoring one or the other feedstock could seriously affect feedstock choices. Thus, early support programs in the United States favored the use of first-use refined soybean oil as a feedstock. Conversely, although Brazil is the world's second largest producer of soybeans, an effort is being made by its government to foster a castor oil-based biodiesel industry because it is felt that adequate markets for soy oil exist, whereas the sale of castor oil into the biodiesel market would provide income to impoverished regions of the country where soy cannot be grown.

Alcohol Reactant

Methanol is the prevalent alcohol, globally, for the production of FA esters for use as biodiesel. Fatty acid methyl esters (FAME) are employed in the vast majority of laboratories, engine test stands, field tests, and field demonstrations conducted on biodiesel. The reason for this choice is that methanol is by far the least expensive of alcohols; in the United States it is half as expensive of ethanol, its nearest cost competitor. In some regions, most notably Brazil, the raw material and technology available allow the economical production of ethanol by fermentation, resulting in a product that is less expensive than methanol. In such areas, ethyl ester biodiesel is a potential product. Ethanol also was used in biodiesel production in test situations in the United States where it was available from the fermentation of starch-rich feed streams (16). However, a detailed economic analysis of this process has not been made, and it is unclear whether the operation was economically viable. The chemical technology described below for the use of methanol can be used for the production of ethyl FA esters, although there are anecdotal reports that the ethyl ester product can be more difficult to recover after purification by water washing.

The use of even longer-chain alcohols, either straight- or branched chain, in biodiesel production was described, and it was determined that the FA esters of these alcohols offer the advantage of exhibiting lower freezing points than methyl ester biodiesel (17,18). The esters produced included the isopropyl and isobutyl esters of tallow; the methyl esters of which are solid at ambient temperatures. The low-temperature properties of these new esters approached those of neat soy methyl esters and were comparable to soy esters at the 20% blend level in petrodiesel. This improvement of properties is desirable because it could facilitate the use of tallow-based fuels at lower temperatures without the danger of fuel solidification and engine failure. The matter of fuel solidification, however, may be more economically addressed with available commercial fuel additives (19). In addition, the higher prices of the longer-chain alcohols render biodiesel made from them impractical as a commercial fuel. Because it appears unlikely that methanol will be generally displaced as the preferred alcohol for biodiesel production, the use of alternate alcohols will not be discussed further here.

Chemical Technology

An attractive feature of the use of refined triacylglycerols as feedstocks, and another factor driving their selection as the predominant biodiesel feedstock, is the relative ease with which they are converted to simple alkyl esters (biodiesel) by chemical transesterification. Freedman *et al.* (20,21) published seminal articles characterizing this reaction, which is readily catalyzed under mild conditions by sodium hydroxide in alcohol, or by sodium methoxide (methylate) produced by the dissolution of metallic sodium in alcohol. Their optimal reaction conditions became the blueprint or at least the starting point for the bulk of contemporary biodiesel production technology. The batch reactions typically involve the use of a sixfold molar excess of alcohol over lipid, sodium hydroxide or sodium methylate as catalysts, reaction times of 2–4 h, reaction temperatures of 60–65°C, ambient pressure, and vigorous mixing to convert soy oil to methyl esters (21,22). Because water catalyzes the hydrolysis of FA esters, the substrate should in all cases be nearly anhydrous (<0.1–0.3% water). Under such conditions, the transesterification reaction is an equilibrium process, in which the yield of ester is only ~75% of theoretical. Typically, the glycerol layer, which contains unreacted alcohol and catalyst, is removed, fresh methanol and catalyst are added, and transesterification repeated. This two-step protocol typically gives high degrees of transesterification (>98%), with negligible amounts of remaining unreacted (complete or partial) acylglycerols. The final ester product separates readily from the polar liquid phase, which contains unreacted alcohol, the glycerol coproduct, and the catalyst. In a subsequent study (23), the alkali-catalyzed methanolysis of refined sunflower oil was optimized by the application of factorial design and response surface methodology. Temperature and catalyst concentration were positively correlated with ester yields. Optimum conditions for methyl ester production in a one-step reaction were identified, but occur only at high-catalyst concentrations. Batch reaction formats were originally used in the industry and remain in use. However, continuous reaction systems, which are easier and more economical to operate, were described (24–26) and are in operation, especially in facilities with annual capacities in the millions of gallons. To see contact information for firms that market biodiesel production equipment see nbb.org/resources/links/providers.shtm.

Potassium hydroxide can also be employed as the catalyst in transesterification. It is rarely used in the U.S. industry, but is reported to be more common than sodium hydroxide in Europe (27). The advantage of this catalyst is that the waste stream may have economic value as a soil fertilizer, due to its potassium content. The major disadvantage is the high cost of potassium hydroxide compared with that of sodium-based catalysts.

For smaller production facilities (6 million gallons of product annually), these metal hydroxide solutions in methanol are acceptable catalysts. Metal alkoxide solutions, such as sodium or potassium methylate, also catalyze the transesterification of lipid-linked FA. These catalysts cost more than the hydroxides, but offer advantages in terms of greater safety and convenience in handling, and a purer glycerol coproduct. These are reported to be the catalysts of choice in larger (>5 million gal/yr) European and some American production plants (27). In these methods, unreacted alcohol, residual acylglycerols, trace glycerol, and catalyst can be removed fairly easily from the crude product, resulting in a fuel able to meet the accepted quality specifications pertaining to the region in which it is produced (28,29). In all considerations of commercial biodiesel production, it is imperative that the product meet these specifications.

Biodiesel Production: Drivers for Change in Feedstock and Catalyst

Even using the least expensive refined oil as feedstock, biodiesel has a difficult or impossible time competing economically with petroleum-based diesel fuel. To our knowledge, all published calculations concluded that biodiesel produced from edible-grade vegetable oils is not economically competitive with petroleum-based fuel (30–32, Haas *et al.*, unpublished data). The primary reason is the relatively high cost of the lipid feedstock, which constitutes between 70 and 85% of the overall production costs when even the least expensive refined vegetable oils are used. This results in an overall production cost that exceeds the price of the petroleum fuel that the biodiesel is designed to replace. This price gap can be as large as fourfold when petroleum prices are low. Sentiment among commercial fleet operators and individual consumers in favor of renewable, domestically produced, low-pollution fuels is not generally strong enough to support the use of alternative fuels at these prices. In Europe, high tax rates on petroleum serve to lower the differential between fossil and renewable fuels and promote more biodiesel use. In the United States, the driving forces promoting biodiesel are mainly environmental and energy security concerns and resulting legislation and regulations (33). These concerns are rarely a sufficient motivator to stimulate widespread use. Legislative approaches, such as the institution of a payment to producers or the elimination of fuel taxes on biodiesel, were taken in some countries to induce renewable fuel usage.

Production-oriented approaches to improving the economics of biodiesel have included the investigation of lower-cost lipids as feedstocks. The composition of these alternate feedstocks, however, can require modification of existing technologies for their conversion into acceptable biodiesel fuels. In addition, a desire to reduce the waste streams of spent catalyst and other by-products resulting from the traditional alkali-catalyzed transesterification reaction has stimulated investigations into alternate means of conducting and catalyzing biodiesel synthesis. Acid-catalyzed alcoholysis of triglycerides to produce alkyl esters for production of biodiesel was examined, but higher reaction temperatures and longer reaction times are required to achieve satisfactory yields (34). In a recent study, a series of Bronsted acids was investigated for the conversion of soy oil to methyl esters at high temperature by conducting the methanolysis reactions in sealed vessels (35,36). Only sulfuric acid, however, was effective in producing high yields of methyl esters. This process, though effective on a laboratory scale, has not been adapted to the large-scale synthesis of biodiesel.

Alternate Feedstocks

In the context of biodiesel production, the consideration of feedstocks is most straightforward if they are grouped according to their degrees of purity, especially with regard to the level of free fatty acids (FFA). The latter are not converted to esters by alkaline-catalyzed transesterification, the conventional method for the production of FAME from triacylglycerols. Because of this, feedstocks containing significant levels of FFA require different processing to biodiesel than do refined oils and fats. It is vital that the FFA be esterified or removed because they can be detrimental to fuel systems and engines. All current biodiesel quality specifications impose strict limits on the levels of FFA allowed (28,29).

Refined vegetable oils (37) and high-quality animal fats (38) can be transesterified directly with both high chemical efficiency and good product yields. However, the efforts expended to obtain and maintain high purity cause them to have relatively high prices. Of the two, animal fats are typically less expensive than refined vegetable oils, because they are a by-product rather than a primary product of animal agriculture, and because demand is lower than for the more common vegetable oils. Animal fats also contain a higher content of saturated fatty acids (SFA) than do vegetable oils. These have relatively high melting points, a trait that may lead to precipitation and poor engine performance in cold weather (19). On the positive side, animal fat-derived biodiesel fuels, because of their higher saturated fatty ester content, generally have higher cetane values than vegetable oil-derived biodiesel (39). There are numerous grades of tallow (beef fat) (40,41), differentiated solely or largely on the basis of FFA content. Only the grades with lowest FFA levels are suitable for successful direct alkali-catalyzed transesterification as described here. Although their transesterification is accomplished by methods analogous to those employed for vegetable oils, some unique considerations necessary to obtain acceptably high degrees of reaction were identified (42,43).

Of potential concern for the use of animal fats, especially bovine lipids, in biofuel production is the possibility of exposure to prions, the infectious proteins responsible for bovine spongiform encephalopathy (BSE) (mad cow) disease in cattle and variant Creutzfeldt-Jacob (vCJD) disease in humans (44). The Scientific Steering Committee of the European Commission examined normal industrial tallow production processes and concluded that the resulting product is free of detectable BSE infectivity, even if the source material was highly infective (45). The U.S. Food and Drug Administration has ruled that tallow and other rendered fats are safe, and specifically omitted them from regulations prohibiting rendered products in feeds for cattle and other ruminants (46). The World Health Organization (WHO) examined the issue and concluded that because prions are proteinaceous, they would partition with the cellular residues of meat and bone, rather than the nonpolar lipid fraction during processing. The tallow fraction was therefore judged not a risk to human or animal health (47). Cummins *et al.* (48) assessed the danger of a human contracting vCJD due to the use of tallow as a fuel in diesel engines. These authors concluded that the risk was several orders of magnitude less than the rate of spontaneous appearance of CJD. Thus, scientific analysis indicates that processed (i.e., rendered) animal fat is not an agent of transmission of BSE.

No special considerations are required when using feedstocks with low FFA levels (≤0.5%) in biodiesel production. The production of oils and fats with such low FFA levels requires special attention to handling and processing, and the resulting material is sufficiently high in quality to be classified as edible grade. During alkaline transesterification, sodium (or potassium) salts of the FFA (soaps) will form, and these will be removed during subsequent purification of the product. The dissociation of the protonated FFA to form a free acid that can react with cations to form soap releases a proton, which will combine with hydroxyl ions in the solution to form water. However, because the FFA levels are low, the resulting reduction in available hydroxyl catalyst, and the accumulation of inhibitory water, are small and do not negatively affect reaction efficiency. The reduction of biodiesel yield due to the loss of FFA to the soap fraction is also insignificant at low FFA levels.

Unrefined vegetable oils from which the phospholipids ("lecithin," "gum") were removed also are acceptable, and can be 10–15% cheaper than highly refined oils. Non-degummed oils can be low in FFA, and on the basis of the discussion above, one would expect them to transesterify well. However, gums can complicate the washing of the crude biodiesel produced by transesterification, leading to increased expense. Thus, the application of degumming is essential with vegetable oil feedstocks, although bleaching and deodorization of the oil, two other common steps in producing edible oils, need not be conducted to produce an acceptable biodiesel feedstock (49). Animal fats do not contain sufficient amounts of phospholipids to require degumming.

There is generally a direct relationship between lipid quality, measured as the inverse of the FFA content, and cost. Thus, there are economic drivers for the choice of feedstocks with higher FFA levels. However, their conversion to biodiesel is less straightforward than in the case of low-FFA lipids. For lipids with FFA levels between ~0.5 and 4%, the loss of catalyst accompanying soap formation during alkali-catalyzed transesterification is sufficient to lower transesterification efficiencies if not compensated by the addition of make-up alkali at transesterification. The approach in such cases is to conduct a pretreatment with alkali to precipitate the FFA as their soaps before beginning transesterification. This increases overall alkali costs, but converts the FFA to a form that can be removed and sold in other commercial outlets. The FFA-depleted lipid is then subjected to alkaline transesterification as for low-FFA feedstocks. Feedstock lipids amenable to this approach are those that are off-specification due to elevated FFA levels, intermediate grades of animal fats [top white, all beef packer, extra fancy, fancy, bleachable according to the U.S. classification scheme (40), and no. 1 tallow by the British scheme (41)] and lightly used deep-fat fryer greases.

At feedstock FFA levels in excess of ~4%, the approach of removing these as their soaps becomes impractical due to excessive alkali consumption and the considerable loss of potential biodiesel product to the soap fraction. Among the lipids in this category are lower-quality animal fats, such as prime tallow, special tallow, "A" tallow, and poultry fat according to U.S. standards (40), and tallows no. 3–6 of the British classification scheme (41). Greases also fall into this feedstock category. In the United States, yellow (FFA ≤15%) and brown grease (FFA >15%) are available (40). British Standards identify one category, "grease," with a maximum FFA content of 20% (41). The greases are sometimes referred to as "recovered vegetable oil" and are typically partially hydrogenated vegetable oils disposed of after use in deep-fat frying. Their cost is one half to one third that of refined oils.

The strategy with these feedstocks is to convert both the FFA and the acylglycerol fractions to biodiesel. Typically, with these feedstocks, two types of reactions are conducted sequentially. The first is the esterification of the FFA to esters, followed by a conventional alkali-catalyzed transesterification step to produce simple alkyl esters from the acylglycerols. Alkali is a poor catalyst for FFA esterification, but mineral acids are efficient in this capacity. Therefore, multistep protocols involving acid-catalyzed esterification followed by alkali-catalyzed transesterification are employed (50). This multistep process is required because the exposure of feedstocks with a high FFA content to the alkaline conditions of the standard transesterification reaction causes the production of soaps. These emulsify and solubilize other lipophilic materials, increasing the difficulty of separating the biodiesel and glycerol streams (50). When greases are used as feedstocks and treated using the two-step sequential approach, it is estimated that the savings in feedstock cost can result in an overall cost reduction of 25–40% relative to the use of virgin soy oil (35). A recently described alternate approach employing only acid-catalyzed ester synthesis is allegedly more economical (25). However, this method is relatively new, and reports of its general use in biodiesel production have not yet appeared.

Due to its low cost, sulfuric acid is the typical catalyst used in the FFA esterification step of the two-step process (35,36). Water, a by-product of esterification, prevents quantitative ester synthesis. By conducting two sequential acid-catalyzed esterifications, with the removal of accumulated water after the first of these, acceptably high degrees of FA esterification can be achieved. Final FFA levels <0.5–1.0% are desired. It also should be mentioned that the oil or fat substrate is partially converted to partial glycerides and FAME during the esterification steps, which facilitates final conversion to FAME in the transesterification step. A one-step conversion of FFA-containing feedstocks to methyl esters for use as biodiesel was reported using a mixed calcium/barium acetate catalyst (51). The process, however, was carried out at temperatures of 200–220°C and pressures of 400–600 psi (2.76–4.14 MPa), and the ester product contained residual levels of soaps and monoacylglycerols. An alternative approach reported for utilizing FFA-containing feedstocks involves a preesterification with glycerol followed by alkali-catalyzed transesterification (52).

Specifications for the contents of greases are broader than those for edible oils, and there can be substantial variations from source to source. This can jeopardize the consistency of biodiesel production. Attention must be paid not only to the FFA level but also to the fatty acyl composition of the grease, especially the SFA content. Animal fats, hydrogenated vegetable oils, and frying fats that contain added chicken fat have higher levels of SFA than do most temperate climate vegetable oils. As discussed above, these FA can cause low-temperature feedstock and fuel performance problems. The use of waste greases and high FFA animal fats lags behind the use of vegetable oils in biodiesel production. Austria and Germany (53) are most active in this area, with some feeds of this type also being used in the United States, particularly by small volume producers

Other low-quality feedstocks that contain mixtures of FFA, acylglycerols, and other components are available. One of these is soapstock, a by-product of vegetable oil refining. Annual U.S. production exceeds 100 million pounds. Soapstock is a rich source of FA, consisting of ~12% acylglycerols, 10% FFA, and 8% phospholipids. It also contains nearly 50% water and is quite alkaline (pH typically >9). Due to the high pH and substantial content of polar lipids, the lipids and water in soapstock are thoroughly emulsified, forming a dense, stable, viscous mass that is solid at room temperature. Recovering any of the components of soapstock is not straightforward; as a consequence, the uses for this material are limited. In the not too distant past, it was disposed of in landfills. Presently the largest use is in animal feeds. Thus, there is interest in finding higher-value applications for soapstock.

Using an inorganic catalyst at pressures of at least 400 psi (2.76 MPa), the production from acidulated soapstock of a simple alkyl FA ester preparation said to be suitable for use as biodiesel was reported (51). To our knowledge, this technology has not been adopted by industry, perhaps because of the requirement for high-pressure processing equipment.

We also investigated the production of simple FA alkyl esters for use as biodiesel from soapstock (54). The approach adopted was first to exploit the already alkaline pH of soapstock to facilitate the complete hydrolysis of all FA ester bonds in soapstock. This was readily achieved by further alkalinization of the material with sodium hydroxide, followed by a 2-h reaction at 95°C. As a result, all acylglycerol and phosphoacylglycerol entities were hydrolyzed. The then necessary removal of the water initially present in the soapstock, which inhibits the esterification of FFA, was achieved by evaporation, with subsequent ready conversion of the FFA to methyl esters by sulfuric acid-catalyzed esterification in the presence of methanol. The resulting ester product met the ASTM Provisional Specifications for Biodiesel that were in effect at that time, and gave emissions and performance in a heavy-duty diesel engine comparable to those of biodiesel produced from soybean oil (55).

One undesirable side effect of this approach to biodiesel production from soapstock, however, was that the sodium added at the saponification step combined with the sulfate added as sulfuric acid during the subsequent esterification reaction, producing substantial amounts of solid sodium sulfate. This precipitated from solution during the esterification reaction. The necessity and cost of disposing of this solid waste constituted a significant disadvantage of the process. To overcome this difficulty, an alternate approach was taken in the removal of water from the saponified soapstock preparation (56). The soapstock processing industry routinely applies a technology known as acidulation. In this process, steam and sulfuric acid are introduced into the soapstock *via* a sparger. The acid protonates the FA soaps, converting them to FFA, which greatly reduces their emulsifying capabilities. When sparging is discontinued, two phases separate in the reactor. The upper phase, known as "acid oil," is rich in lipids and the lower phase consists of the aqueous components of the soapstock. When this technology is applied to saponified soapstock, the acid oil contains >90% FFA and can be subjected to acid-catalyzed esterification as described above for biodiesel production. Sodium sulfate is still formed in this process by the interaction of sodium present in the saponified soapstock with sulfuric acid added during acidulation. However, this is soluble in the aqueous phase generated during acidulation and is discarded along with that phase as liquid waste, which is more readily disposed of than the solid waste produced in the earlier version of this method. This relatively new technology has yet to be adopted on an industrial scale.

Cheaper, and correspondingly more heterogenous, potential feedstocks for biodiesel production can be identified. An example of these is trap grease, the low cost of which suggests that it be considered for biodiesel production. However, there has been only limited use of this material as a feedback to date, with the color and odor of the resulting biodiesel presenting the largest barriers.

Alternate Technologies for FA Ester Synthesis

Alkali-catalyzed transesterification of acylglycerol feedstocks, with the addition of an acid-catalyzed reaction to esterify FFA if they are present, comprises the predominant technologies presently in use for industrial-scale biodiesel production. However, the desire to reduce catalyst costs, waste output, or the need for extensive purification of the product has stimulated the investigation of alternate methods of FA ester synthesis. These methods, described here, are largely in the developmental stage, with little or no actual application in the biodiesel industry to date.

Alkali-Catalyzed Monophasic Transesterification. One feature of conventional alkali-catalyzed transesterification of acylglycerols that reduces the observed rate of transesterification is the fact that the oil substrate is not miscible with the alcohol-catalyst phase. Reaction occurs at the interface between the two phases, resulting in a much lower rate than if the reaction mixture was a single phase. In what has been termed "solvent-assisted methanolysis," the components and ratios of the reaction mixture are altered in order to overcome this limitation (57). Transesterification is conducted in a medium containing oil, methanol, alkali, and an organic solvent such as tetrahydrofuran (THF). In addition to the use of solvent to promote the miscibility of methanol and oil, a high-methanol:oil molar ratio (27:1) is employed, raising the polarity of the medium sufficiently to allow a one-phase system, thereby increasing the transesterification rate. The advantages of this approach are the use of a one-step transesterification process, methyl ester yields >98%, reaction times of <10 min, and lower reaction temperatures. The disadvantages are the necessity of recovering the THF and the large molar excess of unreacted methanol, and the inherent hazards associated with flammable solvents. Nonetheless, adoption of this technology for commercial biodiesel production was reported recently (58). Another nontraditional approach to facilitate the transesterification of intact oils involves conducting the reaction in supercritical methanol (59,60). Although reported conversions are high, it remains to be seen whether this approach can be economically viable.

Enzymatic Conversion of Oils and Fats to Alkyl Esters. Although biodiesel is at present successfully produced chemically, there are several associated problems that impede its continued growth, such as glycerol recovery and the need to use refined oils and fats as primary feedstocks (49). The disadvantages of using chemical catalysts can be overcome by using lipases as the catalysts for ester synthesis (61). Advantages cited for lipase catalysis over chemical methods in the production of simple alkyl esters include: the ability to esterify both acylglycerol-linked and FFA in one step; production of a glycerol side stream with minimal water content and little or no inorganic material; and catalyst reuse. Bottlenecks to the use of enzymatic catalysts include the high cost of lipases compared with inorganic catalysts (in the absence of effective schemes for multiple enzyme use), inactivation of the lipase by contaminants in the feedstocks, and inactivation by polar short-chain alcohols.

Early work on the application of enzymes for biodiesel synthesis was conducted using sunflower oil as the feedstock (62) and various lipases to perform alcoholysis reactions in petroleum ether. Of the lipases tested, only three were found to catalyze alcoholysis, with an immobilized lipase preparation of a *Pseudomonas* sp. giving the best ester yields. Maximum conversion (99%) was obtained with ethanol. When the reaction was repeated without solvent, only 3% product was produced with methanol as alcohol, whereas with absolute ethanol, 96% ethanol, and 1-butanol, conversions ranged between 70 and 82%. Reactions with a series of homologous alcohols showed that reaction rates, with or without the addition of water, increased with increasing chain length of the alcohol. For methanol, the highest conversion was obtained without the addition of water, but for other alcohols, the addition of water increased the esterification rate two to five times.

In a subsequent study, Linko *et al.* (63) reported the lipase-catalyzed alcoholysis of low-erucic acid rapeseed oil without organic solvent in a stirred batch reactor. The best results were obtained with a *Candida rugosa* lipase and, under optimal conditions, a nearly complete conversion to ester was obtained. Other studies (64) reported the ethanolysis of sunflower oil with Lipozyme™ (a commercial immobilized *Rhizomucor meihei* lipase) in a medium totally composed of sunflower oil and ethanol. Conditions studied for the conversion of the oil to esters included substrate molar ratio, reaction temperature and time, and enzyme load. Ethyl ester yields, however, did not exceed 85% even under the optimized reaction conditions. The addition of water (10 wt%), in addition to that associated with the immobilized enzyme, decreased ester yields significantly. The affect of added water in this instance is to be contrasted to the result obtained for reactions in organic solvent. These authors also reported that ester yields could be improved by adding silica to the medium. The positive effect of silica on yield was attributed to the adsorption of the polar glycerol coproduct onto the silica, which reduced glycerol deactivation of the enzyme. Enzyme reuse also was investigated, but ester yields decreased significantly with enzyme recycle, even in the presence of added silica.

In other studies (65,66), mixtures of soybean and rapeseed oils were treated with various immobilized lipase preparations in the presence of methanol. The lipase from *C. antarctica* was the most effective in promoting methyl ester formation. To achieve high levels of conversion to methyl ester, it was necessary to add three equivalents of methanol. Because this level of methanol resulted in lipase deactivation, it was necessary to add the methanol in three separate additions. Under these conditions, >97% conversion of oil to methyl ester was achieved. It was also reported that merely allowing the reaction mixture to stand separated the methyl ester and glycerol layers. In another study (67), it was reported that the lipase of *Rhizopus oryzae* catalyzed the methanolysis of soybean oil in the presence of 4–30% water in the starting materials but was inactive in the absence of water. Methyl ester yields of >90% could be obtained with step-wise additions of methanol to the reaction mixture. Recently, the conversion of soy oil to biodiesel in a continuous batch operation catalyzed by an immobilized lipase of *Thermomyces lanuginose* was reported (68). These authors also used a step-wise addition of methanol to the reaction and in this manner obtained complete conversion to ester. Repeated reuse of the lipase was made possible by removing the bound glycerol by washing with isopropanol. When crude soy oil was used as substrate, a much lower yield of methyl ester was obtained compared with that using refined oil (69). The decrease in ester yields was directly related to the phospholipid content of the oil, which apparently deactivated the lipase. Efficient esterification activity could be attained by preimmersion of the lipase in the crude oil before methanolysis.

Several commercially available lipases were screened for their abilities to transesterify tallow with short-chain alcohols (70). An immobilized lipase from *R. miehei* was the most effective in converting tallow to its corresponding methyl esters, resulting in >95% conversion. The efficiencies of esterification with methanol and ethanol were sensitive to the water content of the reaction mixtures, with water reducing ester yields. *n*-Propyl, *n*-butyl, and isobutyl esters also were prepared at high conversion efficiencies (94–100%). Minor amounts of water did not affect ester production in these instances.

In the transesterification of tallow with secondary alcohols, the lipases from *C. antarctica* (trade name SP435) and *Pseudomonas cepacia* (PS30) gave the best conversions to esters (70). Reactions run without the addition of water were sluggish for both lipases, and conversions of only 60–84% were obtained overnight (16 h). The addition of small amounts of water improved the yields. The opposite effect was observed in the case of methanolysis, which was extremely sensitive to the presence of water. For the branched-chain alcohols, isopropanol and 2-butanol better conversions were obtained when the reactions were run without solvent (71). Reduced yields when using the normal alcohols methanol and ethanol, in solvent-free reactions were attributed to enzyme deactivation by these more polar alcohols. Similar conversions also could be obtained for both the methanolysis and isopropanolysis of soybean and rapeseed oils (71). Engine performance and low-temperature properties of ethyl and isopropyl esters of tallow were comparable to those of the methyl esters of tallow and soy oil (72). The enzymatic conversion of lard to methyl and ethyl esters was reported (15) using a three-step addition of the alcohol to the substrate in solvent-free medium as described (73). The conversion of Nigerian palm oil and the lauric oils, palm kernel and coconut, to simple alkyl esters for use as biodiesel fuels was also reported (74). The best conversions (~85%) were to ethyl esters, and the authors reported several properties of these esters for use as biodiesel fuels.

Enzymatic Conversion of Greases to Biodiesel. Research on low-temperature properties and diesel engine performance of selected mono-alkyl esters derived from tallow and spent restaurant grease strongly suggested that ethyl esters of grease (ethyl greasate) might be an excellent source of biodiesel (72). Ethyl esters of grease have low-temperature properties, including cloud point, pour point, cold filter plugging point, and low-temperature flow test, that closely resemble those of methyl soyate, the predominant form of biodiesel currently marketed in the United States. Diesel engine performance and emissions data were obtained for 20% blends of ethyl greasate or isopropyl tallowate in No. 2 diesel fuel in a matched dual-cylinder diesel engine. Data from the test runs indicated adequate performance, reduced fuel consumption, and no apparent difference in carbon build-up characteristics, or in CO, CO_2, O_2, and NO_x emissions compared to No. 2 diesel (72).

The biodiesel used in these tests was synthesized enzymatically. Low-value lipids, such as waste deep fat fryer grease, usually have relatively high levels of FFA (≥8%). Lipases are of particular interest as catalysts for the production of fatty esters from such feedstocks because they accept both free and glyceride-linked FA as substrates for ester synthesis. In contrast, biodiesel production from such mixed feedstocks (e.g., spent rapeseed oil) using inorganic catalysts requires multistep processing (53). To exploit these attractive features of lipase catalysis, studies were conducted using a lipase from *P. cepacia* and recycled restaurant grease with 95% ethanol in batch reactions (66). Subsequent work (15) showed that methyl and ethyl esters of lard could be obtained by lipase-catalyzed alcoholysis. The methanolysis and ethanolysis of restaurant greases using a series of immobilized lipases from *T. lanuginosa*, *C. antarctica*, and *P. cepacia* in solvent-free media employing a one-step addition of alcohol to the reaction system werre reported (75). The continuous production of ethyl esters of grease using a phyllosilicate sol-gel immobilized lipase from *Burkholderia* (formerly *Pseudomonas*) *cepacia* (IM BS-30) as catalyst was investigated (76). Enzymatic transesterification was carried out in a recirculating packed column reactor using IM BS-30 as the stationary phase and ethanol and restaurant grease as the substrates, without solvent addition. The bioreactor was operated at various temperatures (40–60°C), flow rates (5–50 mL/min), and times (8–48 h) to optimize ester production. Under optimum operating conditions (flow rate, 30 mL/min; temperature, 50°C; mole ratio of substrates, 4:1, ethanol:grease; reaction time, 48 h) the ester yields were >96%.

Other low-cost feedstocks for biodiesel production include the residual oil present in spent bleaching earths and in soapstocks produced during the refining of crude vegetable oils. These contain ~40 and 50% oil by weight, respectively. Residual oil present in the waste bleaching earths from soy, rapeseed, and palm oil refining was extracted with hexane, recovered, and the oils subjected to methanolysis by *R. oryzae* lipase in the presence of a high water content, with a single addition of methanol (77). The highest conversion to methyl esters was 55% with palm oil after 96-h reaction. Adverse viscosity conditions were cited as a possible cause for the low conversions, but inactivation of the lipase by residual phospholipids in the recovered oil, as reported for unrefined oils (69), may also have caused the low yields from soybean and rapeseed oils. The use of enzymes immobilized on solid supports as biocatalysts for the production of simple FA esters of the FFA and acyl lipids present in soapstock was studied (54). However, only low degrees of ester production were achieved. This was likely because the soapstock/lipase/alcohol mixture, was nearly solid during the reaction, resulting in poor mixing between catalyst and substrates.

A two-step enzymatic approach for the conversion of acid oils, a mixture of FFA and partial glycerides obtained after acidulation of soapstock, to fatty esters was used (78). In the first step, the acyl lipids in the acid oil were hydrolyzed completely using *C. cylindracea* lipase. In the second step, the high-acid oils were esterified to short- and long-chain esters using an immobilized *Mucor* (now *Rhizomucor*) *miehei* lipase.

Heterogeneous Catalysts. As noted previously, commonly used methods of producing biodiesel from refined oils and fats rely on the use of soluble metal hydroxide or methoxide catalysts. Removal of these catalysts from the glycerol/alcohol phase is technically difficult; it imparts extra cost to the final product and complicates glycerol purification. With these homogeneous catalysts, high conversions are easy to achieve at temperatures from 40 to 65°C within a few hours of reaction. Higher temperatures are typically not used, to avoid system pressures greater than atmospheric, which would require the use of pressure vessels.

It is possible to conduct transesterification in the absence of added catalyst, an approach that requires high pressures (20 mPa) and temperatures (350°C). This approach is used in some production plants, especially in Europe, but is not widely practiced because of the high pressures necessary to increase ester yields to acceptable levels. In some of these cases, it was determined that esterifications that had been considered to be noncatalyzed were actually catalyzed by the metal surfaces of the reactor (79). Such insoluble catalyst systems are termed "heterogeneous." Compared with the typical homogeneous catalyst reactions, these offer the advantages of greatly simplified product cleanup and a reduction in waste material requiring disposal.

Other research on the alcoholysis of triacylglycerols with heterogeneous catalysts was described. For example, zinc oxide supported on aluminum was employed as catalyst for the alcoholysis of oils and fats with a series of alcohols higher than methanol (80). Another patented procedure (51) used a binary mixture of calcium and barium acetates to catalyze the methanolysis of degummed soy oil, yellow grease, turkey fat, and mixtures of partial acylglycerols oil at 200°C. The application of this technology to the transesterification of soybean soapstock is noted above. The alcoholysis of triacylglycerols with glycerol using basic solid catalysts such as Cs-MCM-41, Cs-sepiolite, and hydrotalcites was evaluated (81). The reaction was carried out at 240°C for 5 h. Hydrotalcite gave a good conversion of 92% followed by Cs-sepiolite (45%) and Cs-MCM-41 (26%). The alcoholysis of rapeseed oil in the presence of Cs-exchanged NaX faujasites and commercial hydrotalcite (KW2200) catalysts was studied (82). At a high methanol to oil ratio of 275 and 22 h of reaction at methanol reflux, the cesium exchanged NaX faujasites gave a conversion of 70%, whereas a 34% conversion was obtained using hydrotalcite. The use of ETS-4 and ETS-10 catalysts to provide conversions of 85.7 and 52.6%, respectively, at 220°C and 1.5 h reaction time was patented (83). Ethyl ester production efficiencies of 78% at 240°C and >95% at 260°C with 18-min reactions were achieved using calcium carbonate rock as the catalyst (84). All these studies required temperatures in excess of 200°C to achieve >90% conversion within the time scales of the experiments. The same group (85) recently reported the use of zeolites, modified by ion exchange by alkali cations or by decomposition of occluded alkali metal salt followed by calcinations at 500°C, as the solid bases. The zeolites faujasite NaX and titanosilicate structure-10 (ETS-10), after modification in this manner, were used for the alcoholysis of soybean oil with methanol. With the faujasite catalyst, yields >90% were reported at 150°C within 24 h, whereas with the calcined ETS-10 catalyst gave an ester yield of 94% at 100°C in 3 h.

In Situ *Transesterification.* Rather than working with isolated, refined oils, an alternate approach is to conduct the transesterification of the oil present in intact oil seeds. This method may serve essentially to reduce substrate costs in biodiesel production. Using sulfuric acid as catalyst, the *in situ* transesterification of homogenized whole sunflower seeds with methanol was explored (86,87); ester yields up to 20% greater than that obtained for extracted oil were reported, which was attributed to the transesterification of the seed hull lipids. In parallel studies, a range of acid and methanol levels in the *in situ* transesterification of homogenized sunflower seeds was studied (88), and ester yields as high as 98% of theoretical based on the oil content of the seeds were reported. The *in situ* acid-catalyzed transesterification of rice bran oil was studied, and it was found that with ethanol 90% of the oil was converted to ester but the product contained high levels of FFA (89,90). These authors also applied this technique to the transesterification of the oil in ground soybeans (91). *In situ* acid-catalyzed methanolysis, however, resulted in only 20–40% of the oil being removed from the seeds.

Recently, the production of simple alkyl FA esters by direct alkali-catalyzed *in situ* transesterification of commercially produced soy flakes by mild agitation of the flakes with alcoholic sodium hydroxide at 60°C was reported (92). Methyl, ethyl, and isopropyl esters were produced. Statistical methods and response surface regression analysis were used to optimize reaction conditions using methanol as the alcohol. At 60°C, the highest yield of methyl esters was predicted at a molar ratio of methanol/acylglycerol/NaOH of 226:1:1.6 at 8 h reaction. At 23°C, maximal methanolysis was predicted at molar ratio of 543:1:2.0. Of the lipid in soy flakes, 95% was removed and transesterified under such conditions [this recovery value is greater than that stated in the original publication, the result of a faulty original value for the lipid content of the substrate (Haas, unpublished data)]. The methyl ester fraction contained minor amounts of FA (<1%) and no acylglycerols. Of the glycerol released by transesterification, >90% was in the alcohol ester phase, with the remainder located in the treated flakes. In recent work, it was shown that by drying the flakes prior to transesterification, the methanol and sodium hydroxide requirements can be reduced by 55–60%, greatly improving the economics of the process (93).

References

1. Harold, S., Industrial Vegetable Oils: Opportunities Within the European Biodiesel and Lubricant Markets. Part 2. Market Characteristics, *Lipid Technol. 10:* 67–70 (1997).
2. Sii, H.S., H. Masjuki, and A.M. Zaki, Dynamometer Evaluation and Engine Wear Characteristics of Palm Oil Diesel Emulsions, *J. Am. Oil Chem. Soc. 72:* 905–909 (1995).
3. Masjuki, H.H., and S.M. Sapuan. Palm Oil Methyl Esters as Lubricant Additives in Small Diesel Engines, *J. Am. Oil. Chem. Soc. 72:* 609–612 (1995).
4. Jewett, B., Biodiesel Powers Up, *inform 14:* 528–530 (2003).
5. Solly, R.K., Coconut Oil and Coconut Oil-Ethanol Derivatives as Fuel for Diesel Engines, *J. Fiji Agric. 42:* 1–6 (1980).
6. Kamini, N.R., and H. Iefuji, Lipase Catalyzed Methanolysis of Vegetable Oils in Aqueous Medium by *Cryptococcus* spp. S-2, *Process. Biochem. 37:* 405–410 (2001).
7. Özgül-Yücel, S., and S. Türkay, FA Monoalkylesters from Rice Bran Oil by *in situ* Esterification, *J. Am. Oil Chem. Soc. 80:* 81–84 (2003).
8. Isigigür, A., F. Karaosmanoğlu, and H.A. Aksoy, Methyl Ester from Safflower Seed Oil of Turkish Origin as a Biofuel for Diesel Engines, *Appl. Biochem. and Biotechnol. 45–46:* 103–112 (1994).
9. Choo, Y.M., K.Y. Cheah, A.N. Ma, and A. Halim, Conversion of Crude Palm Kernel Oil into Its Methyl Esters on a Pilot Plant Scale, in *Proceedings, World Conference on Oleochemicals in the 21st Century*, edited by T.H. Applewhite, AOCS Press, Champaign, IL, 1991, pp. 292–295.

10. Foidl, N., G. Foidl, M. Sanchez, M. Mittelbach, and S. Hackel, *Jatropha curcas* L. as a Source for the Production of Biofuel in Nicaragua, *Bioresource Technol. 58:* 77–82 (1996).
11. Cardone, M., M. Mazzoncini, S. Menini, V. Rocco, A. Senatore, M. Seggiani, and S. Vitolo, *Brassica carinata* as an Alternative Crop for the Production of Biodiesel in Italy: Agronomic Evaluation, Fuel Production by Transesterification and Characterization, *Biomass Bioenergy 25:* 623–636 (2003).
12. Geise, R., Biodiesel's Bright Future Deserves Equality, *Render Mag. 31:* 16–17 (2002).
13. Nautusch, D.F.S., D.W. Richardson, and R.J. Joyce, Methyl Esters of Tallow as a Diesel Extender, in *Proceedings: VI International Symposium on Alcohol Fuels Technology Conference*, May 21–25, 1984, pp. 340–346.
14. Richardson, D.W., R.J. Joyce, T.A. Lister, and D.F.S. Natusch, Methyl Esters of Tallow as a Diesel Component, in *Proceedings 3rd International Conference on Energy from Biomass*, edited by W. Pulz, J. Coombs, and D.O. Hall, Elsevier, New York, 1985, pp. 735–743.
15. Lee, K.-T., T.A. Foglia, and K.-S. Chang, Production of Alkyl Esters as Biodiesel Fuel from Fractionated Lard and Restaurant Grease, *J. Am. Oil Chem. Soc. 79:* 191–195 (2002).
16. Lowe, G.A., C.L. Peterson, J.C. Thompson, J.S. Taberski, P.T. Mann, and C.L. Chase, Producing HySEE Biodiesel from Used French Fry Oil and Ethanol for an Over-the-Road Truck, *ASAE Paper No. 98–6081* (1998).
17. Lee, I., L.A. Johnson, and E.G. Hammond, Use of Branched-Chain Esters to Reduce the Crystallization Temperature of Biodiesel, *J. Am. Oil Chem. Soc. 72:* 1155–1160 (1995).
18. Foglia, T.A., L.L. Nelson, R.O. Dunn, and W.N. Marmer, Low-Temperature Properties of Alkyl Esters of Tallow and Grease, *J. Am. Oil Chem. Soc. 74:* 951–955 (1997).
19. Dunn, R.O., M.W. Shockley, and M.O. Bagby, Improving the Low-Temperature Properties of Alternative Diesel Fuels: Vegetable Oil-Derived Methyl Esters, *J. Am. Oil Chem. Soc. 73:* 1719– 1728 (1996).
20. Freedman, B., E.H. Pryde, and T.L. Mounts, Vairables Affecting the Yields of Fatty Esters from Transesterified Vegetable Oils, *J. Am. Oil Chem. Soc. 61:* 1638–1643 (1984).
21. Freedman, B., R.O. Butterfield, and E. Pryde, Transesterification Kinetics of Soybean Oil, *J. Am. Oil Chem. Soc. 63:* 1375–1380 (1986).
22. Noureddini, H. and D. Zhu, Kinetics of Transesterification of Soybean Oil, *J. Am. Oil Chem. Soc. 74:* 1457–1463 (1997).
23. Vicente, G., A. Coteron, M. Martinez, and J. Aracil, Application of Factorial Design of Experiments and Response Surface Methodology to Optimize Biodiesel Production, *Ind. Crops Prod. 8:* 29–35 (1998).
24. Peterson, C.L., J.L. Cook, J.C. Thompson, and J.S. Taberski, Continuous Flow Biodiesel Production, *Appl. Eng. Agric. 18:* 5–11 (2002).
25. Zhang, Y., M.A. Dube, D.D. McLean, and M. Kates, Biodiesel Production from Waste Cooking Oil: 2. Economic Assessment and Sensitivity Analysis, *Bioresource Technol. 90:* 229–240 (2003).
26. Harvey, A.P., M.R. Mackley, and T. Seliger, Process Intensification of Biodiesel Production Using a Continuous Oscillatory Flow Reactor, *J. Chem. Technol. Biotechnol. 78:* 338–341 (2003).
27. Markolwitz, M., Consider Europe's Most Popular Catalyst, *Biodiesel Mag. 1:* 20–22 (2004).

28. Anonymous, Standard Specification for Biodiesel Fuel (B100) Blend Stock for Distillate Fuels, Designation D 6751–02, American Society for Testing and Materials, West Conshohocken, PA, 2002.
29. Anonymous, European Biodiesel Standard DIN EN 14214, Beuth-Verlag, Berlin, 2003 [www.beuth.de].
30. Bender, M., Economic Feasibility Review for Community-Scale Farmer Cooperatives for Biodiesel, *Bioresour. Technol. 70:* 81–87 (1999).
31. Noordam, M., and R.V. Withers, Producing Biodiesel from Canola in the Inland Northwest: An Economic Feasibility Study, Idaho Agricultural Experiment Station, Bulletin No. 785, University of Idaho College of Agriculture, Moscow, Idaho, 1996.
32. Reining, R.C., and W.E. Tyner, Comparing Liquid Fuel Costs: Grain Alcohol Versus Sunflower Oil, *Am. J. Agric. Econ. 65:* 567–570 (1983).
33. Piazza, G.J., and T.A. Foglia, Rapeseed Oil for Oleochemical Usage, *Eur. J. Lipid Sci. Technol. 103:* 450–454 (2001).
34. Schwab, A.W., M.O. Bagby, and B. Freedman, Preparation and Properties of Diesel Fuels from Vegetable Oils, *Fuel 66:* 1372–1378 (1987).
35. Canakci, M., and J. Van Gerpen, Biodiesel Production *via* Acid Catalysis, *Trans. ASAE 42:* 1203–1210 (1999).
36. Goff, M.J., N.S. Bauer, S. Lopes, W.R. Sutterlin, and G.J. Suppes, Acid-Catalyzed Alcoholysis of Soybean Oil, *J. Am. Oil Chem. Soc. 81:* 415–420 (2004).
37. Ali, Y., and M.A. Hanna, Alternative Diesel Fuels from Vegetable Oils, *Bioresour. Technol. 50:* 153–163 (1994).
38. Ali, Y., and M.H. Hanna, Physical Properties of Tallow Ester and Diesel Fuel Blends, *Bioresour. Technol. 47:* 131–134 (1994).
39. Knothe, G., A.C. Matheaus, and T.W. Ryan, Cetane Numbers of Branched and Straight-Chain Fatty Esters Determined by an Ignition Quality Tester, *Fuel 82:* 971–975 (2003).
40. Anonymous, Specifications for Commercial Grades of Tallows, Animal Fats and Greases, American Fats & Oils Association, Columbia, SC, 2004 [www.afoaonline.org].
41. Anonymous, Specification for Technical Tallow and Animal Grease, *British Standard 3919,* BSI British Standards, London, 1987 [www.bsi-global.com].
42. Ma, F., L.D. Clements, and M.A. Hanna, The Effects of Catalysts, Free Fatty Acids, and Water on Transesterification of Beef Tallow, *Trans. ASAE 41:* 1261–1264 (1998).
43. Ma, F., L.D. Clements, and M.A. Hanna, The Effect of Mixing on Transesterification of Beef Tallow, *Bioresour. Technol. 69:* 289–293 (1999).
44. Erdtmann, R., and L.B. Sivitz, eds., *Advancing Prion Science: Guidance for the National Prion Research Program*, The National Academies Press, Washington, 2004.
45. Anonymous, European Commission, Preliminary Report on Quantitative Risk Assessment on the Use of the Vertebral Column for the Production of Gelatine and Tallow, Submitted to the Scientific Steering Committee at its Meeting of 13–14 April, Brussels, 2000.
46. Anonymous, Department of Health and Human Service, U.S. Food and Drug Administration, 21 CFR Part 589, Substances Prohibited from Use in Animal Food or Feed, *U.S. Fed. Regist. 62:* 30935–30978.
47. Anonymous, World Heath Organization, Report of a WHO Consultation on Medicinal and Other Products in Relation to Human and Animal Transmissible Spongiform Encephalo-pathies, *inform 12:* 588 (2001).

48. Cummins, E.J., S.F. Colgan, P.M. Grace, D.J. Fry, K.P. McDonnell, and S.M. Ward, Human Risks from the Combustion of SRM-Derived Tallow in Ireland, *Hum. Ecol. Risk Assess. 8:* 1177–1192 (2002).
49. Kramer, W., The Potential of Biodiesel Production, *Oils and Fats Int. 11:* 33–34 (1995).
50. Canakci, M., and J. Van Gerpen, Biodiesel Production from Oils and Fats with High Free Fatty Acids, *Trans. ASAE 44:* 1429–1436 (2001).
51. Basu, H.N., and M.E. Norris, Process for the Production of Esters for Use as a Diesel Fuel Substitute Using Non-Alkaline Catalyst, U.S. Patent 5,525,126 (1996).
52. Turck, R., Method of Producing Fatty Acid Esters of Monovalent Alkyl Alcohols and Use Thereof, U.S. Patent 6,538,146 B2 (2003).
53. Mittelbach, M., and H. Enzelsberger, Transesterification of Heated Rapeseed Oil for Extending Diesel Fuel, *J. Am. Oil Chem. Soc. 76:* 545–550 (1999).
54. Haas, M.J., S. Bloomer, and K. Scott, Simple, High-Efficiency Synthesis of Fatty Acid Methyl Esters from Soapstock, *J. Am. Oil Chem. Soc. 77:* 373–379 (2000).
55. Haas, M.J., K.M. Scott, T.L. Alleman, and R.L. McCormick, Engine Performance of Biodiesel Fuel Prepared from Soybean Soapstock: A High Quality Renewable Fuel Produced from a Waste Feedstock, *Energy Fuels 15:* 1207–1212 (2001).
56. Haas, M.J., P.J. Michalski, S. Runyon, A. Nunez, and K.M. Scott, Production of FAME from Acid Oil, a By-Product of Vegetable Oil Refining, *J. Am. Oil Chem. Soc. 80:* 97–102 (2003).
57. Boocock, D.G.B., S.K. Konar, L. Mao, C. Lee, and S. Buligan, Fast Formation of High-Purity Methylesters from Vegetable Oils, *J. Am. Oil Chem. Soc. 75:* 1167–1172 (1998).
58. Caparella, T., Biodiesel Plants Open in Germany, *Render Mag. 37:* 16 (2002).
59. Saka S., and D. Kusdiana, Biodiesel Fuel from Rapeseed Oil as Prepared in Supercritical Methanol, *Fuel 80:* 225–231 (2001).
60. Kusdiana, D., and S. Saka, Kinetics of Transesterification in Rapeseed Oil to Biodiesel Fuel as Treated in Supercritical Methanol, *Fuel 80:* 693–698 (2001).
61. Haas, M.J., G.J. Piazza, and T.A. Foglia, Enzymatic Approaches to the Production of Biodiesel Fuels, in *Lipid Biotechnology*, edited by T.M. Kuo and H.W. Gardner, Marcel Dekker, New York, 2002, pp. 587–598.
62. Mittelbach, M., Lipase-Catalyzed Alcoholysis of Sunflower Oil, *J. Am. Oil Chem. Soc. 61:* 168–170 (1990).
63. Linko, Y-Y., M. Lamsa, X. Wu, E. Uosukainen, J. Seppala, and P. Linko, Biodegradable Products by Lipase Biocatalysis, *J. Biotechnol. 66:* 41–50 (1998).
64. Selmi, B., and D. Thomas, Immobilized Lipase-Catalyzed Ethanolysis of Sunflower Oil in a Solvent-Free Medium, *J. Am. Oil Chem. Soc. 75:* 691–695 (1998).
65. Shimada, Y., Y. Watanabe, T. Samukawa, A. Sugihara, H. Noda, H. Fukuda, and Y. Tominaga, Conversion of Vegetable Oil to Biodiesel Using Immobilized *Candida antarctica* Lipase, *J. Am. Oil Chem. Soc. 76:* 789–793 (1999).
66. Wu, W.H., T.A. Foglia, W.M. Marmer, and J.G. Phillips, Optimizing Production of Ethyl Esters of Grease Using 95% Ethanol by Response Surface Methodology, *J. Am. Oil Chem. Soc. 76:* 517–521 (1999).
67. Kaieda, M., T. Samukawa, T. Matsuumoto, K. Ban, A. Kondo, Y. Shimada, H. Noda, F. Nomoto, K. Ohtsuka, E. Izumoto, and H. Fukada, Biodiesel Fuel Production from Plant Oil Catalyzed by *Rhizopus oryzae* Lipase in a Water-Containing System Without an Organic Solvent, *J. Biosci. Bioeng. 88:* 627–631 (1999).

68. Du, W., Y Xu, J. Zing, and D. Liu, Novozyme 435-Catalyzed Transesterification of Crude Soybean Oils for Biodiesel Production in a Solvent-Free Medium, *Biotechnol. Appl. Biochem. 40:* 187–190 (2004).
69. Du, W., Y. Xu, and D. Liu, Lipase-Catalyzed Transesterification of Soya Bean Oil for Biodiesel Production During Continuous Batch Operation, *Biotechnol. Appl. Biochem. 38:* 103–106 (2003).
70. Nelson, L.L., T.A. Foglia, and W.N. Marmer, Lipase-Catalyzed Production of Biodiesel, *J. Am. Oil Chem. Soc. 73:* 1191–1195 (1996).
71. Foglia, T.A., L.L. Nelson, and W.N. Marmer, Production of Biodiesel, Lubricants, and Fuel and Lubricant Additives, U.S. Patent 5,713,965 (1998).
72. Wu, W.-H., T.A. Foglia, W.N. Marmer, R.O. Dunn, C.E. Goring, and T.E. Briggs, Low-Temperature Properties and Engine Performance Evaluation of Ethyl and Isopropyl Esters of Tallow and Grease, *J. Am. Oil Chem. Soc. 75:* 1173–1178 (1998).
73. Watanabe, Y., Y. Shimada, A. Sugihara, H. Noda, H. Fukuda, and Y. Tominaga, Continuous Production of Biodiesel Fuel from Vegetable Oil Using Immobilized *Candida antarctica* Lipase, *J. Am. Oil Chem. Soc. 77:* 355–359 (2000).
74. Abigor, R.D., P.O. Uadia, T.A. Foglia, M.J. Haas, J.E. Okpefa, and J.U. Obibuzor, Lipase-Catalyzed Production of Biodiesel Fuel from Nigerian Lauric Oils, *Biochem. Soc. Trans. 28:* 979–981 (2000).
75. Hsu, A.-F., K. Jones, T.A. Foglia, and W.N. Marmer, Immobilized Lipase-Catalyzed Production of Alkyl Esters of Restaurant Grease as Biodiesel, *Biotechnol. Appl. Biochem. 36:* 181–186 (2002).
76. Hsu, A.-F., K.C. Jones, T.A. Foglia, and W.N. Marmer, Continuous Production of Ethyl Esters of Grease Using an Immobilized Lipase, *J. Am. Oil Chem. Soc. 81:* 749–752 (2004).
77. Pizarro, A.V.L., and E.Y. Park, Lipase-Catalyzed Production of Biodiesel Fuel from Vegetable Oils Contained in Waste Activated Bleaching Earth, *Process Biochem. 38:* 1077–1082 (2003).
78. Ghosh, S., and D.K. Bhattacharyya, Utilization of Acid Oils in Making Valuable Fatty Products by Microbial Lipase Technology, *J. Am. Oil Chem. Soc. 77:* 1541–1544 (1995).
79. Dasari, M., M.J. Goff, and G.J. Suppes, Noncatalytic Alcoholysis of Soybean Oil, *J. Am. Oil Chem. Soc. 80:* 189–192 (2003).
80. Stern, R., G. Hillion, J.-J. Rouxel, and S. Leporq, Process for the Production of Esters from Vegetable Oils or Animal Oils Alcohols, U.S. Patent 5,908,946 (1999).
81. Corma, A., S. Iborra, S. Miquel, and J. Primo, Catalysts for the Production of Fine Chemicals: Production of Food Emulsifiers, Monoglycerides, by Glycerolysis of Fats with Solid Base Catalysts, *J. Catal. 173:* 315–321 (1998).
82. Leclercq, E., A. Finielsand, and C. Moreau, Transesterification of Rapeseed Oil in the Presence of Basic Zeolites and Related Catalysts, *J. Am. Oil Chem. Soc. 78:* 1161–1165 (2001).
83. Bayense, C.R., H. Hinnekens, and J. Martens, Esterification Process, U.S. Patent 5,508,457 (1996).
84. Suppes, G.J., M.A. Dasari, E.J. Doskocil, P.J. Mankidy, and M.J. Goff, Transesterification of Soybean Oil with Zeolite and Metal Catalysts, *Appl. Catal. A Gen. 257:* 213–223 (2004).
85. Suppes, G.J., K. Bockwinkel, S. Lucas, J.B. Botts, M.H. Mason, and J.A. Heppert, Calcium Carbonate Catalyzed Alcoholysis of Fats and Oils, *J. Am. Oil Chem. Soc. 78:* 139–145 (2001).

86. Harrington, K.J., and C.D'Arcy-Evans, Transesterification *in situ* of Sunflower Seed Oil, *Ind. Eng. Chem. Prod. Res. Dev. 24:* 314–318 (1985).
87. Harrington, K.J., and C.D'Arcy-Evans, A Comparison of Conventional and *in situ* Methods of Transesterification of Seed Oil from a Series of Sunflower Cultivars, *J. Am. Oil Chem. Soc. 62:* 1009–1013 (1985).
88. Siler-Marinkovic, S., and A. Tomasevic, Transesterification of Sunflower Oil *in situ*, *Fuel 77:* 1389–1391 (1998).
89. Özgül, S., and S. Türkay, *In Situ* Esterification of Rice Bran Oil with Methanol and Ethanol, *J. Am. Oil Chem. Soc. 70:* 145–147 (1993).
90. Özgül-Yücel, S., and S. Türkay, Variables Affecting the Yields of Methyl Esters Derived from *in situ* Esterification of Rice Bran Oil, *J. Am. Oil Chem. Soc. 79:* 611–613 (2002).
91. Kildiran, G., S. Özgül-Yücel, and S. Türkay, *In situ* Alcoholysis of Soybean Oil, *J. Am. Oil Chem. Soc. 73:* 225–228 (1996).
92. Haas, M.J., K.M. Scott, W.N. Marmer, and T.A. Foglia, *In situ* Alkaline Transesterification: An Effective Method for the Production of Fatty Acid Esters from Vegetable Oils, *J. Am Oil Chem. Soc. 81:* 83–89 (2004).
93. Haas, M.J., A. McAloon, and K. Scott, Production of Fatty Acid Esters by Direct Alkaline Transesterification: Process Optimization for Improved Economics, *Abstracts of the 95th Annual Meeting & Expo*, AOCS Press, Champaign, IL, 2004, p. 76.

5

Analytical Methods for Biodiesel

Gerhard Knothe

Introduction

As described in previous chapters, during the transesterification process, intermediate glycerols, mono- and diacylglycerols, are formed, small amounts of which can remain in the final biodiesel (methyl or other alkyl ester) product. In addition to these partial glycerols, unreacted triacylglycerols as well as unseparated glycerol, free fatty acids (FFA), residual alcohol, and catalyst can contaminate the final product. The contaminants can lead to severe operational problems when using biodiesel, including engine deposits, filter clogging, or fuel deterioration. Therefore, standards such as those in Europe (EN 14214; EN 14213 when using biodiesel for heating oil purposes) and the United States (ASTM D6751) limit the amount of contaminants in biodiesel fuel (see Appendix B). Under these standards, restrictions are placed on the individual contaminants by inclusion of items such as free and total glycerol for limiting glycerol and acylglycerols, flash point for limiting residual alcohol, acid value for limiting FFA, and ash value for limiting residual catalyst. A more detailed discussion of the rationale for quality parameters in biodiesel fuel standards is given in parts of this book and the literature (1–3). Some methods used in the analysis of biodiesel, including procedures for determination of contaminants such as water and phosphorus, which will not be dealt with here, were also described briefly (3). The determination of fuel quality is therefore an issue of great importance to the successful commercialization of biodiesel. Continuously high fuel quality with no operational problems is a prerequisite for market acceptance of biodiesel.

The major categories of analytical procedures for biodiesel discussed here comprise chromatographic and spectroscopic methods; however, papers dealing with other methods, including those based on physical properties, also have appeared. Categories can overlap due to the advent of hyphenated techniques such as gas chromatography-mass spectrometry (GC-MS), gas chromatography-infrared spectrometry (GC-IR), or liquid chromatography-mass spectrometry (LC-MS). However, few reports exist on the use of hyphenated techniques in biodiesel analysis. The main reasons are likely the higher equipment costs and the higher investment in technical skills of personnel required to interpret the data. This is the case despite the fact that hyphenated techniques could aid in resolving ambiguities remaining after analysis by stand-alone chromatographic methods.

To meet the requirements of biodiesel standards, the quantitation of individual compounds in biodiesel is not necessary but the quantitation of *classes of compounds* is. For example, for the determination of mono-, di-, or triacylglycerol (in European standards), it does not matter which fatty acid(s) is (are) attached to the glycerol backbone. For the determination of total glycerol, it does not matter which kind of acylglycerol (mono-, di-, or tri-) or free glycerol the glycerol stems from as long as the limits of the individual acylglycerol species or free glycerol are observed. That acylglycerols are quantifiable as classes of compounds by GC is a result of the method.

Virtually all methods used in the analysis of biodiesel are suitable (if necessary, with appropriate modifications) for all biodiesel feedstocks even if the authors report their method(s) on one specific feedstock. Also, the ideal analytical method would reliably and inexpensively quantify all contaminants even at trace levels with experimental ease in a matter of seconds at the most, or even faster for on-line reaction monitoring. No current analytical method meets these extreme demands. Therefore, compromises are necessary when selecting (a) method(s) for analyzing biodiesel or monitoring the transesterification reaction.

Due to the increasing use of blends of biodiesel with conventional, petroleum-based diesel fuel, the detection of blend levels is rapidly becoming another important aspect of biodiesel analysis. Different methods for various situations have now been developed. This includes detection of the blend level during use in an engine. The last section of this chapter will therefore deal with the subject of blend level detection.

Chromatographic Methods

Both GC and high-performance liquid chromatography (HPLC) analyses and combinations thereof were reported for biodiesel. Gel permeation chromatography (GPC) as an analytical tool for analysis of transesterification products was reported also. To date, most chromatographic analyses have been applied to methyl esters and not higher esters such as ethyl or *iso*-propyl, for example. Most methods would likely have to be modified for proper analysis of the higher esters. For example, when conducting GC analyses, changes in the temperature programs or other parameters may be necessary. The original work (4) on GC analysis reported the investigation of methyl and butyl esters of soybean oil. Apparently not all individual components could be separated in the analysis of butyl soyate, but classes of compounds could be analyzed. HPLC analysis has been applied to some ethyl, *iso*-propyl, 2-butyl, and *iso*-butyl esters of soybean oil and tallow (5). If an analytical method was applied to esters higher than methyl, it is noted accordingly.

The first report on chromatographic analysis of the transesterification used thin-layer chromatography with flame-ionization detection (TLC/FID; Iatroscan instrument) (6). In another report (7), TLC/FID was used to correlate bound glycerol content to acyl conversion determined by GC. It was found in this work that if conversion to methyl esters was >96%, then the amount of bound glycerol was <0.25 wt%. Although the TLC/FID method is easy to learn and use (6), it was abandoned largely because of lower accuracy, material inconsistencies, as well as sensitivity to humidity (6), and the relatively high cost of the instrument (7).

To date, GC has been the most widely used method for the analysis of biodiesel due to its generally higher accuracy in quantifying minor components. However, the accuracy of GC analyses can be influenced by factors such as baseline drift and overlapping signals. It is not always clear that such factors are compensated for in related reports on biodiesel analysis. The first report on the use of capillary GC discussed the quantitation of esters as well as mono-, di-, and triacylglycerols (4). The samples were reacted with *N,O-bis*(trimethylsilyl)trifluoracetamide (BSTFA) to give the corresponding trimethylsilyl (TMS) derivatives of the hydroxy groups. This kind of derivatization was reported in subsequent papers on GC quantitation of biodiesel. Preparation of TMS derivatives is important because it improves the chromatographic properties of the hydroxylated materials and, in case of coupling to a mass spectrometer, facilitates interpretation of their mass spectra. Although a short (1.8-m) fused silica (100% dimethylpolysiloxane) capillary column was used originally (4), in other work, typically fused silica capillary columns coated with a 0.1-μm film of (5%-phenyl)-methylpolysiloxane of 10–12/15-m length were used. An analysis of rapeseed ethyl esters was carried out on a GC instrument equipped with an FID and a 1.8 m × 4 mm i.d. packed column (8).

Most reports on the use of GC for biodiesel analysis employ FID. The use of mass spectrometric detectors (MSD) would eliminate any ambiguities about the nature of the eluting materials because mass spectra unique to individual compounds would be obtained, although quantitation may be affected. There are two papers in the literature that describe the use of MSD (9,10).

Most papers on GC analysis discuss the determination of a specific contaminant or class of contaminants in methyl esters. The original report on biodiesel GC analysis (4) quantified mono-, di-, and triacylglycerols in methyl soyate on a short 100% dimethylpolysiloxane column (1.8 m × 0.32 mm i.d.). Similar reports on the quantitation of acylglycerols exist (11,12). The main differences are in the column specifications [both (5%-phenyl)-methylpolysiloxane, differences in parameters such as column length] and temperature programs as well as standards.

Other papers deal with the individual or combined determination of other potential contaminants such as free glycerol or methanol. A paper describing the use of a mass selective detector deals with the determination of glycerol (9) and, in an extension thereof, a second paper describes the simultaneous quantitation of glycerol and methanol (10). Other authors have also reported the determination of glycerol (13) or methanol (14). Using the same GC equipment as in the previous determination of glycerol (13) with only a modification of the oven temperature program, methanol could be determined. Ethanol was used as a standard for response factor determination. The flash point of biodiesel from palm oil and the methanol content were correlated. Underivatized glycerol was detected with 1,4-butanediol as a standard on a short 2-m glass column (i.d., 4 mm) loaded with Chromosorb 101 (13), whereas the other method used derivatization and a 60 m × 0.25 mm i.d., film 0.25 μm (5%-phenyl)-methylpolysiloxane column and was reported to be more sensitive (9). The temperature program varied (lower starting temperature when determining methanol) (9,10); otherwise the column was the same.

Two papers (9,10) discussed the use of MS as a detection method in addition to flame ionization. In the determination of free glycerol in biodiesel by GC-MS, selected ion monitoring (SIM) mode was used to track the ions *m/z* 116 and 117 of *bis-O*-trimethylsilyl-1,4-butanediol (from silylation of the 1,4-butanediol standard) and *m/z* 147 and 205 of *tris-O*-trimethylsilyl-1,2,3-propanetriol (from silylation of glycerol). The detection limit was also improved for rapeseed methyl ester (RME) when using MS in SIM mode (10^{-5} %) compared with the FID detector (10^{-4} %) (9). In an extension of this work, the simultaneous detection of methanol and glycerol by MS in SIM mode was reported (10). For detection of (silylated) methanol (trimethylmethoxysilane), peaks at *m/z* 59 and 89 were monitored as were peaks at *m/z* 75 and 103 of the additional (silylated) standard ethanol (trimethylethoxysilane). MS in SIM mode has the additional advantage that interfering signals can be avoided, thus allowing the use of shorter columns (10).

A further extension of the aforementioned papers is the simultaneous determination of glycerol as well as mono-, di-, and triglycerols by GC (15). Figure 1 depicts a GC analysis of biodiesel for glycerol and acylglycerol contaminants from the literature (15). The simultaneous determination of glycerol and acylglycerols in biodiesel led to the development of corresponding standards such as ASTM D6584 and EN 14105, which in turn are included as parameters in full biodiesel standards. Here (15) and in previous work (12), 10-m (5%-phenyl)-methylpoly-siloxane columns with 0.1-μm film [0.25 mm i.d. in (12), 0.32 mm i.d. in (15)] were used. Major differences were the lower starting point of the temperature program (15) and the addition of a standard (1,2,4-butanetriol) for the glycerol analysis. In these works (12,15), a cool on-column injector/inlet was used.

Nonglyceridic materials that can be present in biodiesel also were analyzed by GC. Thus the determination of sterols and sterol esters in biodiesel (16) was reported. The stated reason is that the influence of these compounds, which remain in vegetable oils after processing (and thus in biodiesel after transesterification because they are soluble in methyl esters), on biodiesel fuel quality has not been determined (16). Detection was carried out with an FID as in other GC papers, although in this case, MS detection would appear especially desirable. The method for detection of sterols (16) is virtually identical to the other GC method reported by the authors (12). The only differences were the use of sterol standards and a slight modification of the GC temperature program to spread the sterol peaks under condensation or overlapping of the peaks of the other classes of compounds. Derivatization was carried out again with BSTFA (with 1% trimethylchlorosilane), and the column again was a fused silica capillary column coated with a 0.1-μm film of (5%-phenyl)-methylpolysiloxane.

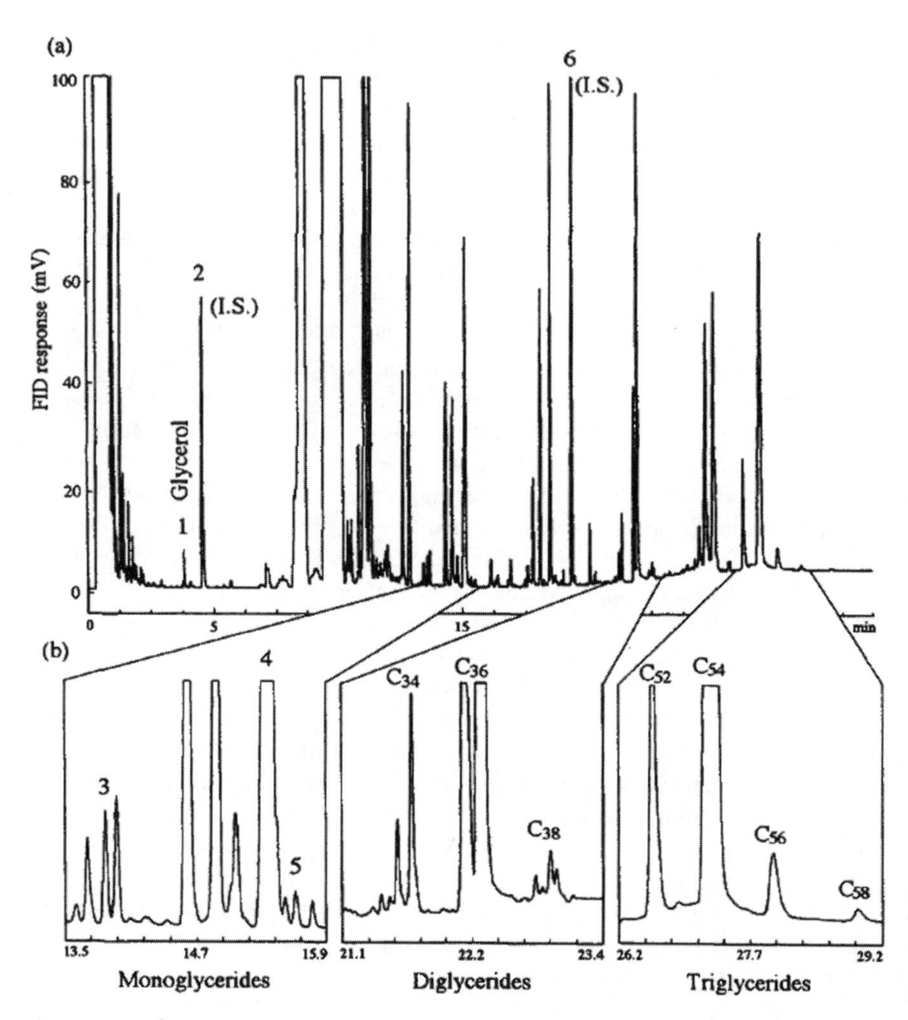

Fig. 1. Gas chromatogram of biodiesel (rapeseed oil methyl esters; free hydroxy groups derivatized by silylation) for quantitative determination of glycerol and acylglycerol contaminants. The inscribed numbers refer to assigned peaks: 1 = glycerol; 2 = 1,2,4-butanetriol (internal standard); 3= monopalmitin; 4 = monoolein, monolinolein, monolinolenin; 5 = monostearin; and 6 = tricaprin (internal standard). In the enlarged portions in b), di- and triacylglycerols containing specific numbers of carbons are also inscribed. Reprinted from Reference 15 with permission from Elsevier.

The total concentration of sterols in rapeseed methyl ester was reported to be 0.339–0.500% and sterol esters were 0.588–0.722%. In another paper on the analysis of sterol content in rapeseed methyl ester (17), the same authors reported a sterol content of 0.70–0.81%. Other authors (11) also pointed out the presence of sterols and sterol esters in biodiesel.

HPLC. A general advantage of HPLC compared with GC is that usually time- and reagent-consuming derivatizations are not necessary, which reduces the time for analyses. Nevertheless, there are fewer reports of HPLC applied to biodiesel than GC analysis. The first report on the use of HPLC (18) described the use of an isocratic solvent system (chloroform with an ethanol content of 0.6%) on a cyano-modified silica column coupled to two GPC columns with density detection. This system allowed for the detection of mono-, di-, and triacylglycerols as well as methyl esters as classes of compounds. The system was useful for quantitating various degrees of conversion of the transesterification reaction.

HPLC with pulsed amperometric detection (detection limit usually 10–100 times lower than for amperometric detection; detection limit 1 μg/g) was used to determine the amount of free glycerol in vegetable oil esters (19). The major advantage of this detection method was its high sensitivity. The simultaneous detection of residual alcohol also is possible with this technique.

Reaction mixtures obtained from lipase-catalyzed transesterification were analyzed by HPLC using an evaporative light scattering detector (ELSD) (5). This method is able to quantitate product esters, FFA, and the various forms of acylglycerols. A solvent system consisting of hexane and methyl *tert*-butyl ether (each with 0.4% acetic acid) with a gradient elution profile was used. It can be applied to esters higher than methyl as discussed above.

In an extensive study (20), reversed-phase HPLC was used with different detection methods [ultraviolet (UV) detection at 205 nm, evaporative light scattering detection and atmospheric pressure chemical ionization-MS (APCI-MS) in positive-ion mode]. Two gradient solvent systems, one consisting of mixing methanol (A) with 5:4 2-propanol/hexane (B) from 100% A to 50:50 A:B (a non-aqueous reversed-phase solvent system), the other of mixing water (A), acetonitrile (B), and 5:4 2-propanol/hexane (C) in two linear gradient steps (30:70 A:B at 0 min, 100% B in 10 min, 50:50 B:C in 20 min, and finally, isocratic 50:50 B:C for 5 min), were applied. The first solvent system was developed for rapid quantitation of the transesterification of rapeseed oil with methanol by comparing the peak areas of methyl esters and triacylglycerols. The contents of individual acids (using normalized peak areas) were subject to error, and the results differed for the various detection methods. The sensitivity and linearity of each detection method varied with the individual triacylglycerols. APCI-MS and ELSD had decreased sensitivity with increasing number of double bonds in the fatty acid methyl esters, whereas UV will not quantify the saturates. APCI-MS was stated to be the most suitable detection method for the analysis of rapeseed oil and biodiesel. The HPLC-MS-APCI method was reviewed briefly with respect to its applicability to biodiesel analysis (21).

GPC. One report exists that describes the use of GPC (which is similar to HPLC in instrumentation except for the nature of the column and the underlying separation principle, namely, molecular weight of the analytes for GPC) for the analysis of transesterification products (22). Using a refractive index detector and tetrahydrofuran as mobile phase, mono-, di-, and triacylglycerols as well as the methyl esters and glycerol could be analyzed. The method was tailored for palm oil, and standards were selected accordingly. Reproducibility was good with an SD of 0.27–3.87% at different rates of conversion.

Hyphenated Method: LC-GC. The combination of LC with GC was also reported. The purpose of the combination of the two separation methods is to reduce the complexity of the gas chromatograms and to obtain more reliable peak assignments (23). A fully automated LC-GC instrument was employed in the determination of acylglycerols in vegetable oil methyl esters (23). Hydroxy groups were acetylated, and then the methyl esters (sterols and esterified sterols elute with methyl esters) and acylglycerols were preseparated by LC (variable wavelength detector). The solvent system for LC was hexane/methylene chloride/acetonitrile 79.97:20:0.05. GC (FID) was performed on a 10-m (5%-phenyl)-methylpolysiloxane column. One LC-GC run required 52 min.

LC-GC was also applied to the analysis of sterols in biodiesel derived from rapeseed oil (24,25). Five different types of methyl esters were analyzed for sterols by on-line LC-GC (25). The vegetable oil methyl esters were those of rapeseed oil, soybean oil, sunflower oil, high-oleic sunflower oil, and used frying oil. The sterols were silylated before analysis with *N*-methyl-*N*-trimethylsilyltrifluoracetamide (MSTFA). No saponification or off-line preseparation was required. The methyl esters were separated from the sterols by LC with a hexane/methylene chloride/acetonitrile 79.9:20:0.1 solvent system. GC was again carried out with a 12-m (5%-phenyl)-methylpolysiloxane column and FID detection. Total concentrations of free sterols were 0.20–0.35 wt% for the five samples, whereas sterol esters displayed a range of 0.15–0.73 wt%. Soybean oil methyl ester was at the lower end (0.20 and 0.15%, respectively), whereas rapeseed oil methyl ester was at the higher end (0.33 and 0.73%, respectively). In a comparison of two methods, saponification and isolation of the sterol fraction with subsequent GC analysis and LC-GC analysis of sterols in rapeseed oil methyl ester (25), despite the sophisticated instrumentation required, LC-GC was recommended because of additional information, short analysis time, and reproducibility. The total sterol content in RME was 0.70–0.81 wt%.

Spectroscopic Methods

Spectroscopic methods evaluated for the analysis of biodiesel and/or monitoring of the transesterification reaction are ^{1}H as well as ^{13}C nuclear magnetic resonance (NMR) spectroscopy and near infrared (NIR) spectroscopy. The first report on spectroscopic determination of the yield of a transesterification reaction utilized ^{1}H NMR (26). Figure 2 depicts the ^{1}H NMR spectrum of a progressing transesterification reaction. These authors used the protons of the methylene group adjacent to the ester moiety in triglycerols and the protons in the alcohol moiety of the product methyl esters to monitor the yield. A simple equation given by the authors (terminology slightly modified here) is as follows:

$$C = 100 \times (2A_{ME}/3A_{\alpha\text{-CH2}})$$

in which C is the conversion of triacylglycerol feedstock (vegetable oil) to the corresponding methyl ester, A_{ME} is the integration value of the protons of the methyl esters (the strong singlet peak), and $A_{\alpha\text{-CH2}}$ is the integration value of the methylene protons. The factors 2 and 3 derive from the fact that the methylene carbon possesses two protons and the alcohol (methanol-derived) carbon has three attached protons.

Turnover and reaction kinetics of the transesterification of rapeseed oil with methanol were studied by ^{13}C NMR (27) with benzene-d_6 as solvent. The signals at ~14.5 ppm of the terminal methyl groups unaffected by the transesterification were used as an internal quantitation standard. The methyl signal of the product methyl esters registered at ~51 ppm and the glyceridic carbons of the mono-, di-, and triacylglycerols at 62–71 ppm. Analysis of the latter peak range allowed the determination of transesterification kinetics showing that the formation of partial acylglycerols from the triglycerols is the slower, rate-determining step.

NIR spectroscopy was used to monitor the transesterification reaction (28). The basis for quantitation is differences in the NIR spectra at 6005 and at 4425– 4430 cm^{-1}, where methyl esters display peaks, whereas triacylglycerols exhibit only shoul-

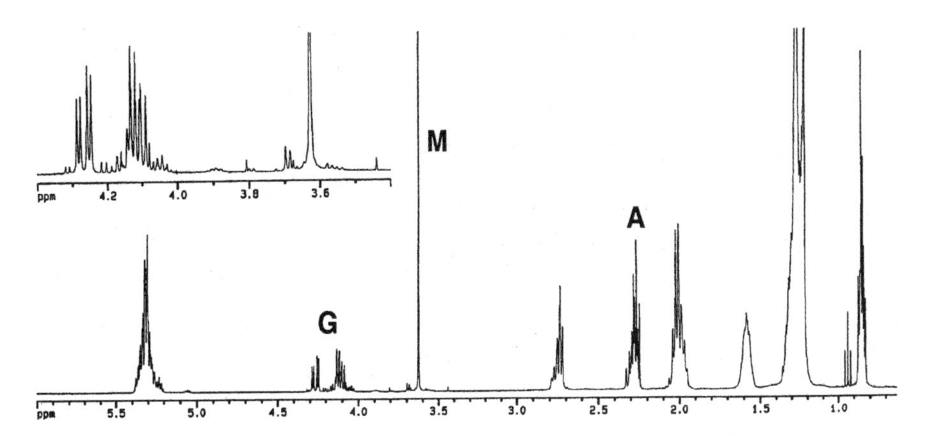

Fig. 2. ^1H NMR spectrum of a progressing transesterification reaction. The signals at 4.1–4.3 ppm are caused by the protons attached to the glycerol moiety of mono-, di, or triacylglycerols. The strong singlet at 3.6 ppm indicates methyl ester (-CO_2CH_3) formation. The signals at 2.3 ppm result from the protons on the CH_2 groups adjacent to the methyl or glyceryl ester moieties (–$CH_2CO_2CH_3$ for methyl esters). These signals can be used for quantitation. *Source:* Reference 29.

ders (see Fig. 3). Ethyl esters could be distinguished in a similar fashion (28). Using the absorption at 6005 cm^{-1} rather than the one at 4425 cm^{-1} gave better quantitation results. It appears that ethyl esters, and perhaps even higher esters, may be distinguished similarly by NIR from triacylglycerols but no results have been reported to date. NIR spectra were obtained with the aid of a fiber-optic probe coupled to the spectrometer, thus rendering their acquisition particularly easy and time-efficient.

Contaminants of biodiesel cannot be fully quantified by NIR at the low levels called for in biodiesel standards. The accuracy of the NIR method in distinguishing triacylglycerols and methyl esters is in the range of 1–1.5%, although in most cases, better results are achieved. To circumvent this difficulty, an inductive method can be applied. The inductive method consists of verifying by GC, for example, that a biodiesel sample meets standards. The NIR spectrum of this sample would be recorded. The NIR spectrum of the feedstock would also be recorded as well as the spectra of intermediate samples at conversions of 25, 50, and 75%, for example. A quantitative NIR evaluation method could then be established. If while conducting another transesterification reaction, the NIR spectrum indicates that the reaction with the same parameters has attained conversion to a product that

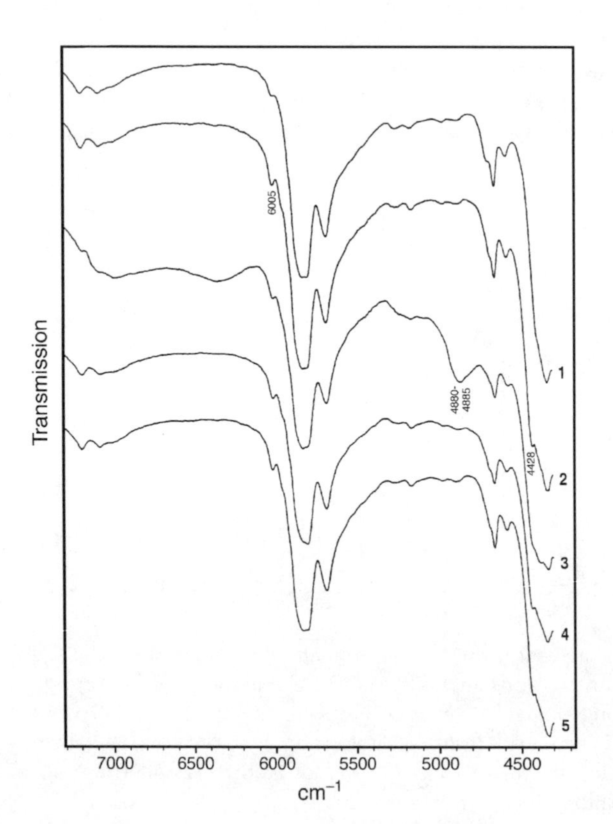

Fig. 3. NIR spectra of soybean oil (SBO) and methyl soyate (SME) and SME containing significant amounts of methanol. The inscribed wave numbers highlight the possibilities for distinguishing the spectra and thus quantifying the components. *Source:* Reference 28.

(within experimental error of NIR) conforms to standards, it can be safely assumed that this result is correct, even if not all potential contaminants have been fully analyzed. Only if a significant deviation is indicated by NIR would a detailed investigation by a more complex method such as GC be necessary. The NIR procedure is considerably less labor-intensive, faster, and easier to perform than GC.

Although the first NIR paper used a model system to describe monitoring of transesterification and to develop quantitation methods, a second paper applied the method to a transesterification reaction in progress on a 6-L scale. Here, spectroscopic results were obtained not only by NIR but also by ^{1}H NMR and NIR (29). The results of both spectroscopic methods, which can be correlated by simple equations, were in good agreement. Two NMR approaches were used; one was the use of the methyl ester protons (peak at 3.6 ppm in Fig. 2) and the protons on the carbons next to the glyceryl moiety (α-CH_2; peaks at 2.3 ppm in Fig. 2) (29). The second approach was the use of the methyl ester protons and the protons of the glyceryl moiety (peaks at 4.1–4.3 ppm in Fig. 2) in the triacylglycerols (29).

In related work, determining the amount of biodiesel in lubricating oil (30) by mid-IR spectroscopy with a fiber-optic probe was reported. The problem is significant because biodiesel can cause dilution of the lubricant, which may ultimately result in engine failure. The dilution is attributed to the higher boiling range of biodiesel (30,31) compared with conventional diesel fuel, whose more volatile components have less chance to dilute the lubricant. The mid-IR range used was 1820–1680 cm^{-1}, which is typical for carbonyl absorption and is not observed in conventional diesel fuel or in the lubricating oil. Previous to this work, other authors had used IR spectroscopy (without the aid of a fiber-optic probe) in the range 1850–1700 cm^{-1} to analyze biodiesel in lubricating oil (31). The carbonyl absorption at 1750 cm^{-1} was not disturbed by the absorption of oxidation products at 1710 cm^{-1}. However, the carbonyl absorptions in the mid-range IR spectra of triacylglycerols and fatty acid methyl esters are almost identical, and care must be taken to know whether triacylglycerols or methyl esters are being analyzed.

Other Methods

Viscometry. The viscosity difference between component triacylglycerols of vegetable oils and their corresponding methyl esters resulting from transesterification is approximately one order of magnitude (see Chapter 6.2), tables in Appendix A). The viscosity difference forms the basis of an analytical method, viscometry, applied to determining the conversion of a vegetable oil to methyl ester (32). Viscosities determined at 20 and 37.8°C were in good agreement with GC analyses conducted for verification purposes. The viscometric method, especially results obtained at 20°C, is reported to be suitable for process control purposes due to its rapidity (32). Similar results were obtained from density measurements (32). However, it appears that the viscosity of the final product, which depends on the fatty acid composition, must be known.

Titration for Determining FFA. Titration methods for determining the neutralization number (NN) of biodiesel were described (33). Two methods for determining strong acids and FFA in one measurement were developed. One method, of particular interest, used potentiometry, whereas the other used two acid-base indicators (neutral red, phenolphthalein). The potentiometric method is more reliable and even with the use of two indicators, the NN values derived from the titration method are 10–20% greater relative to the real acidity of the sample.

Other Methods. An enzymatic method for analyzing glycerol in biodiesel was described to test for completeness of the transesterification reaction (34). Solid-phase extraction of the reaction mixture with subsequent enzymatic analysis was applied. This method was originally intended as a simple method for glycerol determination, but reproducibility and complexity concerns exist (13,19). Recently, an enzymatic method for determining free and total glycerol became available commercially (35).

Biodiesel Blends

As mentioned above, mid-IR spectroscopy was used for determining biodiesel in lubricating oil (30) for direct determination of blend levels of biodiesel with petroleum-based diesel fuel (36). The peak utilized is that of the carbonyl moiety at ~1740 cm^{-1}. Blend detection by IR using this peak is the basis of the European standard EN 14078 [Determination of fatty acid methyl esters (FAME) in middle distillates—IR spectroscopy method].

In NIR spectroscopy, the peaks used for quantitation, like those for monitoring transesterification and fuel quality, appear to be suitable for purposes of blend level determination (37). The use of the NIR range may permit using a spectrometer without any changes in instrument settings for reaction and fuel quality monitoring as well as determination of blend levels. Also, some characteristic peaks of triacylglycerols in vegetable oils or animal fats and methyl esters occur at nearly the same wavenumber (1740 cm^{-1}) in mid-IR, whereas NIR utilizes differences in the spectra of methyl esters and triacylglycerols. Thus, NIR may be able to detect simultaneously whether the petroleum diesel fuel was blended with biodiesel or triacylglycerol-containing oil or fat, with the latter being unacceptable.

Silica cartridge chromatography with hexane/diethyl ether as solvents was employed to separate biodiesel from conventional diesel fuel, which was then analyzed by GC (38). Acetylation of the contaminants in a blend was carried out relatedly, the blend separated by means of a silica cartridge with hexane as solvent, and then the biodiesel fraction analyzed by GC (39).

Although no reports yet exist, GC of a blend would likely cause very complex chromatograms due to the numerous components of conventional diesel fuel. In HPLC, as done previously, the classes of compounds may elute and be analyzable, thus reducing complexity.

Another method for blend level detection of biodiesel utilizes the saponification value (38). Relatedly, the ester number, defined as the difference between the saponification value and the acid value, of blended fuels was determined and the methyl ester fraction determined using an average molecular weight of methyl esters (37). If the average molecular weight of the biodiesel is unknown, methyl oleate can be used as a reference (37). The ester number method yielded results comparable to those of IR (37).

On-Vehicle Blend Sensors. In addition to these analytical methods, on-vehicle analysis of biodiesel blends may be required to adjust engine settings such as fuel injection timing for improving performance and emissions (40,41) due to fueling with different blend levels of biodiesel or when refueling with neat biodiesel or petroleum-based diesel fuel in alternating fashion. For this purpose, a commercial sensor originally developed for detecting the level of alcohol (methanol or ethanol) in gasoline-alcohol blends (40) was used. The average frequency difference of ~7 Hz suffices for use in blend level detection (40). Another suitable sensor was developed (41,42), which functioned better than sensors originally developed for testing soil humidity and salinity. The frequency output of the sensors, which monitors the dielectric constant of the mixture, is linearly proportional to the blend level of biodiesel.

References

1. Mittelbach, M., Analytical Aspects and Quality Criteria for Biodiesel Derived from Vegetable Oils, *Proceedings of an Alternative Energy Conference: Liquid Fuels, Lubricants, and Additives from Biomass*, ASAE, St. Joseph, MI, 1994, pp. 151–156.
2. Mittelbach, M., Diesel Fuel Derived from Vegetable Oils, VI: Specifications and Quality Control of Biodiesel, *Bioresour. Technol. 56:* 7–11 (1996).
3. Komers, K., R. Stloukal, J. Machek, F. Skopal, and A. Komersová, Biodiesel Fuel from Rapeseed Oil, Methanol, and KOH. Analytical Methods in Research and Production, *Fett/Lipid 100:* 507–512 (1998).
4. Freedman, B., W.F. Kwolek, and E.H. Pryde, Quantitation in the Analysis of Transesterified Soybean Oil by Capillary Gas Chromatography, *J. Am. Oil Chem. Soc. 63:* 1370–1375 (1986).
5. Foglia, T.A., and K.C. Jones, Quantitation of Neutral Lipid Mixtures Using High-Performance Liquid Chromatography with Light Scattering Detection, *J. Liq. Chromatogr. Relat. Technol. 20:* 1829–1838 (1997).
6. Freedman, B., E.H. Pryde, and W.F. Kwolek, Thin-Layer Chromatography/Flame-Ionization Analysis of Transesterified Vegetable Oils, *J. Am. Oil Chem. Soc. 61:* 1215–1220 (1984).
7. Cvengroš, J., and Z. Cvengrošová, Quality Control of Rapeseed Oil Methyl Esters by Determination of Acyl Conversion, *J. Am. Oil Chem. Soc. 71:* 1349–1352 (1994).
8. Cvengrošová, Z., J. Cvengroš, and M. Hronec, Rapeseed Oil Ethyl Esters as Alternative Fuels and Their Quality Control, *Petrol. Coal 39:* 36–40 (1997).
9. Mittelbach, M., Diesel Fuel Derived from Vegetable Oils, V [1]: Gas Chromatographic Determination of Free Glycerol in Transesterified Vegetable Oils, *Chromatographia 37:* 623–626 (1993).

10. Mittelbach, M., G. Roth, and A. Bergmann, Simultaneous Gas Chromatographic Determination of Methanol and Free Glycerol in Biodiesel, *Chromatographia 42:* 431–434 (1996).
11. Mariani, C., P. Bondioli, S. Venturini, and E. Fedeli, Vegetable Oil Derivatives as Diesel Fuel. Analytical Aspects. Note 1: Determination of Methyl Esters, Mono-, Di-, and Triglycerides, *Riv. Ital. Sostanze Grasse 69:* 549–551 (1991).
12. Plank, C., and E. Lorbeer, Quality Control of Vegetable Oil Methyl Esters Used as Diesel Fuel Substitutes: Quantitative Determination of Mono-, Di-, and Triglycerides by Capillary GC, *J. High Resolut. Chromatogr. 16:* 609–612 (1992).
13. Bondioli, P., C. Mariani, A. Lanzani, and E. Fedeli, Vegetable Oil Derivatives as Diesel Fuel Substitutes. Analytical Aspects. Note 2: Determination of Free Glycerol, *Riv. Ital. Sostanze Grasse 69:* 7–9 (1992).
14. Bondioli, P., C. Mariani, E. Fedeli, A.M. Gomez, and S. Veronese, Vegetable Oil Derivatives as Diesel Fuel Substitutes. Analytical Aspects. Note 3: Determination of Methanol, *Riv. Ital. Sostanze Grasse 69:* 467–469 (1992).
15. Plank, C., and E. Lorbeer, Simultaneous Determination of Glycerol, and Mono-, Di-, and Triglycerides in Vegetable Oil Methyl Esters by Capillary Gas Chromatography, *J. Chromatogr. A 697:* 461–468 (1995).
16. Plank, C., and E. Lorbeer, Analysis of Free and Esterified Sterols in Vegetable Oil Methyl Esters by Capillary GC, *J. High Resolut. Chromatogr. 16:* 483–487 (1993).
17. Plank, C., and E. Lorbeer, Minor Components in Vegetable Oil Methyl Esters I: Sterols in Rapeseed Oil Methyl Ester, *Fett Wiss. Technol. 96:* 379–386 (1994).
18. Trathnigg, B., and M. Mittelbach, Analysis of Triglyceride Methanolysis Mixtures Using Isocratic HPLC with Density Detection, *J. Liq. Chromatogr. 13:* 95–105 (1990).
19. Lozano, P., N. Chirat, J. Graille, and D. Pioch, Measurement of Free Glycerol in Biofuels, *Fresenius J. Anal. Chem. 354:* 319–322 (1996).
20. Holčapek, M., P. Jandera, J. Fischer, and B. Prokeš, Analytical Monitoring of the Production of Biodiesel by High-Performance Liquid Chromatography with Various Detection Methods, *J. Chromatogr. A. 858:* 13–31 (1999).
21. Holčapek, M., P. Jandera, and J. Fischer, Analysis of Acylglycerols and Methyl Esters of Fatty Acids in Vegetable Oils and Biodiesel, *Crit. Rev. Anal. Chem. 31:* 53–56 (2001).
22. Darnoko, D., M. Cheryan, and E.G. Perkins, Analysis of Vegetable Oil Transesterification Products by Gel Permeation Chromatography, *J. Liq. Chrom. Rel. Technol. 23:* 2327–2335 (2000).
23. Lechner, M., C. Bauer-Plank, and E. Lorbeer, Determination of Acylglycerols in Vegetable Oil Methyl Esters by On-Line Normal Phase LC-GC, *J. High Resolut. Chromatogr. 20:* 581–585 (1997).
24. Plank, C., and E. Lorbeer, Minor Components in Vegetable Oil Methyl Esters I: Sterols in Rapeseed Oil Methyl Ester, *Fett Wiss. Technol. 96:* 379–386 (1994).
25. Plank, C., and E. Lorbeer, On-Line Liquid Chromatography—Gas Chromatography for the Analysis of Free and Esterified Sterols in Vegetable Oil Methyl Esters Used as Diesel Fuel Substitutes, *J. Chromatogr. A 683:* 95–104 (1994).
26. Gelbard, G., O. Brès, R.M. Vargas, F. Vielfaure, and U.F. Schuchardt, ^{1}H Nuclear Magnetic Resonance Determination of the Yield of the Transesesterification of Rapeseed Oil with Methanol, *J. Am. Oil Chem. Soc. 72:* 1239–1241 (1995).
27. Dimmig, T., W. Radig, C. Knoll, and T. Dittmar, ^{13}C-NMR-Spektroskopie zur Bestimmung von Umsatz und Reaktionskinetik der Umesterung von Triglyceriden zu

Methylestern (^{13}C-NMR Spectroscopic Determination of the Conversion and Reaction Kinetics of Transesteri-fication of Triglycerols to Methyl Esters), *Chem. Tech. (Leipzig) 51:* 326–329 (1999).

28. Knothe, G., Rapid Monitoring of Transesterification and Assessing Biodiesel Fuel Quality by NIR Spectroscopy Using a Fiber-Optic Probe, *J. Am. Oil Chem. Soc. 76:* 795–800 (1999).
29. Knothe, G., Monitoring the Turnover of a Progressing Transesterification Reaction by Fiber-Optic NIR Spectroscopy with Correlation to ^{1}H-NMR Spectroscopy, *J. Am. Oil Chem. Soc. 77:* 489–493 (2000).
30. Sadeghi-Jorabchi, H., V.M.E. Wood, F. Jeffery, A. Bruster-Davies, N. Loh, and D. Coombs, Estimation of Biodiesel in Lubricating Oil Using Fourier Transform Infrared Spectroscopy Combined with a Mid-Infrared Fibre Optic Probe, *Spectroscopy Eur. 6:* 16,18,20–21 (1994).
31. Siekmann, R.W., G.H. Pischinger, D. Blackman, and L.D. Carvalho, The Influence of Lubricant Contamination by Methyl Esters of Plant Oils on Oxidation Stability and Life, in *Proceedings of the International Conference on Plant and Vegetable Oils as Fuels*, ASAE Publication 4-82, ASAE, St. Joseph, MI, 1982, pp. 209–217.
32 De Filippis, P., C. Giavarini, M. Scarsella, and M. Sorrentino, Transesterification Processes for Vegetable Oils: A Simple Control Method of Methyl Ester Content, *J. Am. Oil Chem. Soc. 72:* 1399–1404 (1995).
33. Komers, K., F. Skopal, and R. Stloukal, Determination of the Neutralization Number for Biodiesel Fuel Production, *Fett/Lipid 99:* 52–54 (1997).
34. Bailer, J., and K. de Hueber, Determination of Saponifiable Glycerol in "Bio-Diesel," *Fresenius J. Anal. Chem. 340:* 186 (1991).
35. Anonymous, Glycerine, *Chem. Market Reporter 263*, No. 21 (May 26, 2003), p. 12.
36. Bírová, A., E. Švajdlenka, J. Cvengroš, and V. Dostálíková, Determination of the Mass Fraction of Methyl Esters in Mixed Fuels, *Eur. J. Lipid Sci. Technol. 104:* 271–277 (2002).
37. Knothe, G., Determining the Blend Level of Mixtures of Biodiesel with Conventional Diesel Fuel by Fiber-Optic NIR Spectroscopy and [1]H Nuclear Magnetic Resonance Spectroscopy, *J. Am. Oil Chem. Soc. 78:* 1025–1028 (2001).
38. Bondioli, P., A. Lanzani, E. Fedeli, M. Sala, and S. Veronese, Vegetable Oil Derivatives as Diesel Fuel Substitutes. Analytical Aspects. Note 4: Determination of Biodiesel and Diesel Fuel in Mixture, *Riv. Ital. Sostanze Grasse 71:* 287–289 (1994).
39. Bondioli, P., and L. Della Bella, The Evaluation of Biodiesel Quality in Commercial Blends with Diesel Fuel, *Riv. Ital. Sostanze Grasse 80:* 173–176 (2003).
40. Tat, M.E., and J.H. Van Gerpen, Biodiesel Blend Detection with a Fuel Composition Sensor, *Appl. Eng. Agric. 19:* 125–131 (2003).
41. Munack, A., J. Krahl, and H. Speckmann, A Fuel Sensor for Biodiesel, Fossil Diesel Fuel, and Their Blends, presented at the 2002 ASAE Annual Meeting/CIGR XVth World Congress, Chicago, ASAE Paper No. 02-6081, 2002.
42. Munack, A., and J. Krahl, Erkennung des RME-Betriebs mittels eines Biodiesel-Kraftstoffsensors (Identifying Use of RME with a Biodiesel Fuel Sensor), *Landbauforschung Völkenrode, Sonderheft 257* (Special Issue) (2003).

6

Fuel Properties

6.1

Cetane Numbers–Heat of Combustion–Why Vegetable Oils and Their Derivatives Are Suitable as a Diesel Fuel

Gerhard Knothe

Cetane Number

Generally, the cetane number (CN) is a dimensionless descriptor of the ignition quality of a diesel fuel (DF). As such, it is a prime indicator of DF quality. For the following discussion, it is appropriate to briefly discuss conventional DF first.

Conventional DF (petrodiesel) is a product of the cracking of petroleum. Petrodiesel is a fraction boiling in the mid-range of cracking products; thus, it is also termed "middle distillates" (1). Petrodiesel is further classified as No. 1, No. 2, and No. 4 DF in the United States by the standard ASTM D975. No. 1 (DF1) is obtained from the 170–270°C boiling range (as are kerosene and jet fuel) (2) and is applicable to high-speed engines whose operation involves frequent and relatively wide variations in engine load and speed. It is required for use at abnormally low temperatures. No. 2 (DF2) is in the 180–340°C boiling range (2). This grade is suitable for use in high-speed engines under relatively high loads and uniform speeds. DF2 can be used in engines not requiring fuels that have the greater volatility and other properties specified for DF1 (1). DF2 is the transportation fuel to which biodiesel is usually compared. DF2 contains many *n*-alkanes, cycloalkanes, as well as alkylbenzenes and various mono- and polyaromatic compounds (2). No. 4 DF (DF4) comprises more viscous distillates and their blends with residual fuel oils and usually is satisfactory only for low-speed and medium-speed engines operated under sustained load at nearly constant speed (1). Today, many "clean" DF containing significantly reduced amounts of aromatics and/or sulfur are in use.

A scale, the CN, conceptually similar to the octane scale used for gasoline (British term: petrol), was established for describing the ignition quality of petrodiesel or its components. Generally, a compound that has a high octane number tends to have a low CN and *vice versa*. Thus, 2,2,4-trimethylpentane (*iso*-octane), a short, branched alkane, is the high-quality standard (a primary reference fuel; PRF) for the octane scale of gasoline (and also gives it its name); it has an octane number of 100, whereas *n*-heptane is the low-quality PRF with an octane number of 0

(3). For the cetane scale, a long, straight-chain hydrocarbon, hexadecane ($C_{16}H_{34}$; trivial name cetane, giving the cetane scale its name), is the high-quality standard (and a PRF); it was assigned a CN of 100. At the other end of the scale, a highly branched compound, 2,2,4,4,6,8,8-heptamethylnonane (HMN, also $C_{16}H_{34}$), a compound with poor ignition quality in a diesel engine, was assigned a CN of 15 and it also is a PRF. Thus, branching and chain length influence CN with that number becoming smaller with decreasing chain length and increasing branching. Aromatic compounds, as mentioned above, occur in significant amounts in conventional DF. They have low CN but these increase with increasing size of *n*-alkyl side chains (4,5). The CN of a DF is determined by the ignition delay time, i.e., the time that passes between injection of the fuel into the cylinder and the onset of ignition. The shorter the ignition delay time, the higher the CN and *vice versa*. The cetane scale is arbitrary and compounds with CN >100 (although the cetane scale does not provide for compounds with CN >100) or CN <15 have been identified. Too high or too low a CN can cause operational problems. If a CN is too high, combustion can occur before the fuel and air are properly mixed, resulting in incomplete combustion and smoke. If a CN is too low, engine roughness, misfiring, higher air temperatures, slower engine warm-up, and also incomplete combustion occur. Most engine manufacturers in the United States designate a range of required CN, usually 40–50, for their engines.

The cetane scale clarifies why triacylglycerols as found in vegetable oils, animal fats, and derivatives thereof are suitable as alternative DF. The key is the long, unbranched chains of fatty acids, which are similar to those of the *n*-alkanes of good conventional DF.

Standards have been established worldwide for CN determination, for example, ASTM D613 in the United States and internationally, the International Organization for Standardization (ISO) standard ISO 5165. In the ASTM standard, hexadecane and HMN are the reference compounds. The standard ASTM D975 for conventional DF requires a minimum CN of 40, whereas the standards for biodiesel prescribe a minimum of 47 (ASTM D6751) or 51 (European standard pr EN14214). Due to the high CN of many fatty compounds, which can exceed the cetane scale, the term “lipid combustion quality number” for these compounds was suggested (6).

For petrodiesel, higher CN were correlated with reduced nitrogen oxides (NO_x) exhaust emissions (7). This correlation led to efforts to improve the CN of biodiesel fuels by means of additives known as cetane improvers (8). Despite the inherently relatively high CN of fatty compounds, NO_x exhaust emissions usually increase slightly when operating a diesel engine with biodiesel. The connection between the structure of fatty esters and exhaust emissions was investigated (9) by studying the exhaust emissions caused by enriched fatty acid alkyl esters as fuel. NO_x exhaust emissions reportedly increase with increasing unsaturation and decreasing chain length, which can also lead to a connection with the CN of these compounds. Particulate emissions, on the other hand, were only slightly influenced by the aforementioned structural factors. The relation between the CN and engine emissions is complicated by many factors, including the technology level of the engine. Older, lower injection pressure engines are generally very sensitive to CN, with increased CN causing significant reductions in the NO_x emissions due to shorter ignition delay times and the resulting lower average combustion temperatures. More modern engines that are equipped with injection systems that control the rate of injection are not highly sensitive to CN (10–12). Exhaust emissions from the operation of diesel engines on biodiesel are discussed in more detail in Chapter 7.

Historically, the first CN tests were carried out on palm oil ethyl esters (see Chapter 2), that had a high CN, a result confirmed by later studies on many other vegetable oil-based diesel fuels and individual fatty compounds. The influence of compound structure on the CN of fatty compounds was discussed recently (13); the predictions made in that report were confirmed by practical cetane tests (6,8,14–16). The CN of neat fatty compounds are given in Table A-1 of Appendix A. In summary, the results are that CN is lower with increasing unsaturation and higher with increasing chain length, i.e., uninterrupted CH_2 moieties. However, branched esters derived from alcohols such as *iso*-propanol have CN that are competitive with methyl or other straight-chain alkyl esters (14,17). Thus, one long straight chain suffices to impart a high CN even if the other moiety is branched. Branched esters are of interest because they exhibit improved low-temperature properties (see Chapter 6.3). Boiling point is the physical property that correlates best with CN for saturated fatty esters (16). The lower CN of unsaturated fatty compounds may be explained in part by the enhanced formation of intermediary precombustion species such as aromatics, which have low CN (18).

Cetane studies on fatty compounds were conducted using an Ignition Quality Tester™ (IQT™) (14). The IQT™ is a further, automated development of a constant volume combustion apparatus (CVCA) (19,20). The CVCA was originally developed for determining CN more rapidly, with greater experimental ease, better reproducibility, reduced use of fuel and therefore less cost than the ASTM method D613 utilizing a cetane engine. The IQT™ method, which is the basis of ASTM D6890, was shown to be reproducible and the results competitive with those derived from ASTM D613. Some results from the IQT™ are included in Table A-1 of Appendix A. For the IQT™, ignition delay (ID) and CN are related by Equation 1:

$$CN_{IQT} = 83.99 \times (ID - 1.512)^{-0.658} + 3.547 \quad [1]$$

In the recently approved method ASTM D6890, which is based on this technology, only ID times of 3.6–5.5 ms [corresponding to 55.3 to 40.5 derived cetane number (DCN)] are covered because it is stated that precision may be affected outside that range. However, the results for fatty compounds with the IQT™ are comparable to those obtained by other methods (14). Generally, the results of cetane testing for compounds with lower CN, such as the more unsaturated fatty compounds, show better agreement over the various related literature references than the results for compounds with higher CN. The reason is the nonlinear relation (see Eq. 1) between the ID time and the CN. This nonlinear relation between the ID time and CN was observed previously (21). Thus, small changes at shorter ID times result in greater changes in CN than at longer ID times. This would indicate a leveling-off effect on emissions such as NO_x as discussed above once a certain ID time with corresponding CN was reached because the formation of certain species depends on the ID time. However, for newer engines, this aspect must be modified as discussed above.

Heat of Combustion

In addition to CN, gross heat of combustion (HG) is a property proving the suitability of using fatty compounds as DF. The heat content of vegetable oils and their alkyl esters is nearly 90% that of DF2 (see Chapter 5 and Tables A-3 and A-4 in Appendix A). The heats of combustion of fatty esters and triacylglycerols (22,23) are in the range of ~1300–3500 kg-cal/mol for C_8–C_{22} fatty acids and esters (see Table A-1 in Appendix A). HG increases with chain length. Fatty alcohols possess heats of combustion in the same range (24). For purposes of comparison, the literature value (23) for the heat of combustion of hexadecane (cetane) is 2559.1 kg-cal/mol (at 20°C); thus it is the same range as fatty compounds.

References

1. Lane, J.C., Gasoline and Other Motor Fuels, in *Kirk-Othmer, Encyclopedia of Chemical Technology*, 3rd edn., edited by M. Grayson, D. Eckroth, G.J. Bushey, C.I. Eastman, A. Klingsberg, and L. Spiro, John Wiley & Sons, New York, Vol. 11, 1980, pp. 682–689.
2. Van Gerpen, J. and R. Reitz, Diesel Combustion and Fuels, in *Diesel Engine Reference Book*, edited by B. Challen and R. Baranescu, Society of Automotive Engineers, Warrendale, PA, 1998, pp. 89–104.
3. Hochhauser, A.M., Gasoline and Other Motor Fuels, in *Kirk-Othmer, Encyclopedia of Chemical Technology*, 4th edn., edited by J.I. Kroschwitz and M. Howe-Grant, John Wiley & Sons, New York, Vol. 12, 1994, pp. 341–388.
4. Puckett, A.D., and B.H. Caudle, *U.S. Bureau of Mines Information Circular No. 7474*, 1948.
5. Clothier, P.Q.E., B.D. Aguda, A. Moise, and H. Pritchard, How Do Diesel-Fuel Ignition Improvers Work? *Chem. Soc. Rev. 22:* 101 (1993).
6. Freedman, B., M.O. Bagby, T.J. Callahan, and T.W. Ryan III, Cetane Numbers of Fatty Esters, Fatty Alcohols and Triglycerides Determined in a Constant Volume Combustion Bomb, SAE Technical Paper Series 900343, SAE, Warrendale, PA, 1990.
7. Ladommatos, N., M. Parsi, and A. Knowles, The Effect of Fuel Cetane Improver on Diesel Pollutant Emissions, *Fuel 75:* 8–14 (1996).
8. Knothe, G., M.O. Bagby, and T.W. Ryan, III, Cetane Numbers of Fatty Compounds: Influence of Compound Structure and of Various Potential Cetane Improvers, SAE Technical Paper Series 971681, in *State of Alternative Fuel Technologies*, SAE Publication SP-1274, SAE, Warrendale, PA, 1997, pp. 127–132.
9. McCormick, R.L., M.S. Graboski, T.L. Alleman, and A.M. Herring, Impact of Biodiesel Source Material and Chemical Structure on Emissions of Criteria Pollutants from a Heavy-Duty Engine, *Environ. Sci. Technol. 35:* 1742–1747 (2001).
10. Mason, R.L., A.C. Matheaus, T.W. Ryan, III, R.A. Sobotowski, J.C. Wall, C.H. Hobbs,

G.W. Passavant, and T.J. Bond, EPA HDEWG Program—Statistical Analysis, SAE Paper 2001–01–1859, also in *Diesel and Gasoline Performance and Additives*, SAE Special Publication SP-1551, SAE, Warrendale, PA, 2001.

11. Matheaus, A.C., G.D. Neely, T.W. Ryan, III, R.A. Sobotowski, J.C. Wall, C.H. Hobbs, G.W. Passavant, and T.J. Bond, EPA HDEWG Program—Engine Test Results, SAE Paper 2001–01–1858, also in *Diesel and Gasoline Performance and Additives*, SAE Special Publication SP-1551, SAE, Warrendale, PA, 2001.
12. Sobotowski, R.A., J.C. Wall, C.H. Hobbs, A.C. Matheaus, R.L. Mason, T.W. Ryan, III, G.W. Passavant, and T.J. Bond, EPA HDEWG Program—Test Fuel Development, SAE Paper 2001–01–1857, also in *Diesel and Gasoline Performance and Additives*, SAE Special Publication SP-1551, SAE, Warrendale, PA, 2001.
13. Harrington, K.J., Chemical and Physical Properties of Vegetable Oil Esters and Their Effect on Diesel Fuel Performance, *Biomass 9:* 1–17 (1986).
14. Knothe, G., A.C. Matheaus, and T.W. Ryan, III, Cetane Numbers of Branched and Straight-Chain Fatty Esters Determined in an Ignition Quality Tester, *Fuel 82:* 971–975 (2003).
15. Klopfenstein, W.E., Effect of Molecular Weights of Fatty Acid Esters on Cetane Numbers as Diesel Fuels, *J. Am. Oil Chem. Soc. 62:* 1029–1031 (1985).
16. Freedman, B., and M.O. Bagby, Predicting Cetane Numbers of *n*-Alcohols and Methyl Esters from Their Physical Properties, *J. Am. Oil Chem. Soc. 67:* 565–571 (1990).
17. Zhang, Y., and J.H. Van Gerpen, Combustion Analysis of Esters of Soybean Oil in a Diesel Engine. Performance of Alternative Fuels for SI and CI Engines, SAE Technical Paper Series 960765, also in *Performance of Alternative Fuels for SI and CI Engines*, SAE Special Publication SP-1160, SAE, Warrendale, PA, 1996, pp. 1–15.
18. Knothe, G., M.O. Bagby, and T.W. Ryan, III, Precombustion of Fatty Acids and Esters of Biodiesel. A Possible Explanation for Differing Cetane Numbers, *J. Am. Oil Chem. Soc. 75:* 1007–1013 (1998).
19. Ryan, III, T.W., and B. Stapper, Diesel Fuel Ignition Quality as Determined in a Constant Volume Combustion Bomb, SAE Technical Paper Series 870586, SAE, Warrendale, PA, 1987.
20. Aradi, A.A., and T.W. Ryan, III, Cetane Effect on Diesel Ignition Delay Times Measured in a Constant Volume Combustion Apparatus, SAE Technical Paper Series 952352, also in SAE Special Publication SP-1119, *Emission Processes and Control Technologies in Diesel Engines*, SAE, Warrendale, PA, 1995, p. 43.
21. Allard, L.N., G.D. Webster, N.J. Hole, T.W. Ryan, III, D. Ott, and C.W. Fairbridge, Diesel Fuel Ignition Quality as Determined in the Ignition Quality Tester (IQT), SAE Technical Paper Series 961182, SAE, Warrendale, PA, 1996.
22. Freedman, B., and M.O. Bagby, Heats of Combustion of Fatty Esters and Triglycerides, *J. Am. Oil Chem. Soc. 66:* 1601–1605 (1989).
23. Weast, R.C., M.J. Astle, and W.H. Beyer, *Handbook of Chemistry and Physics*, 66th edn., CRC Press, Boca Raton, FL, 1985–1986, pp. D-272–D-278.
24. Freedman, B., M.O. Bagby, and H. Khoury, Correlation of Heats of Combustion with Empirical Formulas for Fatty Alcohols, *J. Am. Oil Chem. Soc. 66:* 595–596 (1989).

6.2

Viscosity of Biodiesel

Gerhard Knothe

Viscosity, which is a measure of resistance to flow of a liquid due to internal friction of one part of a fluid moving over another, affects the atomization of a fuel upon injection into the combustion chamber and thereby, ultimately, the formation of engine deposits. The higher the viscosity, the greater the tendency of the fuel to cause such problems. The viscosity of a transesterified oil, i.e., biodiesel, is about an order of magnitude lower than that of the parent oil (see tables in Appendix A). High viscosity is the major fuel property explaining why neat vegetable oils have largely been abandoned as an alternative diesel fuel (DF). Kinematic viscosity (ν), which is related to dynamic viscosity (η) by density as a factor, is included as a specification in biodiesel standards (see tables in Appendix B). It can be determined by standards such as ASTM D445 or ISO 3104. Values for kinematic viscosity of numerous fatty acid compounds, including methyl esters, were reported (1–4). Data on the dynamic viscosity of fatty materials are also available in the literature (5–10). Values for η of fatty acid alkyl esters are compiled in Table A-1 and of fats and oil as well as their alkyl esters in Tables A-3 and A-4 of Appendix A. Fatty acid methyl esters are Newtonian fluids at temperatures above 5°C (11).

The viscosity of petrodiesel fuel is lower than that of biodiesel, which is also reflected in the kinematic viscosity limits (all at 40°C) of petrodiesel standards, which are 1.9–4.1 mm^2/s for DF2 (1.3–2.4 mm^2/s for DF1) in the ASTM petrodiesel standard D975 and 2.0–4.5 mm^2/s in the European petrodiesel standard EN 590.

The difference in viscosity between the parent oil and the alkyl ester derivatives can be used to monitor biodiesel production (12); (see also Chapter 5). The effect on viscosity of blending biodiesel and conventional petroleum-derived diesel fuel was also investigated (13), and an equation was derived that allows calculation of the viscosity of such blends.

Prediction of the viscosity of fatty materials has received considerable attention in the literature. Viscosity values of biodiesel/fatty ester mixtures were predicted from the viscosities of the individual components by a logarithmic equation for dynamic viscosity (5). Viscosity increases with chain length (number of carbon atoms) and with increasing degree of saturation. This holds also for the alcohol moiety because the viscosity of ethyl esters is slightly higher than that of methyl esters. Factors such as double-bond configuration influence viscosity (*cis* double-bond configuration giving a lower viscosity than *trans*), whereas double-bond position affects viscosity less (unpublished results). Branching in the ester moiety, however, has little or no influence on viscosity (unpublished results).

References

1. Gouw, T.H., J.C. Vlugter, and C.J.A. Roelands, Physical Properties of Fatty Acid Methyl Esters. VI. Viscosity, *J. Am. Oil Chem. Soc. 43:* 433–434 (1966).
2. Valeri, D., and A.J.A. Meirelles, Viscosities of Fatty Acids, Triglycerides, and Their Binary Mixtures, *J. Am. Oil. Chem. Soc. 74:* 1221–1226 (1997).
3. Bonhorst, C.W., P.M. Althouse, and H.O. Triebold, Esters of Naturally Occurring Fatty Acids, *Ind. Eng. Chem. 40:* 2379–2384 (1948).
4. Formo, M.W., Physical Properties of Fats and Fatty Acids, in *Bailey's Industrial and Oil Products,* Vol. 1, 4th edn., John Wiley & Sons, New York, 1979, pp. 177–232.
5. Allen, C.A.W., K.C. Watts, R.G. Ackman, and M.J. Pegg, Predicting the Viscosity of Biodiesel Fuels from Their Fatty Acid Ester Composition, *Fuel 78:* 1319–1326 (1999).
6. Noureddini, H., B.C. Teoh, and L.D. Clements, Viscosities of Vegetable Oils and Fatty Acids, *J. Am. Oil Chem. Soc. 69:* 1189–1191 (1992).
7. Fernandez-Martin, F., and F. Montes, Viscosity of Multicomponent Systems of Normal Fatty Acids: Principle of Congruence, *J. Am. Oil Chem. Soc. 53:* 130–131 (1976).
8. Shigley, J.W., C.W. Bonhorst, C.C. Liang, P.M. Althouse, and H.O. Triebold, Physical Characterization of a) a Series of Ethyl Esters and b) a Series of Ethanoate Esters, *J. Am. Oil Chem. Soc. 32:* 213–215 (1955).
9. Gros, A.T., and R.O. Feuge, Surface and Interfacial Tensions, Viscosities, and Other Physical Properties of Some *n*-Aliphatic Acids and Their Methyl and Ethyl Esters, *J. Am. Oil Chem. Soc. 29:* 313–317 (1952).
10. Kern, D.Q., and W. Van Nostrand, Heat Transfer Characteristics of Fatty Acids, *Ind. Eng. Chem. 41:* 2209–2212 (1948).
11. Srivastava, A., and R. Prasad, Rheological Behavior of Fatty Acid Methyl Esters, *Indian J. Chem. Technol. 8:* 473–481 (2001).
12. De Filippis, P., C. Giavarini, M. Scarsella, and M. Sorrentino, Transesterification Processes for Vegetable Oils: A Simple Control Method of Methyl Ester Content, *J. Am. Oil Chem. Soc. 71:* 1399–1404 (1995).
13. Tat, M.E., and J.H. Van Gerpen, The Kinematic Viscosity of Biodiesel and Its Blends with Diesel Fuel, *J. Am. Oil Chem. Soc. 76:* 1511 (1999).

6.3

Cold Weather Properties and Performance of Biodiesel

Robert O. Dunn

Introduction

In spite of biodiesel's many advantages, performance during cold weather may affect its year-round commercial viability in moderate temperature climates. Although field studies for biodiesel performance in cooler weather are scarce, there is evidence that using the soybean oil methyl ester (SME) form of biodiesel (made by transesterification of soybean oil and methanol) raises performance issues when ambient temperatures approach 0–2°C. As overnight temperatures fall into this range, saturated methyl esters within SME nucleate and form solid crystals. These crystals plug or restrict flow through fuel lines and filters during start-up the next morning and can lead to fuel starvation and engine failure.

This chapter opens by examining pertinent fuel characteristics for assessing the effects of biodiesel on cold weather performance as a neat (100%) fuel and in blends with petrodiesel. A review of relevant standard test methods for measuring cold flow characteristics is presented followed by a discussion of the available technology for improving cold flow properties of biodiesel from several feedstocks. This chapter closes with an examination of research and development on the cold weather performance of biodiesel as an alternative fuel and fuel extender.

Cold Flow Properties of Diesel Fuels

All diesel fuels are susceptible to start-up and performance problems when vehicles and fuel systems are subjected to cold temperatures. As ambient temperatures cool toward their saturation temperature, high-molecular-weight paraffins (C_{18}–C_{30} *n*-alkanes) present in petrodiesel begin to nucleate and form wax crystals suspended in a liquid phase composed of shorter-chain *n*-alkanes and aromatics (1–5). If the fuel is left unattended in cold temperatures for a long period of time (e.g., overnight), the presence of solid wax crystals may cause start-up and performance problems the next morning. The tendency of a fuel to solidify or gel at low temperatures can be quantified by several experimental parameters as defined below.

Cloud Point and Pour Point

Initially, cooling temperatures cause the formation of solid wax crystal nuclei that are submicron in scale and invisible to the human eye. Further decreases in temperature cause these crystals to grow. The temperature at which crystals become visible [diameter (d) $\geq$ 0.5 μm] is defined as the cloud point (CP) because the crystals usually form a cloudy or hazy suspension (1,3,6–8). Due to the *orthorhombic* crystalline structure, unchecked crystalline growth continues rapidly in two dimensions forming large platelet lamellae (2,4,5,8–12). At temperatures below CP, larger crystals (d ~ 0.5–1 mm × 0.01 mm thick) fuse together and form large agglomerates that can restrict or cut off flow through fuel lines and filters and cause start-up and performance problems the next morning (1,2,4,5,8,11–15). The temperature at which crystal agglomeration is extensive enough to prevent free pouring of fluid is determined by measurement of its pour point (PP) (3,6–8). Some petrodiesel fuels can reach their PP with as little as 2% wax out of solution (12).

Method numbers and short descriptions corresponding to ASTM standard methods for determining CP, PP, and other relevant cold flow properties are summarized in Table 1. Also listed in Table 1 are two examples of automated test methods that employ static light-scattering (LS) technology to measure CP (determined by particle counting) or PP (surface movement). In general, measurement requires a maximum time of 12 min due to very low sample quantity (<150 mL) required to perform an analysis. Other standard automated methods include D 5771 (CP), D 5950 (PP), D 5772 (CP), and D 5985 (PP) (7). These particular methods demonstrate very little bias (<0.03°C for CP) with respect to "manual" methods D 2500 (7,16). No studies comparing results from manual and automated methods have been reported for biodiesel or biodiesel/conventional diesel fuel blends.

Thermal analytical methods such as subambient differential scanning calorimetry (DSC) were successfully applied to neat petrodiesel (14,17–21) and engine oils (18,22–24). DSC has the advantages of rapid and accurate determination of melting characteristics, analysis of samples that are solid at room temperature, and relatively small sample sizes (<20 mg). Heating and cooling DSC scans were analyzed to determine crystallization onset temperature (CP), melting points (PP), and glass transition temperatures.

The application of DSC in the analysis of the melting characteristics of neat biodiesel was also reported. The crystallization onset temperature (T_{Cryst}) was determined by rapidly cooling the sample at 100°C/min, equilibration at –70°C, then heating to 60°C at 5°C/min (25,26). Heating curves were analyzed at the high-temperature end of the highest melting peak for biodiesel made from soybean oil and low-palmitic soybean oil transesterified with various straight and branched-chain alcohols. The same protocol was applied in an analysis of esters derived from tallow and waste grease. Studies on correlating CP, PP, and other cold flow properties from various parameters obtained from analysis of heating (27) and cooling DSC scans were also reported (28). Peak temperatures and freezing points determined for the highest freezing peak in the cooling curves yielded the most accurate means for determining the CP and PP of SME.

Wax Appearance Point and Wax Precipitation Index

Although CP and PP are relatively easy to measure in bench scale, neither parameter is very useful for predicting cold flow performance of diesel fuels under field conditions. CP data consistently overpredict the cold temperature limit at which start-up or performance problems may be expected to occur, whereas PP data tend to be optimistic (1,3,12,29,30). Although the wax appearance point (WAP, ASTM D 3117) shows better precision than CP, some studies have shown that these two parameters are essentially equivalent within 1–2°C (3). The wax precipitation index (WPI) for predicting minimum vehicle operating temperature during cold weather was introduced (30). WPI is determined by an empirical relation based on CP and PP as shown in Table 1. Although good correlation was demonstrated for petrodiesel, the WPI has seen little application by industry (3,30).

Low-Temperature Filterability Tests

In the mid-1960s, attention focused on developing laboratory bench-scale tests independent of CP or PP to predict minimum operability temperatures for diesel fuels. In Western Europe, this work resulted in the development of the cold filter plugging point (CFPP) test method (1,3,11–13,15). This method (ASTM standard D 6371) calls for cooling an oil sample at a specified rate and drawing it under vacuum through a wire mesh filter screen (see Table 1). CFPP is then defined as the lowest temperature at which 20 mL of oil safely passes through the filter within 60 s (1,3,6,8).

Although CFPP is acceptable nearly worldwide as a standard bench test method, field studies in the early 1980s showed that more stringent test conditions were needed to adequately correlate results to fuels and equipment prevalent in North America. That work resulted in the development of the less user-friendly Low-Temperature Flow Test (LTFT) (1,3,5,13,15,30). This method (ASTM D 4539) is nearly identical to CFPP except for the larger sample volume, generally slower cooling rates (1°C/h), smaller pore-size wire mesh filter, and stronger vacuum force applied to draw the sample (see Table 1). Like CFPP, LTFT is defined as the lowest temperature at which 180 mL of oil safely passes through the filter within 60 s (1,3,6,29). One recent study (13) confirmed that LTFT was the best predictor for cold weather performance for applications of fuels treated with cold flow improver (CFI) additives (discussed below) in North America.

Viscosity

Viscosity (see also Chapter 6.2) is defined as the resistance by one portion of a material moving over another portion of the same material. Dynamic viscosity (η) is defined as the ratio of shear stress existing between layers of moving fluid and

TABLE 1
Cold Flow Properties of Petroleum Middle Distillate (Petrodiesel) Fuels[a,b]

Parameter	Test methods[c]	Description
Cloud point (CP)	ASTM D 2500 (IP 219, ISO 3015, DIN 51597, JIS K 2269, AFNOR T60-105)	Cool at specified rate, examine at 1°C intervals; CP = temperature at which haziness is observed
CP (automated method)	ASTM D 5773 (IP 446)	Sample cooled at 1.5°C/min, examine continuously under LS; CP = temperature at which particles are detected
Cold filter plugging point (CFPP)	ASTM D 6371 (IP 309, EN 116)	Cool at specified rate, examine at 1°C intervals; CFPP = lowest temperature at which a 20-mL sample passes through 45-µm wire mesh under 0.0194 atm vacuum within 60 s
Freezing point (FP)	ASTM D 2386 (IP 16, ISO 3013, DIN 51421, JIS K 2276, AFNOR M07-048)	Cool in double-walled, jacketed, clear glass tube with agitation until haziness is observed, then reheat, examine at 0.5°C intervals; FP = temperature at which haziness disappears completely
FP (automated method)	ASTM D5972 (IP 435)	Sample is cooled at 15°C/min[d] until crystals are detected and then reheated at 10°C/min; FP = temperature at which particles cannot be detected
Kinematic viscosity (ν)	ASTM D 445 (IP 71-1, ISO 3104, DIN 51562, JIS K 2283, AFNOR T60-100)	Cool in calibrated viscometer in constant-temperature bath, measure time (t) necessary for fixed volume to flow under gravity through capillary; $\nu = k(t)$, k = viscometer calibration constant

Low-temperature flow test (LTFT)	ASTM D 4539	Cool at 1°C/h, examine at 1°C intervals; LTFT = lowest temperature at which 180-mL sample passes through a 17-µm wire-mesh filter under 0.197 atm vacuum within 60 s
Pour point (PP)	ASTM D 97 (IP 15, ISO 3016, DIN 51597, JIS K 2269, AFNOR T60-105)	Cool at specified rate; examine at 3°C intervals; PP = lowest temperature at which movement is detected
PP (automated method)	ASTM D 5949	Sample cooled at 1.5°C/min[d]; apply moving force (pressurized N_2 gas) at 1, 2, or 3°C intervals; examine by LS detection; PP = lowest temperature at which surface movement is detected during pulse
Wax appearance point (WAP)	ASTM D 3117	Cool in jacketed, double-walled clear glass tube with agitation, examine at 1°C intervals; WAP = temperature at which "swirl" of crystals is observed
Wax precipitation index (WPI)	None	Two-parameter correlation for predicting minimum vehicle operating temperature: $WPI = CP + x(CP - PP - y)^z$ where CP and PP are in °C and x, y, and z are constants[e]

[a]*Sources:* References 3,7,30,31,50,51,107.
[b]AFNOR, Association Francaise de Normalisation (Paris); ASTM, American Society for Testing and Materials (USA); DIN, Deutsche Institut Fur Normung (Germany); IP, Institute of Petroleum (UK); ISO, International Organization for Standardization (Switzerland); JIS, Japan Industrial Standards (Tokyo); LS, light scattering.
[c]Equivalent methods in parentheses (AFNOR, IP, and DIN are currently being integrated into ISO/EN-ISO series standards as mandated by European Union legal requirements).
[d]Samples cooled by Peltier device.
[e]For conventional diesel fuels (D-2, D-2/1 winter blends) where $(CP - PP) > 1.1$, $x = 1.3$, $y = 1.1$, and $z = 0.5$; if $(CP - PP) \leq 1.1$, $WPI = CP$.

the rate of shear between the layers. The resistance to flow of a liquid under gravity (kinematic viscosity, ν) is the ratio of η to the density (ρ) of the fluid (31).

Most fluids such as petrodiesel and biodiesel increase in viscosity with decreasing temperature. Biodiesel and petrodiesel fuel standards limit ν (32; see Appendix B). Decreasing the temperature from 40 to –3°C increased ν from 2.81 to 10.4 mm^2/s for D-2 and 1.59 to 4.20 mm^2/s for D-1 (33).

Significant increases in ν may be accompanied by transition into non-Newtonian behavior, defined as a fluid that does not exhibit constant viscosity at all shear rates. As a result, changes in rheological flow properties may provide a way to restrict the flow through the wire mesh screens in addition to blockage from large wax crystals, during CFPP or LTFT testing. Refined oils without polymeric additives such as petrodiesel are typically Newtonian (31). However, as will be discussed in the following section, studies on biodiesel suggest that a transition to non-Newtonian behavior will affect viscosity and other flow properties at low temperatures.

Cold Flow Properties of Biodiesel

Cold flow properties of methyl and ethyl ester forms of biodiesel derived from several feedstocks are summarized in Table 2. Transesterification does not alter the fatty acid composition of the feedstocks. Therefore, biodiesel made from feedstocks containing higher concentrations of high-melting point saturated long-chain fatty acids tends to have relatively poor cold flow properties.

The fatty acid compositions of many oils and fats commonly used to make biodiesel are presented in Table A-2 in Appendix A. Due to its content of saturated compounds, tallow methyl ester (TME) has CP = 17°C (34). Another example is palm oil methyl ester, whose CP = 13°C (35). In contrast, feedstocks with relatively low concentrations of saturated long-chain fatty acids generally yield biodiesel with much lower CP and PP. Thus, feedstocks such as linseed, olive, rapeseed, and safflower oils tend to yield biodiesel with CP ≤ 0°C (36–40).

Figures 1 and 2 are plots of CP and PP, respectively, vs. blend ratio (vol% esters) for SME in blends with D-1 and D-2 and in blends with JP-8 jet fuel (33,41). The data in these figures demonstrate that SME significantly affects both CP and PP at relatively low blend ratios in D-1 and JP-8 fuel. For blends in D-2, increasing the blend ratio results in a linear increase in CP (R^2 = 0.99) and nearly linear increase in PP (R^2 = 0.96) (33). D-1 blends showed an increase in PP of only 4°C between 0 and 10 vol% SME compared with an increase of 12°C between 10 and 20 vol% SME (see also Table 3). Thus, up to a cut-off blend ratio close to 10 vol% SME, D-1 appears to predominate in determining the cold flow property behavior of blends. This trend is also reflected in CP data in Figure 1. For JP-8 blends, a similar cut-off blend ratio close to 20 vol% SME is noticeable in Figure 2. Overall, these results indicate that diesel engines fueled by biodiesel/petrodiesel blends increase in susceptibility to cold flow start-up and performance problems with an increasing blend ratio.

TABLE 2
Cold Flow Properties of Biodiesel (Methyl and Ethyl Esters) Derived from Oils and Fats[a]

Oil or fat	Alkyl group	CP (°C)	PP (°C)	CFPP (°C)	LTFT (°C)	Reference
Babassu	Methyl	4				39
Canola	Methy	l	−9			41
Canola	Ethyl	−1	−6			41
Coconut	Ethyl	5	−3			42
Cottonseed	Methyl		−4			108
Linseed	Methyl	0	−9			41
Linseed	Ethyl	−2	−6			41
Mustardseed	Ethyl	1	−15			42
Olive	Methyl	−2	−3	−6		40
Palm	Methyl	13	16			39, 109
Palm	Ethyl	8	6			48
Peanut	Methyl	5				39
Rapeseed	Methyl	−2	−9	−8		44, 110
Rapeseed	Ethyl	−2	−15			41
Safflower	Methyl		−6			43
Safflower	Ethyl	−6	−6			42
Soybean	Methyl	0	−2	−2	0	35
Soybean	Ethyl	1	−4			37
Sunflowerseed	Methyl	2	−3	−2		40
Sunflowerseed	Ethyl	−1	−5			41
HO Sunflowerseed[b]	Methyl			−12		109
Tallow	Methyl	17	15	9	20	37
Tallow	Ethyl	15	12	8	13	37
Used hydrogenated soybean[c]	Ethyl	7	6			42
Waste cooking[d]	Methyl			−1		99
Waste grease[e]	Ethyl	9	−3	0	9	37
Waste olive	Methyl	−2	−6	−9		53

[a]Biodiesel from transesterification of "oil or fat" with "alkyl" alcohol; see Table 1 for abbreviations.
[b]High-oleic (77.9 wt%) sunflowerseed oil.
[c]Hydrogenated to iodine value (IV) of ~65.
[d]Total saturated methyl ester content ~19.2 wt%.
[e]Contained ~9 wt% free fatty acids.

Earlier studies (33,42,43) reported a linear correlation between low-temperature filterability (CFPP and LTFT) and CP for biodiesel and its blends with D-1 and D-2. Figures 3 and 4 are plots of CFPP and LTFT vs. CP data reported collectively in the three studies cited above. Data for blends, neat SME, neat TME, SME/TME admixtures, and formulations treated with CFI additives are represented in both figures. Effects of CFI additives are discussed in further detail later in the chapter.

Least-squares regression analysis of filterability vs. CP data for neat oils and blends not treated with CFI additives yielded the following equations:

$$\text{CFPP} = 1.019(\text{CP}) - 2.9 \quad [1]$$

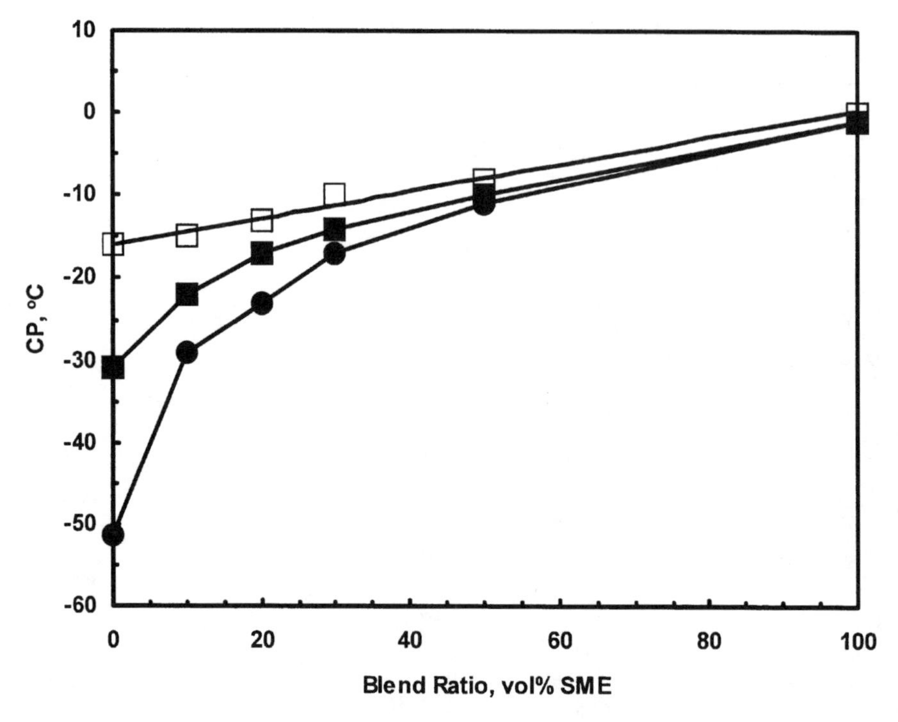

Fig. 1. Cloud point (CP) vs. blend ratio (vol%) of soybean oil fatty acid methyl esters (SME) for blends with No. 1 (D-1) and No. 2 (D-2) diesel fuels and JP-8 jet fuel. Legend: □ = D-2 blends; ■ = D-1 blends; ● = JP-8 blends. Line through D-2 blend data: CP = 0.1618[blend ratio] – 16.0 ($R^2 = 0.99$, $\sigma_y = 0.67$).

$$\text{LTFT} = 1.020(\text{CP}) + 0.4 \qquad [2]$$

with $R^2 = 0.90$ and 0.95 and $\sigma_y = 2.5$ and 1.8, respectively. ANOVA revealed high probabilities that slope = 1 for both equations ($P = 0.81$ and 0.78, respectively), suggesting that a 1°C decrease in CP results in a 1°C decrease in either CFPP or LTFT (33). Hence, this work showed that research on improving the cold flow performance of biodiesel should focus on approaches that significantly decrease CP. Furthermore, this conclusion applied to blends with as little as 10 vol% biodiesel in D-1 or D-2.

Several studies (33,40,44–46) reported that biodiesel in the form of methyl or ethyl esters derived from most feedstocks has $\nu = 4.1–6.7$ mm^2/s at 40°C. In contrast, measurement at 5°C yielded $\nu = 11.4$ mm^2/s for SME (33). In that study, samples were sealed in a Cannon-Fenske viscometer, immersed in a constant temperature bath, and allowed to sit overnight to simulate cooling of the fuel for a sustained period (~16 h). Admixtures of SME with up to 20 vol% TME decreased ν slightly to 10.8 mm^2/s. Results from a similar analysis of neat D-2 (blend ratio = B0) yielded ν

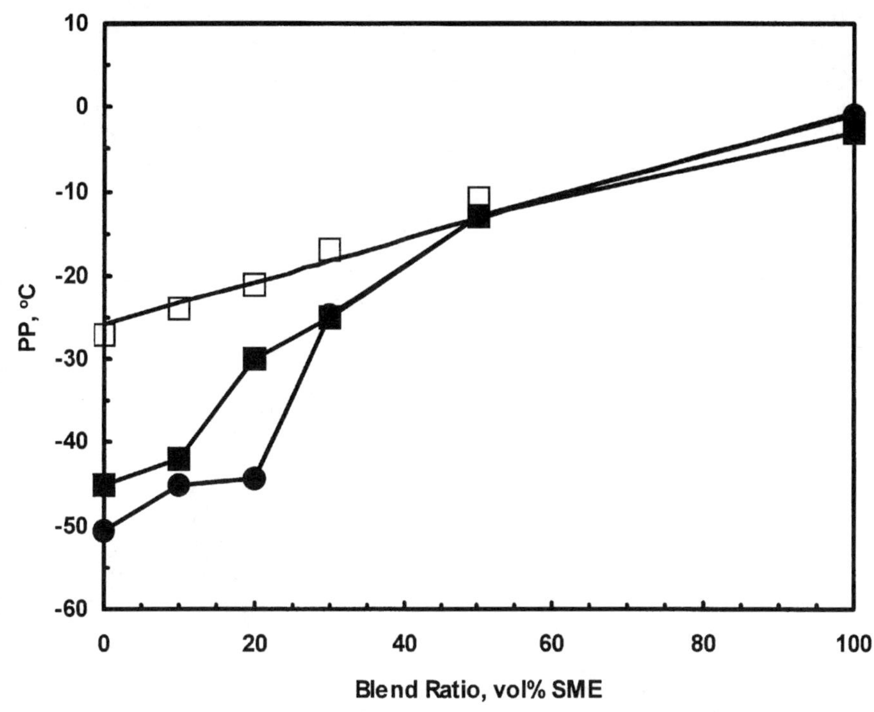

Fig. 2. Pour point (PP) vs. blend ratio (vol%) of soybean oil fatty acid methyl esters (SME) for blends with No. 1 (D-1) and No. 2 (D-2) diesel fuels and JP-8 jet fuel. Legend: □ = D-2 blends; ■ = D-1 blends; ● = JP-8 blends. Line through D-2 blend data: PP = 0.2519[blend ratio] – 25.8 (R^2 = 0.96, σ_y = 1.6).

= 10.4 mm^2/s when cooled overnight at –3°C, in comparison with ν = 2.81 mm^2/s at 40°C.

Attempts to measure ν of SME or SME/TME admixtures at temperatures <5°C were frustrated by the apparent solidification of the sample during the 16-h equilibration period (33). It was reported (47) that at temperatures near 5°C, the rheology of SME (as well as that of mustardseed oil methyl esters) undergoes transition from Newtonian to *pseudoplastic*-type flow. Pseudoplastic fluids experience shear thinning or the property of non-Newtonian fluids exhibiting reductions in viscosity with increasing shear rate (31). Thus, it is possible at temperatures <5°C that a transition to pseudoplastic fluid facilitated the formation of a network of interlocked crystals while SME stood quiescent overnight under zero or very little shear (force), plugging the viscometer tube.

Similar attempts to measure ν of a 7:3 (vol/vol) SME/TME admixture at 5°C were also frustrated (33). Given the relatively high CP and PP of neat TME (see Table 2), if TME undergoes a rheological transition to pseudoplastic fluid, then it is likely to occur at a higher temperature than SME. Apparently, rheology in

TABLE 3
Cold Flow Properties of Biodiesel/Petrodiesel Blends[a]

Oil or fat	Alkyl group	Diesel grade	Blend ratio	CP (°C)	PP (°C)	CFPP (°C)	LTFT (°C)	Ref.
—	—	D-1	B0	–31	–46	–42	–27	35
Soybean	Methyl	D-1	B10	–22	–42			35
Soybean	Methyl	D-1	B20	–17	–30	–27	–19	35
Soybean	Methyl	D-1	B30	–14	–25	–20	–16	35
Soybean/tallow[b]	Methyl	D-1	B20	–21	–29	–21	–18	35
Soybean/tallow[b]	Methyl	D-1	B30	–13	–24	–18	–14	35
—	—	D-2	B0	–16	–27	–18	–14	35
Coconut	Ethyl	D-2	B20	–7	–15			42
Rapeseed	Ethyl	D-2	B20	–13	–15			42
Soybean	Methyl	D-2	B20	–14	–21	–14	–12	35
Soybean	Methyl	D-2	B30	–10	–17	–12	–12	35
HO sunflowerseed[c]	Methyl	D-2	B30		–12			110
Tallow	Methyl	D-2	B20	–5	–9	–8		37
Tallow	Ethyl	D-2	B20	–3	–12	–10	1	37
Soybean/tallow[b]	Methyl	D-2	B20	–12	–20	–13	–10	35
Soybean/tallow[b]	Methyl	D-2	B30	–10	–12	–11	–9	35
Used hydrogenated soybean[d]	Ethyl	D-2	B20	–9	–9			42
Waste grease[e]	Ethyl	D-2	B20	–12	–21	–12	–3	29

[a]Biodiesel from transesterification of "oil or fat" with "alkyl" alcohol; Diesel grade is according to ASTM fuel specification D 975 (see Ref. 32) in which D-1 = grade No. 1 and D-2 = grade No. 2; Blend ratio = "B*X*" where *X* = vol% biodiesel in blend with petrodiesel; see Tables 1 and 2 for other abbreviations.
[b]4:1 vol/vol soybean oil methyl ester/tallow methyl ester.
[c]High-oleic (77.9 wt%) sunflowerseed oil.
[d]Hydrogenated to iodine value = 65.
[e]Contained ~9 wt% free fatty acids before conversion to biodiesel.

SME/TME admixtures at cold temperatures more closely resembles that of neat SME for TME contents up to 20 vol%, whereas higher contents allow TME to influence rheology.

Changes in fluid rheology may also explain why PP exceeds the CFPP of neat TME by at least 6°C (34). Reviewing the summary of ASTM method D 97 shown in Table 1, samples are cooled at a specified rate and checked visually in 3°C intervals (48); thus, for neat TME, failure to flow was actually detected at 12°C, a temperature that exceeds CFPP by 3°C. Although this explains discrepancies between the PP and CFPP data in Table 2 for tallow ethyl esters, olive oil methyl esters, and waste olive oil methyl esters (34,36,49), there remains a 3°C discrepancy for neat TME.

It is possible that the transition of TME to a pseudoplastic fluid also facilitated formation of interlocking crystals after extended residence time in viscometer tubes similar to that noted above for SME. Without shear force other than that induced by gravity as the sample test tube was tilted, it is possible the interlocking

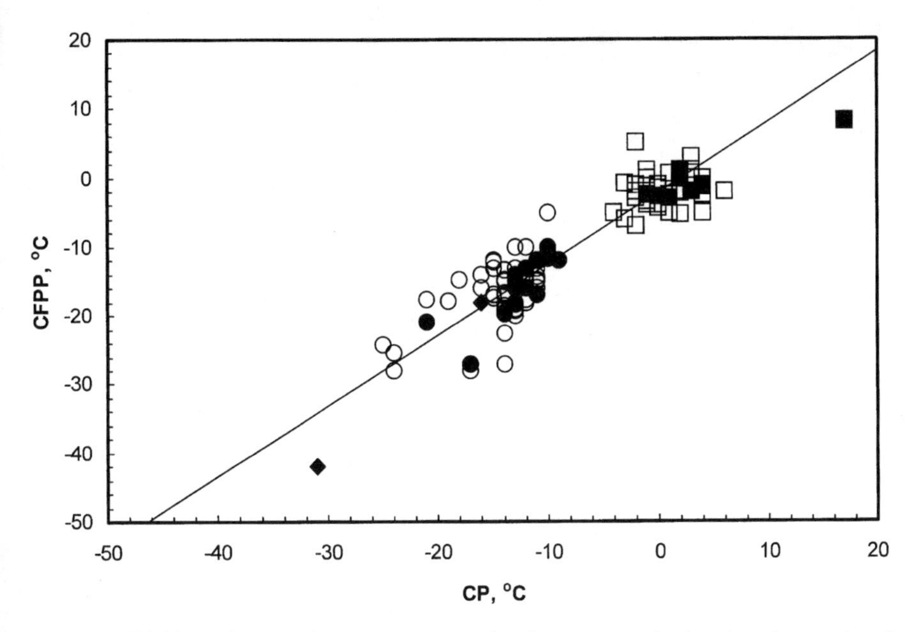

Fig. 3. Cold filter plugging point (CFPP) vs. cloud point (CP) for biodiesel/petrodiesel blends. Legend: ◆ = petrodiesel (B0); ■ = neat biodiesel (B100); □ = B100 + CFI; ● = blends; ○ = blends + CFI. Biodiesel = fatty acid methyl esters; CFI = cold flow improver. Regression line: CFPP = 1.0276[CP] – 2.2 (R^2 = 0.82, σ_y = 3.5).

crystals also prevented movement in the sample at temperatures <12°C. Although fluid movement was not detected under the force of gravity, the interlocked crystals may have retained some pseudoplastic flow character. If so, then application of vacuum shear force may have been sufficient to decrease pseudoplastic fluid viscosity and allow the material to pass safely through the 45-µm wire mesh screen used to measure CFPP in accordance with ASTM D 6371 (50). In the LTFT test, effects of a stronger vacuum shear in forcing the pseudoplastic fluid to flow with reduced viscosity appear to have been countered by the smaller wire mesh pore size (17 µm) employed in the test (51). Therefore, LTFT determinations for TME were more in line with PP (and CP) results shown in Table 2 (34).

Trace concentrations of contaminants can also influence cold flow properties of biodiesel. Studies (52) on the effects of residual contaminants arising from the refining and transesterification processes on cold flow properties of neat SME and SME/D-1 blends showed that, although PP was not affected, CP increased with increasing concentration of both monoglycerides and diglycerides. Concentrations as low as 0.1 wt% (1000 ppm) saturated monoglycerides or diglycerides raised CP, whereas unsaturated monoolein did not affect CP or PP. Unsaponifiable matter increases T_{Cryst}, CP, and PP of SME at concentrations of 3 wt% but had essentially no effect on a 20 vol% SME blend. Other contaminants that may arise from refin-

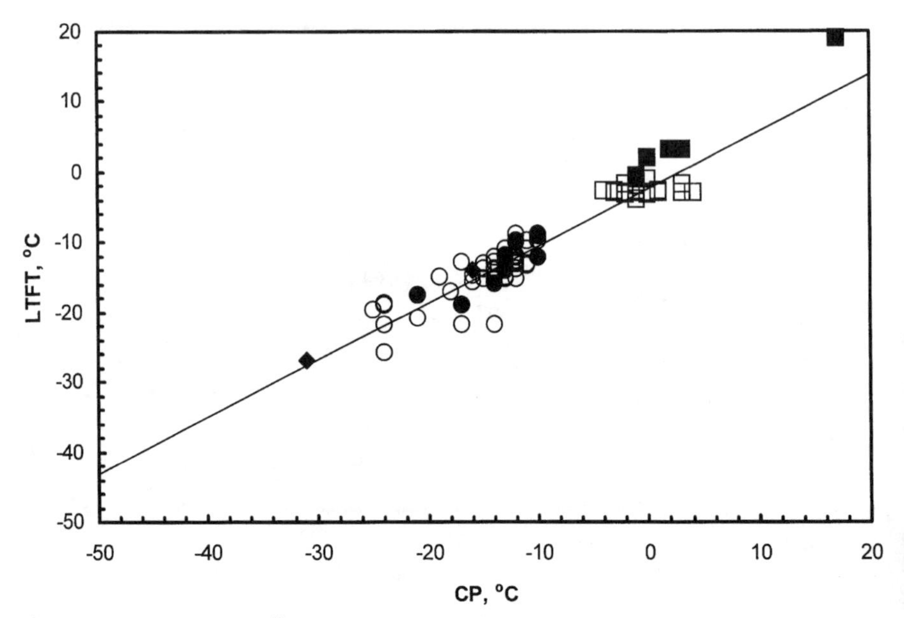

Fig. 4. Low-temperature flow test (LTFT) vs. cloud point (CP) for biodiesel/petrodiesel blends. Legend: ◆ = petrodiesel (B0); ■ = neat biodiesel (B100); □ = B100 + CFI; ● = blends; ○ = blends + CFI. Biodiesel = fatty acid methyl esters; CFI = cold flow improver. Regression line: CFPP = 0.8140[CP] – 2.4 (R^2 = 0.90, σ_y = 2.0).

ing or transesterification processes that can influence cold flow properties include alcohol, free fatty acids, and unreacted triacylglyceride.

Improving the Cold Flow Properties of Biodiesel

Based on available technology, several approaches for reducing CP of biodiesel were investigated. Approaches attracting the most attention include the following: (i) blending with petrodiesel; (ii) treating with petrodiesel fuel CFI additives; (iii) developing new additives designed for biodiesel; (iv) transesterification of vegetable oils or fats with long- or branched-chain alcohols; and (v) crystallization fractionation. The remainder of this section discusses the effects of each of these approaches on improving the cold weather performance of biodiesel.

Blending with Petrodiesel Fuel

One of the most effective ways to reduce CP and improve the pumpability of D-2 during cold weather is to blend it with D-1, kerosene or jet fuel (1,5,8,12,13,29). Each 10 vol% of D-1 decreases the CP and CFPP of the blend by 2°C (8,12). Blending with D-1 also decreases viscosity, decreases energy content, which reduces power output, and increases fuel consumption, wear on injector pumps, and cost (1,5,8,11–13,29).

In terms of improving cold weather performance of petrodiesel, treatment with CFI additives is preferred to blending with D-1, kerosene, or jet fuel because CFI additives generally do not have the deleterious side effects on power, fuel consumption, injector pumps, or cost (1,5,8,13,29). Nevertheless, studies on blending biodiesel with petrodiesel were conducted within the context of improving biodiesel performance during cold weather.

Cold flow property data of methyl and ethyl esters of various feedstocks in blends with D-1 and D-2 are summarized in Table 3. Properties of neat D-1 and D-2 (B0 in both cases) are listed for comparison. Comparing data for blends with corresponding results in Table 2 shows that the CFPP and LTFT of neat biodiesel (B100) occur at temperatures 14–16°C higher than those for D-2. Although cold flow properties are improved with respect to B100, diesel engines fueled by biodiesel blends are more susceptible to start-up and performance problems during cold weather.

An earlier study (33) showed that D-1 blends generally exhibit better CP, PP, CFPP, and LTFT than D-2 blends largely because neat D-1 has lower cold flow properties than D-2. As discussed above, CP and PP of SME/D-2 blends increase nearly linearly with respect to increasing SME blend ratio. Nevertheless, D-2 blends at B20–B30 blend ratios did not greatly increase CFPP or LTFT, although exceptions were reported for LTFT of tallow and waste grease ethyl esters (see Table 3). D-1 blends had significant increases in CP, PP, CFPP and LTFT at nearly all blend ratios, although blend ratios up to B30 might be allowable under certain conditions (i.e., CFPP = –20°C; LTFT = –16°C).

As discussed above, results in Figures 3 and 4 show near-linear correlation between low-temperature filterability and CP. Regression analysis of blends at all blend ratios and omitting data for blends treated with CFI additives yielded Equation 1 and $R^2 = 0.90$ for CFPP and Equation 2 and $R^2 = 0.95$ for LTFT (33, 42). Furthermore, statistical analyses indicated a very good probability ($P = 0.94$) that LTFT = CP, suggesting that measuring CP was essentially equivalent to determining the LTFT of biodiesel/petrodiesel blends (33). This result was important because measuring LTFT is appreciably more demanding and time-consuming than CP.

Finally, blending SME or SME/TME admixtures with petrodiesel decreased ν for blend ratios up to B50 (33). This was expected because neat D-1 and D-2 have lower ν values (4.2 and 10.4 mm^2/s) when measured after equilibration at –3°C overnight (16 h). In contrast to the discussion of ν results for neat SME and SME/TME admixtures earlier, increasing volumetric TME content in D-2 blends increased ν at a constant blend ratio. These results suggested that decreasing the degree of unsaturation in the methyl esters increases the viscosity of biodiesel/petrodiesel blends and were consistent with studies on fatty alcohols and triacylglycerides not blended with petrodiesel (53,54).

Treating with Commercial Petrodiesel CFI Additives

The economic and performance benefits of using CFI additives to improve cold flow properties of petroleum middle distillates have been recognized for more than

TABLE 4
Cold Flow Properties of Biodiesel and Biodiesel/Petrodiesel Blends Treated with Cold Flow Improver (CFI) Additives[a]

Biodiesel	Diesel grade	Blend ratio	CFI additive[b]	Loading (ppm)	CP (°C)	PP (°C)	Ref.
SME		B100	DFI-100	1000	–2	–6	43
SME		B100	DFI-200	1000	–1	–8	43
SME		B100	DFI-200	2000	–1	–16	94
SME		B100	Hitec 672	1000	–2	–6	43
SME		B100	OS 110050	1000	–1	–7	43
SME		B100	Paramins	1000	0	–5	43
SME		B100	Winterflow	1000	0	–5	43
SME		B100	Winterflow	2000	–1	–17	94
SME/TME (4:1 vol/vol)		B100	DFI-100	2000	4	0[c]	42
SME/TME (4:1 vol/vol)		B100	Hitec 672	2000	2	–5[c]	42
SME	D-1	B30	DFI-100	1000	–14	–49	43
SME	D-1	B30	DFI-200	1000	–21	–45	43
SME	D-1	B30	Hitech 672	1000	–13	–44	43
SME	D-1	B30	OS 110050	1000	–17	–46	43
SME	D-1	B30	Paramins	1000	–14	–29	43
SME	D-1	B30	Winterflow	1000	–19	–39	43
SME	D-2	B20	DFI-100	1000	–14	–26	43
SME	D-2	B20	DFI-200	1000	–14	–32	43
SME	D-2	B20	Hitech 672	1000	–14	–27	43
SME	D-2	B20	OS 110050	1000	–15	–18	43
SME	D-2	B20	Paramins	1000	–14	–27	43
SME	D-2	B20	Winterflow	1000	–13	–39	43

[a]Biodiesel from transesterification of "oil or fat" with "alkyl" alcohol; Diesel grade is according to ASTM fuel specification D 975 (34); SME = soybean oil methyl esters; TME = tallow methyl esters; see Tables 1 and 2 for other abbreviations.
[b]Vendors: Du Pont (DFI-100, DFI-200); Ethyl Corp. (Hitec 672); Exxon Chemical (Paramins); SVO/Lubrizol (OS 110050); Starreon Corp. (Winterflow).
[c]CFPP results.

40 years (1,5,8,12,14,29). Results from studies on the effects of additives developed for treating petrodiesel on the cold flow properties of biodiesel, and its blends with D-1 or D-2 are summarized in Table 4. Before discussing these results, it is necessary to examine pertinent background information on the types of commercial CFI additives.

Pour Point Depressants. The first generation of additives, PP-depressants (PPD), were developed initially in the 1950s. These additives are employed in refinery locations to improve pumpability of crude oil and are most effective for after-market applications in heating oils or lubricants (1,3,5,9,17,55). Most PPD do not affect nucleation, and the crystalline growth habit (shape) typically remains orthorhombic

(8,9,56). These additives inhibit crystalline growth and eliminate agglomeration (gelling), reducing sizes to d = 10–100 μm and preventing the formation of large flat plate crystals that clog lines and filters (2,4,5,8,9,11,30,55). PPD are typically composed of low-molecular-weight copolymers similar in structure and melting point to the *n*-alkane paraffin molecules, making it possible for them to adsorb or co-crystallize after nucleation has been initiated (1,55,56). Although a reduction in CFPP is possible at higher concentrations (3,7), the effects on CP and LTFT are negligible (1,29). The predominant PPD currently marketed are ethylene vinyl ester copolymers (1,2,8,9,12,17,56). Other examples include copolymers with long-chain alkyl groups (derived from fatty alcohols) as pendant groups, polymethacrylates, polyalkylacrylates, polyalkylmethacrylates, copolymers containing esterified derivatives of maleic anhydride, chlorinated polyethylenes, and copolymers of styrene-maleate esters and vinyl acetate-maleate esters (9,17,22,30,55).

Wax Crystalline Modifiers. In response to the development of low-temperature predictive filterability tests (see earlier discussion), more advanced wax crystalline modifier (WCM)-type CFI additives were developed (3,9,11,15). Many examples are reported in the scientific and patent literature, including ethylene vinyl ester copolymers, alkenyl succinic amides, long-chain polyalkylacrylates, polyethylenes, copolymers of linear α-olefins with acrylic, vinylic, and maleic compounds, secondary amines, random terpolymers of α-olefin, stearylacrylate and *N*-alkylmaleimide, copolymers of acrylate/methacrylate with maleic anhydride partially amidated with *n*-hexadecylamine, fumarate vinyl acetate copolymers, itaconate copolymers, polyethylene-polypropylene block copolymers, polyamides of linear or branched acids and copolymers of α-olefins, maleic anhydride copolymers, carboxy-containing interpolymers, styrene-maleic anhydride copolymers, and polyoxyalkylene compounds (1–3,9,10,12,14,17,21,57–66).

These additives attack one or more phases of the crystallization process, i.e., nucleation, growth, or agglomeration (1,2,4,5,8,10,12). Their combined effect is to promote formation of a greater number of smaller, more compact wax crystals (1–5,8–10,12). WCM additives are copolymers with chemically structured characteristics tailored to match the wax type and the rate of precipitation of fuels (2,9,11). Some are *comb-* or *brush*-shaped, consisting of a straight-chain backbone and teeth structured with moieties designed to interact with and absorb paraffin molecules (1,2,9,11,14,21,57). They modify the crystal habit to form small, needle-shaped crystals that do not clog fuel lines or plug primary filters (2,4,5,9,11,12,29). A build-up of solids on secondary filters (d = 2–10 μm) results in the formation of a permeable cake layer that allows some flow of liquid fuel to the injectors (1,2,4,5,9,11,12). As long as excess fuel can be recycled back to the fuel tank, warming of the crystals on the filter surface eventually melts the cake layer (1–4,9,11,12). Some additives modify the crystalline cell geometry to form hexagonal or a mixture of hexagonal and rhombic lattices (10). WCM additives are better suited for after-market petrodiesel applications than PPD additives because typically they allow operation of engines at temperatures as low as 10°C below the CP of the fuel (4,5,8,29). In general, WCM were developed and categorized as (i) CFPP improvers, (ii) CP depressants, and (iii) wax antisettling flow improvers.

Among the WCM, CFPP improvers were the first type of the new-generation additives to be developed. They typically provide dual functionality by reducing PP and CFPP and are sometimes referred to as middle distillate flow improvers (1,3,4,11,12). CFPP improvers are capable of reducing CFPP by 10–20°C (3–5,8,29). Although some additives were reported to reduce LTFT (1,5,8,15,29), most CFPP improvers typically do not affect CP (1,9,29). Reductions in CFPP to temperatures more than 12°C below the CP usually result in the CFPP test no longer being an accurate predictor of operability (3,5,12,15).

The development of CP depressants (CPD) began in the late 1970s (3). These additives are typically low-molecular-weight comb-shaped copolymers and work by preferentially interacting with and adsorbing the first paraffin molecules to crystallize in competition with normal nuclei (1,2,8). In contrast to CFPP improvers, CPD are designed with a soluble backbone that allows the additive-paraffin complex to remain soluble at temperatures below the CP (1,9,14,21). Although CPD are capable of reducing CP by a maximum of 3–5°C (1,3,8,9,12,21), many are antagonistic toward CFPP improvers, and combinations thereof may worsen CFPP (3,9,12).

Though not designed specifically for improving flow and pumpability, wax antisettling flow improvers (WAFI) are typically employed to prevent wax build-up at the bottom of storage and fuel tanks (11,12,29). Allowing diesel fuels to stand cold overnight at 10°C or more below CP causes the wax crystals to grow large enough to settle at the bottom of the tank (2,11,13). WAFI are similar in structure to CPD additives and work by co-crystallizing with nuclei and imparting a dispersive effect caused by attached highly polar functional groups based on heteroatoms such as nitrogen, oxygen, sulfur, or phosphorus (9,58). Wax crystals have a slightly higher density than petrodiesel, and settling is prevented by keeping their sizes very small ($d < 5$–$10\ \mu m$) (2,4,11,15,58). The effect of particle size (d) on the rate of settling (R_S) may be determined by employing Stokes' law for spheres suspended in a fluid (3,11,12,15). Stokes' law shows that a fivefold reduction in d results in a 25-fold reduction in R_S. WAFI decrease CFPP and PP but typically do not alter CP (58). Some PPD increase the rate of wax settling (11).

Effects of Petrodiesel CFI Additives on Biodiesel. Several commercial CFI additives developed for application in petrodiesel fuels were studied (42,43); the results are summarized in Table 4. Petrodiesel CFI additives demonstrated the ability to decrease PP by up to 18–20°C for SME/D-1 (B30) and SME/D-2 (B20) blends. Comparing results in Tables 2 and 4, CFI additives decreased PP of neat SME by as much as 6°C. Under similar conditions, additives decreased PP by 7 and 23°C for unblended D-1 and D-2, respectively (43). These results suggested that mechanisms associated with crystalline growth and agglomeration in neat biodiesel were similar to those for petrodiesel fuels.

Data in Table 4 also show that increasing additive loading (concentration) further reduces the PP of blends. This was the case for most of the additives studied, with respect to loadings in the range 0–2000 ppm. Reductions in PP tended to be proportionate to loading, although some additives were more efficient than others. Results also showed that additive effectiveness decreased with increasing blend ratio at constant loading (43).

Overall, PP results were encouraging from the standpoint of utilizing additives to ease biodiesel pumpability operations during cooler weather. Most of the additives listed in Table 4 were also effective in decreasing CFPP (43). The ability to decrease both PP and CFPP suggests that such additives are CFPP improvers as defined above to the extent that they are capable of altering wax crystallization in neat biodiesel and biodiesel/petrodiesel blends. Similar results were reported (67) when studying effects of a CFI improver on rapeseed oil methyl ester/D-2 blends. At present, many biodiesel producers and sellers in the United States are using CFPP improvers during cooler weather.

In contrast to PP results, a comparison of CP data summarized in Tables 2 and 4 for neat SME or SME/TME admixtures shows that none of the CFI additives tested greatly affected CP (43). In terms of wax crystallization in neat biodiesel, CFI additives structurally designed to modify wax formation in neat petrodiesel as discussed in an earlier section did not selectively modify crystal nucleation in biodiesel to significantly affect CP.

Incorporating results for formulations treated with CFI additives modified the regression analyses discussed above for Equations 1 and 2 as follows (42,43):

$$\mathrm{CFPP} = 1.03(\mathrm{CP}) - 2.2 \qquad [3]$$

$$\mathrm{LTFT} = 0.81(\mathrm{CP}) - 2.4 \qquad [4]$$

with $R^2 = 0.82$ and 0.90 and $\sigma_y = 3.5$ and 4.0, respectively. Although ANOVA showed a $P = 0.743$ likelihood that the slope of [3] was close to unity, the slope of [4] was not ($P < 0.001$). The disruption in the apparent 1:1 correlation between LTFT and CP was attributed to effects of additives on neat biodiesel (B100), which showed small decreases in CP coupled with very little effect on LTFT. Factoring in additive-treated formulations decreased the slope of [4], meaning that each 1°C decrease in LTFT now requires at least a 1.25°C decrease in CP. Finally, Figures 3 and 4 show a higher degree of scatter associated with formulations treated with additives (empty symbols) than for those not treated with additives (filled symbols); hence, R^2 values for [3] and [4] decreased relative to those for [1] and [2]. Nevertheless, the results in Figures 3 and 4 showed that the most effective approaches for improving cold flow properties and performance of biodiesel will be those that significantly decrease CP.

Finally, the effects of treating SME/D-2 blends with the most effective PPD/CFPP improvers on ν were also studied (43). Statistical analyses of results showed that increasing the blend ratio from B10 to B50 does not significantly affect ν measured at 40 or –3°C. Similar analyses showed that increasing additive loading from 0 to 2000 ppm also did not significantly affect ν under the same temperature conditions (5°C for neat SME).

Development of New CFI Additives for Biodiesel

Most commercial CFI additives were designed to treat petroleum derivatives. More recent generations of WCM were developed with higher degrees of selectivity based on the concentration of specific high-melting point *n*-alkanes present in refined petrodiesel (1–3,5,8–11,14,15,21,55,57,58). Petrodiesel is typically defined as a middle distillate, generally boiling between 170 and 390°C and comprising 15–30 wt% paraffinic hydrocarbon (wax) plus aromatic and olefinic compounds (5,9,12,15,21). However, most petrodiesel fuels produced in North America and Japan tend to have narrower boiling ranges with a lower final boiling point compared with fuels in Europe, India, or Singapore (3,5,12). Narrow *cut* fuels are more difficult to treat because directionally they have a higher wax precipitation and growth rate (3,5,11,12). Analogous to these complexities, differences in molecular structure, crystal nucleation, and growth mechanisms may also limit the effectiveness of modern petrodiesel CFI additives in treating biodiesel.

Given that most additives studied as CFI additives for biodiesel were initially designed to treat petrodiesel, it is not surprising that many of these same additives were effective in reducing the PP of neat and blended biodiesel (see Tables 2, 3, and 4) because PPD tend to have the lowest degree of structural selectivity. In contrast, CFPP improvers are more selective and tend to promote the formation of smaller wax crystals than additives that exclusively decrease PP (1,2,5,8,9,11,58). CPD and WAFI have the highest selectivity and tend to promote even smaller crystals (1,2,9,11). As noted earlier, when applied to neat biodiesel or biodiesel/petrodiesel blends, the CFI additives listed in Table 4 significantly decreased PP and CFPP, slightly decreased (or increased) CP, and had very little effect on LTFT (43). The failure of these CFI additives to more effectively reduce CP or LTFT suggests that their ability to alter the nucleation and crystalline growth mechanisms present in neat and blended biodiesel may be limited by structural selectivity.

As also outlined above, there is a nearly linear correlation between CP and LTFT (Figs. 3 and 4). This means that improving the performance of biodiesel as a neat fuel and in blends with petrodiesel in North America during cold weather depends primarily on reducing the CP. Therefore, the studies summarized herein suggest the best hope for improving cold flow properties of biodiesel through treatment with CFI additives is to design new compounds whose molecular structures carry a greater degree of selectivity toward high-melting-point alkyl ester compounds to allow modification in the nucleation and crystalline growth mechanisms prevalent in biodiesel.

Reviewing the earlier discussion of wax crystallization in petrodiesel, a decreasing temperature causes the formation of orthorhombic crystal structures in

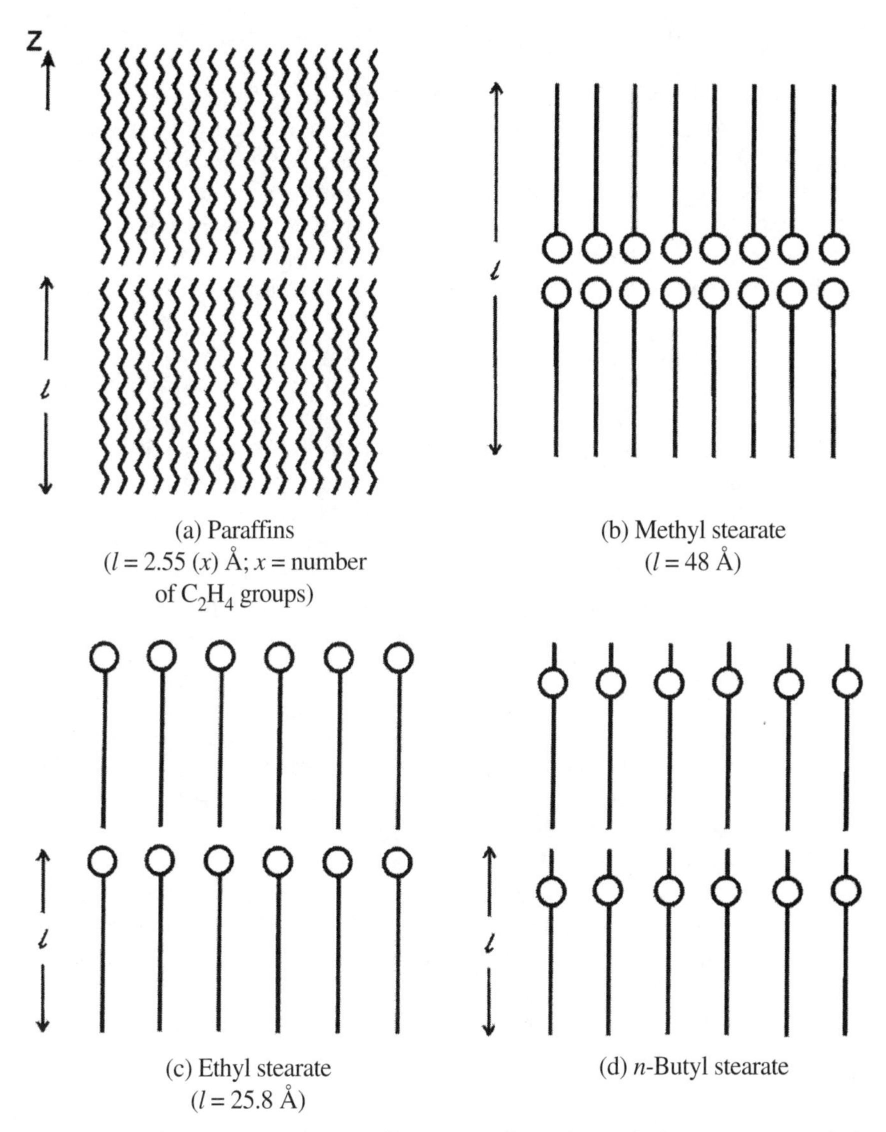

Fig. 5. Crystal structures of (a) *n*-alkane (paraffin); (b) methyl stearate; (c) ethyl stearate; and (d) butyl stearate, viewed along the shortest unit cell axis. Legend: l = long spacing. *Source:* References 4, 74.

which long-chain paraffin molecules are stacked together side-by-side as shown in Figure 5(a) (2,9,21). Unencumbered crystal growth continues in two dimensions (the *XY*-plane) as paraffin molecules continue to stack against each other. Very weak intermolecular forces between the ends of the hydrocarbon chains slow growth in the *Z*-direction, resulting in the formation of large platelet lamellae (2,4,5,8–12).

X-ray diffraction studies of fatty materials containing large quantities of long-chain hydrocarbon compounds revealed the formation of unit crystal cells with triclinic, orthorhombic, or hexagonal chain-packing geometries at lower temperatures (68–70). Slow cooling or crystallization from nonpolar solvent causes saturated alkyl esters such as those present in biodiesel to favor the formation of prisms with orthorhombic chain packing (70,71). These crystals typically have two short and one long spacing and are tilted in the direction of the hydrocarbon endgroup plane (69–72).

Schematic diagrams of crystal lattices for methyl, ethyl, and butyl stearate are shown in Figure 5(b), (c) and (d). For methyl stearate, X-ray diffraction studies revealed long spacings that are nearly twice those of ethyl stearate (70). The methyl stearate molecules possess sufficient polarity in the carboxylic headgroup, giving them an amphiphilic nature and allowing the formation of bilayer structures with headgroups aligned next to each other inside the crystal and away from nonpolar bulk liquid as shown in Figure 5(b) (70,71). Similar structures were observed for long-chain fatty acids (69–72). Ethyl, butyl, and larger alkyl esters have nonpolar chains in the headgroups that are sufficient in size to shield the forces between more polar portions of the headgroup. Hence, these esters orient themselves head-to-tail with hydrocarbon-chain tailgroups parallel to each other as shown in Figure 5(c) and (d) (71). Analogous to long-chain paraffin molecules, crystal growth in alkyl stearates continues predominately in two dimensions, forming large platelet lamellae (70,71). Growth in the *Z*-dimension is generally hindered by relatively weak intermolecular forces between alkyl tailgroups for methyl stearate or between carboxylic headgroups and hydrocarbon tailgroups for other alkyl esters (71).

Under slow cooling conditions such as those experienced during cold weather, evidence suggests that the crystal shapes formed in neat biodiesel are similar to those observed in petrodiesel. In addition, data in Table 2 show that biodiesel from feedstocks such as olive, palm, safflower, and soybean oils as well as tallow has a relatively small CP-PP differential ([CP – PP] ≤ 3°C) relative to petrodiesel (9–15°C, Table 3). This suggests that despite relatively slow crystalline growth in the *Z*-direction, formation of crystalline bridges between larger crystals [sintering (68)] rapidly results in a transition to a semisolid dispersion. Thus, transition to an unpumpable solid occurs at a much faster rate in biodiesel than petrodiesel. As noted above, some CFI additives are effective in slowing the rate of sintering and decreasing PP of neat biodiesel.

Echoing earlier discussion, the next logical step in the development of similar additives to treat biodiesel is to identify compounds with increased selectivity toward modifying nucleation and disrupting crystalline growth in one or two dimensions. Analogous to CFI additives for petrodiesel, such compounds for treating biodiesel should possess some CPD characteristics. The following two hypotheses were suggested: (i) synthesis of fatty compounds similar in structure to saturated esters and containing bulky moieties and (ii) modification of block copolymers similar to those used to treat petrodiesel.

For examining the first hypothesis, several novel fatty diesters were synthesized by *p*-toluene sulfonic acid-catalyzed esterification in toluene solvent by reacting diols with acids and diacids with 2-octanol (73). Testing these products in SME followed the supposition that co-crystallization within saturated ester crystals allows the bulky moieties to disrupt solid crystal formation from otherwise harmonious orientation in one direction as shown in Figure 5(b). However, results from that study showed only a slight effect on CP or PP (≤1°C) at loading = 2000 ppm. Increasing loading to 5000 or 10,000 ppm yielded no significant benefits.

According to the second hypothesis, at least two reports in the patent literature claim the invention of CFI additives specifically targeted to improve the cold weather performance of biodiesel. It was reported (74,75) that block copolymers of long-chain alkyl methacrylates and acrylates were effective as PPD and flow improvers for lubricant oils and biodiesel fuel additives. Similarly, methacrylate copolymers were reported to decrease CFPP of biologically derived fuel oils and biodiesel fuels made from rapeseed oil (76).

Efforts were also made to employ glycerol (see Chapter 11), which is typically yielded as a coproduct from biodiesel production, in the synthesis of agents that effectively improve cold flow properties of biodiesel fuels. Glycerol can be reacted with isobutylene or isoamylene in the presence of a strong acid catalyst to produce glycerol ether derivatives (77). Adding the derivatives back to biodiesel was shown to improve fuel characteristics, although very large quantities (>1%) were generally required to significantly reduce CP.

Medium and Branched-Chain Mono-Alkyl Esters

Transesterification of oils or fats with medium- (C_3–C_8) or branched-chain alkyl alcohols is known to produce biodiesel with improved cold flow properties. As noted earlier, ethyl and larger alkyl esters tend to form thin flat lamellae during nucleation [e.g., Fig. 5(d)]. Large or bulky headgroups also disrupt spacing between individual molecules in the lamellae causing rotational disorder in the hydrocarbon tailgroup chains. This disorder results in the initial formation of crystal nuclei with less stable chain packing followed by transformation to a more stable form at lower temperatures. Thus, melting points for ethyl palmitate and stearate are below those of their corresponding methyl esters (see Table A-1 in Appendix A). Melting points for alkyl palmitate and stearate esters continue to decrease up to an alkyl carbon chain length of 4 (*n*-butyl), then increase with chain length for C_5 (*n*-pentyl) and larger alkyl esters (71).

Comparing data for canola, linseed, and soybean oil alkyl esters summarized in Tables 2 and 5, CP steadily decreases with increasing alkyl chain length in the ester headgroup (33,34,37,46). In addition, PP decreased by 4–7°C between the methyl to *n*-butyl esters of these oils. CP of tallow alkyl esters decreased from

TABLE 5
Cold Flow Properties of Selected Mono-Alkyl Esters[a]

Oil or fat	Alkyl group	CP (°C)	PP (°C)	CFPP (°C)	LTFT (°C)	Reference
Canola	Isopropyl	7	–12			37
Canola	*n*-Butyl	–6	–16			37
Linseed	Isopropyl	3	–12			37
Linseed	*n*-Butyl	–10	–13			37
Soybean	Isopropyl	–9	–12			26, 78
Soybean	*n*-Butyl	–3	–7			48
Soybean	2-Butyl	–12	–15			26
Tallow	*n*-Propyl	12	9	7	18	34
Tallow	Isopropyl	8	0	7	19	34
Tallow	*n*-Butyl	9	6	3	13	34
Tallow	Isobutyl	8	3	8	17	34
Tallow	2-Butyl	9	0	4	12	34

[a]Biodiesel from transesterification of "oil or fat" with "alkyl" alcohol; see Tables 1 and 2 for other abbreviations.

17°C for TME to 15, 12, and 9°C for ethyl, *n*-propyl, and *n*-butyl esters, respectively, with an overall decrease of 9°C in PP (34). That study also reported CFPP = 3°C and LTFT = 13°C for *n*-butyl esters, compared with CFPP = 9°C and LTFT = 20°C for TME.

Comparing data for alkyl esters in Tables 2 and 5 also shows that biodiesel made from branched-chain alkyl alcohols can significantly improve cold flow properties relative to the corresponding methyl esters. Substituting isopropyl for the methyl group in the ester headgroup reduces T_{Cryst} by 11°C (26). A comparison of data for these soybean oil alkyl esters in Tables 2 and 5 shows reductions from 0 to –9°C for CP and –2 to –10°C for PP (26,34,78). A similar comparison of data for soybean oil 2-butyl esters shows reductions of 14°C for T_{Cryst}, 12°C for CP, and –15°C for PP (26,34).

Results for all four major cold flow properties, CP, PP, CFPP, and LTFT, are listed in Tables 2 and 5 only for the tallow alkyl esters. CP decreased from 17°C for TME to 8°C for both isopropyl and isobutyl tallow esters; PP decreased from 15°C for TME to 3 and 0°C for isopropyl and isobutyl esters (34). However, CFPP and LTFT decreased only slightly (1–2°C).

In addition to improving cold flow properties, transesterification with longer-chain alkyl alcohols can improve the ignition quality of biodiesel. Increasing hydrocarbon chain length in saturated methyl esters increases the cetane number (CN), a parameter that can influence combustion quality and emissions (79). However, increased branching in the hydrocarbon chain decreases CN (80). Few data exist in the literature on engine performance and emissions testing of longer or branched-chain alkyl esters. One report (27) showed that a 20 vol% blend of ethyl esters of waste grease in D-2 performed better in engine testing than a similar blend with TME.

Nevertheless, the conversion of vegetable oils or fats using methanol remains the most economical means for producing biodiesel. According to recently published prices (81), substituting ethanol for methanol increases the cost of biodiesel production by \$0.039/L (\$0.147/gal) biodiesel. For a B20 ethyl ester blend, the cost of fuel increases by only \$0.008/L (\$0.029/gal). On the other hand, conversion with longer- or branched-chain alkyl alcohols is significantly more expensive. Substituting isopropyl alcohol for methanol raises biodiesel production costs by at least \$0.211/L (\$0.803/gal). Substituting 2-butanol raises production costs by at least \$0.372/L (\$1.41/gal).

Crystallization Fractionation

The nature of biodiesel suggests that crystallization fractionation may be a useful technique for decreasing CP by reducing total saturated alkyl ester content. Biodiesel made from the most common feedstocks (e.g., Table 2) may be considered a pseudobinary mixture consisting of components that fall into one of two families, those with high melting points and those with low melting points. For example, a relatively large differential exists between melting points of pure methyl palmitate and methyl oleate, 30 and –19.9°C, respectively (34). This large differential typically means that a binary mixture of these two compounds would experience nucleation and crystal growth kinetics similar to precipitation of solute (methyl palmitate) from solvent (methyl oleate). Crystallization does not occur in bulk in these types of mixtures; rather, it is governed by the relative solubility of the solute (82). Applying this reasoning to multicomponent mixtures such as biodiesel suggests that an effective approach for improving cold flow properties is to reduce the total concentration of high-melting point components through crystallization fractionation.

Crystallization fractionation is the separation of the components of lipids (vegetable oils, fats, fatty acids, fatty acid esters, monodiglycerides and other derivatives) based on differences in crystallizing (or melting) temperature (83–90). Crystallization fractionation of oleo-margarine from beef tallow yielding 60% yellow (soft) and 40% white (hard) fractions dates back to ~1869 (83,91). Winterization processing evolved from the observation that storing cottonseed oil in outside tanks during the winter caused the oil to separate into hard and clear fractions (83–86). The clear liquid was decanted and marketed as winter salad oil (86,91), whereas small quantities (2–5 wt%) of the hard fraction (wax) were processed into margarine oil streams (91). Because winterization is typically associated with crystallization during long-term storage in cold temperatures (84,85,87,88,90,91), commercial-scale crystallization fractionation as a practical technique for production of margarine ceased after World War I in favor of the more rapid hydrogenation processes (83,91). However, increased demand for salad oils and other high-quality oil and fat products coupled with an upsurge in palm oil production in the mid-1960s led to the development of faster mechanical crystallization fractionation processes (83,85,

86,91). Commercial processes were applied in the fractionation of canola, coconut, corn, cottonseed, fish, palm, safflower, sesame, and sunflower oils, butter, lard, and esterified lard, milk fats, palm kernel, palm olein, tallow, and yellow grease, as well as partially hydrogenated soybean, cottonseed, and fish oils (83–88,90,91). In addition to salad oils and margarines, fractionation of lipids yields stearic and oleic acids, medium-chain (C_6–C_{10}) saturated triacylglycerols, high-(oxidative) stability liquid oils (Active Oxygen Method >350 h), cocoa butter substitutes, coatings for confections, whipped toppings, dairy products, cooking oils, shortenings, oleochemicals such as soaps, fatty acids and fatty acid derivatives, and specialty fats (83–85,90).

The traditional fractionation process comprises two stages. The crystallization stage consists of selective nucleation and crystal growth under a strictly controlled cooling rate combined with gentle agitation. Once well-defined crystals with narrow distribution of specific size and habit are formed, the resulting slurry is transferred to the second stage for separation into solid and liquid fractions, typically by filtration or centrifugation (83–88,90,91).

Filtering aids or crystallization modifiers may be added to the crystallizer to promote nucleation, modify crystal growth and habit, reduce entrainment of the liquid phase within solid crystal agglomerates, facilitate separation by filtration, or retard crystallization in the final product fractions. Analogous to the role of CFI additives in diesel fuel discussed earlier, crystallization modification of fatty derivatives is imparted either through action as seed agents promoting nucleation or by co-crystallization on crystal surfaces to disrupt otherwise orderly patterns limiting growth and agglomeration. Examples of crystallization modifiers include commercial lecithins, mono- and diglycerides, monodiglycerides esterified with citric acid, free fatty acids, fatty acid esters of sorbitol, fatty acid esters of polyglycerol or other polyhydric alcohols, aluminum stearate, unsplit fat, polysaccharides, or peptide esters (83–88,90).

Crystal fractions in fatty mixtures tend to be voluminous, making complete separation from the mother liquor difficult (89). Hence, product yield as defined by separation of high- and low-melting point fractions depends greatly on maintaining control of both stages of the process (86–88,91). The cooling rate in the crystallizer controls the rate of nucleation, number, size, and habit of crystals and reduces the rate of agglomeration of crystals, thus reducing entrainment of the liquid phase (83–88,91). Agitation prevents build-up of heat transfer-reducing crystals on the walls of the chiller, reduces entrainment of the liquid phase by reducing agglomeration caused by interactions of small crystals, and reduces the effects of viscosity. Viscosity, which increases with decreasing temperature, can affect crystal growth rate by hindering the mass transfer of molecules from bulk liquid to the crystal surface and decreasing heat transfer away from the crystals. Strict control of agitation is necessary to prevent detrimental effects of high shear rates, which can fragment or destroy crystals, and imparting mechanical work into the system (83–87). Other process variables include the composition of the source oil or fat, crystallization temperature, and crystallization time; the last-mentioned two can significantly affect crystalline growth rates (83–85,88).

The separation stage is usually performed by filtration, centrifugation, or decantation. Most commercial filtration equipment employs plate and frame, flat-bed vacuum band, rotary drum vacuum, membrane (polypropylene or synthetic rubber), hydraulic press, or pressure leaf type filters. Conditions are also tightly controlled and generally determine the optimal crystal sizes that should be generated in the crystallization stage (83,85–87,90,91).

The following three unit processes were employed in commercial crystallization fractionation of fatty derivatives: dry, solvent, and detergent fractionation (83–88,91). Two of these processes, dry and solvent fractionation, were applied in studies with biodiesel and are discussed in more detail below.

Dry Fractionation. Dry fractionation, defined as crystallization from a melt without dilution in solvent (85,86,90), is the simplest and least expensive process for separating high- and low-melting point fatty derivatives (83–87). Thus, it is the most widely practiced form of crystallization fractionation and most common of all fat fraction technologies (83,85).

Dewaxing and winterization are narrow forms of dry fractionation (84–88,90, 91). They are often associated with oil refining because the total content of wax removed is relatively small (<2 wt%) and does not affect physical properties other than appearance; this wax arises as residuals from seed hulls after extraction of the oil from seeds, high-melting point triacylglycerols, and other low-solubility components (84,85,87,91).

Dry fractionation more generally refers to a modification process in which drastic changes in composition are accompanied by significantly altered CP [measured by AOCS Cc 6–25 (92)], cold stability test [AOCS Cc 11–53 (93)], and other properties (84–88,91). The crystallization and separation stages are technologically more sophisticated and require a higher degree of control than dewaxing or winterization (84,87). Vegetable oils or fats are separated with a high degree of selectivity (86,87); the collected fractions provide new materials considered more useful than the corresponding natural lipid product (85).

Cold flow properties of liquid fractions obtained from dry fractionation of SME and waste cooking oil methyl esters are summarized in Table 6. Earlier studies (43,94) showed that application of bench-scale dry fractionation to SME decreases the CP to –20°C and LTFT to –16°C, results that were lower than those for neat D-2 (–16 and –14°C, shown in Table 3). Dry fractionation also significantly reduced the PP and CFPP of SME. Although fractionation reduced the total concentration of saturated methyl esters (C_{16}, C_{18}) to as low as 5.6 wt%, liquid yields were only 25–33% relative to the mass of the starting material. Stepwise crystallization in 2- to 3°C-increments was necessary to maintain control over the crystallization stage because the CP was only 2–3°C > PP (see Table 2). The initial cooling rate for each step depended on the temperature differential between the ambient and crystallization temperatures (T_B).

TABLE 6
Cold Flow Properties of Fractionated Methyl Esters[a]

Source: oil or fat	Solvent	CFI additive[b]	B/S (g/g)	Steps	TB (°C)	Yield (g/g)	Total saturates (wt%)	CP (°C)	PP (°C)	CFPP (°C)	Ref.
Soybean[c]	None	None	—	6	–10	0.334	6.3	–20	–21	–19	94, 100
Waste cooking[d]	None	None	—	1	–1	0.25–0.30	14			–5	95
Soybean	None	DFI-200	—	6	–10	0.801	9.8	–11		–12	94, 100
Soybean	None	Winterflow	—	6	–10	0.870	9.3	–11		–11	94, 100
Soybean	Hexane	None	0.284	1	–25	0.784	16.2	–10	–11	–10	94, 100
Soybean[e]	Hexane	DFI-200	0.200	1	–34	0.992	13.5	–5	–12		94
Soybean[e]	Hexane	Winterflow	0.200	1	–34	1.029	11.1	–5	–12		94
Soybean	Isopropanol	None	0.228	1	–15	0.860	10.8	–9	–9	–9	94, 100
Soybean[e]	Isopropanol	DFI-200	0.200	1	–20	0.952	12.8	–6	–9[f]		94
Soybean[e]	Isopropanol	Winterflow	0.199	1	–20	0.989	13.3	–5	–9[f]		94

[a]Biodiesel from transesterification of "oil or fat" with methanol; B/S = biodiesel/solvent mass ratio; T_B = coolant temperature (final step); Yield = mass ratio liquid fraction to starting material; Total saturates = of liquid fraction (determined by gas chromatography); see Tables 1, 2, and 4 for other abbreviations.
[b]Loading = 2000 ppm (before first fractionation step).
[c]LTFT = –16°C.
[d]Cooling rate = 0.1°C/min.
[e]Liquid fractions contained small quantities of residual solvent after evaporation.
[f]For samples fractionated at –15°C.

Total residence time in the bath including cooling and equilibration was ~16 h (overnight) for each step, and 5–6 steps were necessary to significantly affect CP.

Lee *et al.* (25) studied dry fractionation of SME under similar conditions, obtaining nearly identical results. In that work, stepwise crystallization was pursued with incremental equilibration times varying in relation to the time required for significant crystallization to occur. Typically, liquid fraction yields for each step were 84–90%. After 10 steps with total residence time of 84 h, the final liquid fraction had a total saturated methyl esters content = 5.5 wt% and T_{Cryst} = –7.1°C, compared with values of 15.6% and 3.7°C for nonfractionated SME. The liquid yield was also only 25.5% relative to the starting material. Fractionation of low-palmitic (4.0 wt% C_{16}) SME under similar conditions significantly increased yield to 85.7% and required fewer steps (3, total residence time = 40 h) to produce a liquid fraction with T_{Cryst} = –6.5°C.

Contrasting results from studies conducted by Lee *et al.* (25) with Dunn *et al.* (43,94), the former sought to control the nucleation and crystal growth processes during each fractionation step by keeping the mass of crystallized phase nearly constant (84–90%), whereas the latter simply employed a constant residence time (16 h) for each step. Both control schemes resulted in relatively poor yields in terms of liquid fractions. Thus, occlusion of the liquid phase within solid crystals as they grow and agglomerate led to substantial loss in yield during filtration. Contrasting the control schemes also suggests that the rate of crystalline growth and agglomeration was rapid, causing substantial entrainment of the liquid phase during the early stages of crystallization.

The use of dry fractionation to reduce CFPP of waste cooking oil methyl esters by first cooling them to T_B = –1°C with a relatively slow cooling rate = 0.1°C/min was studied (95). Comparing results shown in Tables 3 and 6, this process decreased the total saturated ester content from 19.2 to 14 wt% and CFPP from –1 to –5°C. Results also demonstrated a nearly linear relationship (R^2 = 0.95) between CFPP and total saturation ester content.

Dry fractionation also exhibited potential for improving cold flow properties of biodiesel made from high-melting feedstocks. Crystallization fractionation of TME under controlled conditions resulted in 60–65% yield of liquid fraction in which the iodine value (IV) increased from 41 to 60 and CP (measured by AOCS Cc 6–25) decreased from 11 to –1°C (87).

Crystal Modifiers. Reviewing the earlier discussion on additives, it was noted that several commercial petrodiesel CFI additives were also effective in decreasing the PP and CFPP of neat biodiesel (B100). Comparison of data summarized in Tables 2 and 4 indicates that treating SME with 2000 ppm DFI-200 or Winterflow decreases the PP without significantly affecting the CP, expanding the [CP-PP] differential from 2 to 15–16°C (33,94). These results demonstrated that the CFI additives were also effective in reducing the rates of growth and agglomeration at temperatures below the CP. These observations now suggest that treating biodiesel with commercial diesel fuel CFI additives before crystallization may be an effective means of increasing the liquid fraction yield by reducing entrainment of the liquid phase in solid crystal agglomerates.

Listed in Table 6 are results from stepwise winterization of two SME samples, each treated with 2000 ppm DFI-200 and Winterflow. Similar to stepwise fractionation of neat SME, a series of six crystallization-filtration steps were conducted on succeeding liquid fractions with coolant temperatures for each step decreased incrementally by 2–3°C. Results showed that adding DFI-200 and Winterflow increased yield to 80.1 and 87.0%, respectively. Although yields increased significantly compared with stepwise fractionation of neat SME, the CP results for final liquid fractions were at higher temperatures. For both additives, CP = –11°C, a value very close to the final T_B = –10°C, suggesting that a relatively high separation efficiency occurred after separation of the solid and liquid fractions (94). Overall, these results appear to confirm that hindering crystalline growth and agglomeration by treating biodiesel with crystal modifiers reduces entrainment and increases liquid fraction yields.

Solvent Fractionation. Relative to dry fractionation, crystallization fractionation from dilute solution in organic solvent offers many advantages. Solvent fractionation significantly reduces viscosity and entrainment of the liquid phase in solid crystals, reduces crystallization times, increases ease of separation, provides relatively high separation efficiencies, and improves yield (83–85,87,88,90). It is also the most efficient of all fractionation processes (83,84). Disadvantages include increased costs for safety, handling, and recovery of solvent (83–88,90,91). Decontaminating collected fractions from trace concentrations of residual solvent may also prove difficult. It is the most expensive of the fractionation processes (83) and is usually reserved for production of high-quality products or oils or fats with unique properties (85,88,96).

Factors affecting the selection of solvents include polarity, the relative solubility of compounds to be crystallized, and the presence of unsaturated fatty acids. Solvent polarity affects crystalline morphology, which can affect growth rate and habit (70,71). Alkyl esters have higher solubility in a given organic solvent than corresponding free fatty acids. As a result, alkyl esters will require much lower crystallization temperatures (e.g., linoleic acid is easily crystallized at –75°C, whereas methyl or ethyl linoleate may require temperatures below –100°C). Finally, the presence of unsaturated fatty acids solubilizes (increases the solubility of) saturated acids and esters. Acetone, methanol, Skellysolve B, and ether were employed as solvents for the fractionation of alkyl esters made from cottonseed oil, soybean oil, and other long-chain (C_{18}) fatty acids (89). Other solvents that were employed in the fractionation of lipids include chloroform, ethyl acetate, ethanol (5 wt% aqueous), hexane, isopropanol, 2-nitropropane, and chlorinated and petroleum naphthas (83–85,87,89–91).

Solvent crystallization fraction was applied to the separation of biodiesel. Results from an earlier study (94) are summarized in Table 6. Each fractionation was conducted in one step with residence time of 3.5–6.5 h. For SME/hexane with a biodiesel/solvent (B/S) mass ratio = 0.284 g/g and T_B = –25°C, the liquid fraction was separated with yield = 78.4% and CP = –10°C. Total saturated methyl ester content was relatively high, 16.3 wt% compared with 19.7% in nonfractionated SME. The relatively small decrease in total saturated esters suggests that the improvement in CP may have been caused by the retention of residual hexane solvent after rotary evaporation of the liquid fraction. Although decreasing T_B to –30°C reduced the yield to 59.6%, total saturated esters in the liquid fraction were reduced to 11.3 wt%, a level that allowed a more reasonable explanation for the decreases in CP to –10°C and PP to –11°C.

Solvent fractionation of SME from dilution in isopropanol solvent (B/S = 0.228 g/g) at T_B = –15°C resulted in a separation of the liquid fraction with yield = 86.0% and CP = –9°C (94). The reduction in total saturated ester content to 10.8 wt% in the liquid fraction also appeared reasonable, given the reduction in CP, PP, and CFPP as noted in Table 6.

Fractionation of SME from dilution in hexane employing three crystallization steps was studied (25). The initial concentration of the mixture was 217 g SME in 1 L hexane, residence times for each step were 16, 16, and 5 h, respectively, and the final T_B = –28.4°C. The liquid fraction was separated with yield = 77 wt%, a significant increase relative to dry fractionation of SME in the same study (discussed above). Liquid fractions were characterized by total saturated ester content = 6.0 wt% and T_{Cryst} = –5.8°C. Results also showed the importance of the nature of the solvent. SME/methanol separated into two liquid layers as coolant temperature approached –1.6°C; acetone resulted in no reduction in T_{Cryst} relative to nonfractionated SME; and the study of chloroform was abandoned after crystals failed to form at temperatures below T_B = –25°C.

Crystallization of TME in blends with ethanol, D-2, and ethanol/D-2 solvents was studied (97). Experimental procedures for this study resembled traditional winterization in which samples were stored for 3 wk in walk-in freezers in a series of stepwise increments at T_B = 10, 0, –5, –10, and –16°C. After each step, the samples were filtered to collect the crystals and the liquid fraction placed for the next incremental step. Adding ethanol decreased the formation of crystals in TME and TME/D-2 mixtures. Winterization of 1:9 (vol/vol) TME/D-2 and 16.5/13.5/70 TME/ethanol/D-2 mixtures reduced CP to below –5°C.

Four mixtures listed in Table 6 demonstrated results from CFI additive-modified solvent fractionation of SME (94). For both hexane and isopropanol, lower values for T_B were necessary to maintain short residence times (4–6.5 h). Liquid fractions exhibited very high yields (95.2–103%) relative to solvent fractionation, an indication that residual solvent was likely retained after rotary evaporation. Total saturated esters from isopropanol fractionation were also slightly higher when modified by CFI additives. Although PP results were comparable, CP show less improvement when modified by CFI additives.

Disadvantages of Fractionation. Reducing the total saturated ester concentration in biodiesel affects other fuel properties in addition to cold flow. An earlier study

(98) examined the effects of dry and solvent fractionation on fuel properties of SME. Although dry fractionation increased ν (40°C) slightly, solvent fractionation decreased ν slightly. The latter effect was likely caused by traces of residual solvent in the liquid fraction after rotary evaporation. Neither dry nor solvent fractionation increased ν to exceed limits imposed by ASTM biodiesel fuel specification D 6751 (see Table B-1 in Appendix B). Fractionation increased acid value, although not enough to exceed maximum limitation imposed by D 6751 (see Table B-1 in Appendix B). Dry fractionation slightly increased specific gravity (SG), whereas solvent fractionation caused very little change in SG. The peroxide value increased slightly for both dry and solvent fractionation (98). As expected, fractionation to increase total concentration of unsaturated esters also increased the IV of SME. Liquid SME fractions exhibited larger reductions in CP, and other cold flow properties naturally showed larger increases in IV. The IV of fractionated samples, as inferred from GC analyses, remained within the range characteristic for nonfractionated SME (see Table A-2 in Appendix A for typical IV). Dry and solvent fractionation of SME significantly decreased the oxidation induction period (98). This was determined by comparing oil stability index results determined isothermally for neat SME and 20 vol% blends in D-2 with those for corresponding fractionated SME samples. Again, removing saturated esters increased total unsaturated ester content per unit mass, making the liquid fraction more susceptible to oxidative degradation from contact with oxygen present in ambient air.

Ignition quality (i.e., ignition delay time) as measured by the CN of biodiesel may also be adversely affected by fractionation. Increasing the total degree of unsaturation in the hydrocarbon chain structure decreases CN, relative to constant chain-length (99–101). Ignition delay time generally increases with decreasing CN, an effect that may worsen performance and emissions.

Finally, fractionation may have economic effects on the production of biodiesel. Crystallization and separation of high-melting point components as well as any after-treatments such as solvent or crystallization modifier recovery will increase production costs. Secondary effects may be assumed if special conditions and equipment are required to maintain fuel quality during storage and handling. Another consideration is the disposition of *co-products* generated from solid fractions, although one possibility is conversion to fatty acids or alcohols for use in the oleochemicals industry.

Outlook for Future Studies

In the United States, the Environmental Protection Agency decreed that the sulfur content of petrodiesel for fueling heavy-duty engines must be reduced to 15 ppm by June 2006 (102). Meeting the ultralow sulfur requirement with current refining technology increases paraffin wax content. Consequently, the process of blending biodiesel with petrodiesel will likely be exacerbated by these changes. Thus, developing technology to mitigate or improve the effects of biodiesel on cold flow properties of blends will increase in relevance in the future.

This review demonstrates that the key to improving the cold weather performance of biodiesel as a neat fuel and in blends with petrodiesel is to decrease its CP. Based on this criterion, the most promising approaches involve fractionation to remove high-melting point constituents. Transportation and off-road applications (power generation, heater and boiler fuel) in cooler climates as well as aviation fuel applications may provide driving forces for the development of commercial-scale fractionation processes for improving the cold flow properties of biodiesel.

Surfactant (or detergent) fractionation, a modification of dry fractionation, is a technique that may be explored in future studies. After the crystallization step, the oil is mixed with a cool aqueous solution containing wetting agent plus electrolyte (83–85,87,88,91). Surfactant molecules attach themselves to crystal surfaces, resulting in a suspension of surfactant-coated crystals in the aqueous phase (84,85, 87,88,91). Agitation is heavy at first to eliminate entrainment and promote wetting, then decreased to allow formation of a suspension (87,91). The entire mixture is centrifuged to complete the phase separation. Heating and centrifuging the aqueous phase allow recovery of solids and recycling of surfactant solution. Although surfactant fractionation improves yield and separation efficiency relative to traditional dry fractionation, its disadvantages are higher cost and contamination of end products (83–85,87,88,91).

Another fractionation technique that may merit future investigation is supercritical fluid extraction. Separation by supercritical fluids is based on the relative solubility of the components, which may be affected by chain length and degree of unsaturation (83,87). Although most studies have explored supercritical CO_2, fatty derivatives such as biodiesel demonstrate much higher solubilities in supercritical ethylene or propane (83).

Traditional breeding or gene modification could be employed in designing new oilseeds with modified fatty acid profiles. Such oilseeds would be extracted for oil to be converted into biodiesel with inherently better cold flow properties. One example is low-palmitic SME studied by Lee *et al.* (25) (as discussed earlier). Another example is the development of canola, which was bred from rapeseed to reduce the concentration of erucic acid in its oil (Table A-2 in Appendix A).

Although approaches for developing new additives structured to enhance cold flow of biodiesel were discussed earlier, very little research has been published outside of those studies summarized above. For applications in which small (5°C) reductions in CP or LTFT are sufficient, this approach has potential if polymer additives can be developed that interfere or hinder nucleation mechanisms prevalent for saturated esters in biodiesel.

Admixed alkyl esters tend to exhibit cold flow properties favoring the predominant alkyl ester present in the admixture. Earlier studies (34) with SME/TME admixtures comparing D-1 and D-2 blends with 20 vol% neat SME and a 7:3 (vol/vol) SME/TME admixture showed very little increase in CP or PP (Table 3). A similar comparison demonstrated that blending with a 4:1 (vol/vol) SME/TME admixture did not compromise LTFT with respect to blends with neat SME. Under these conditions, SME appeared to dominate the effects on cold flow properties of blends with SME/TME admixtures. Thus, it is possible that admixing SME with other alkyl esters may impart positive effects relative to neat SME. For example, ethyl laurate weakly depresses PP in admixtures with biodiesel (103).

Another possible avenue for future research is the development of agents that act as freezing point (FP) depressants when mixed with biodiesel. Holder and Winkler (104) employed FP depression theory over the simple assumption of linear mixing rules for solid solutions. Comparing CP and PP data for dewaxed gas oils spiked with up to 8% (C_{20}–C_{28}) *n*-alkanes with results from FP depression theory, they determined that very dissimilar *n*-alkanes (e.g., C_{20} and C_{28}) underwent independent crystallization and formed pure component crystals, whereas those with similar constituents co-crystallized to form solid solutions. Independent crystallization was preferred as differences in molecular weight (i.e., melting point) increased. Finally, they used FP depression theory to explain how a small quantity of heavy wax influences cold flow properties.

FP depression theory was applied to studies on the crystallization behavior in binary mixtures of tripalmitin and palm stearin in sesame seed oil (105). Similarly, the theory was applied to studies on the cold flow properties of Fischer-Tropsch fuels (106). Models were developed for both independent crystallization and co-crystallization in solid solution (106). Results showed that distinguishing between these two types of behavior is an important aspect in controlling cold flow properties by modifying the chemical composition of fuels.

Binary mixtures of compounds that undergo independent crystallization of each component frequently exhibit eutectic transitions in which, at a singular composition, the minimum FP occurs at a temperature that is below both pure components. Mixtures that undergo co-crystallization and form solid solutions generally demonstrate transition temperatures in the range between the FP of the pure components at all compositions (87,90). Thus, future studies should be conducted, perhaps incorporating FP depression theory, to identify agents and diluents that exhibit eutectic transitions when mixed in pseudobinary solution with biodiesel.

References

1. Chandler, J.E., F.G. Horneck, and G.I. Brown, The Effect of Cold Flow Additives on Low Temperature Operability of Diesel Fuels, in *Proceedings of the SAE International Fuels and Lubricants Meeting and Exposition*, San Francisco, CA, SAE Paper No. 922186, Warrendale, PA, 1992.
2. Lewtas, K., R.D. Tack, D.H.M. Beiny, and J.W. Mullin, Wax Crystallization in Diesel Fuel: Habit Modification and the Growth of *n*-Alkane Crystals, in *Advances in Industrial Crystallization*, edited by J. Garside, R.J. Davey, and A.G. Jones, Butterworth-Heinemann, Oxford, 1991, pp. 166–179.
3. Owen, K., and T. Coley, in *Automotive Fuels Handbook,* Society of Automotive Engineers, Warrendale, PA, 1990, pp. 353–403.
4. Brown, G.I., E.W. Lehmann, and K. Lewtas, Evolution of Diesel Fuel Cold Flow: The Next Frontier, in SAE Technical Paper Series, Paper No. 890031, Society of Automotive Engineers, Warrendale, PA, 1989.

5. Zielinski, J., and F. Rossi, Wax and Flow in Diesel Fuels, in *Proceedings of SAE Fuels and Lubricants Meeting and Exposition*, Paper No. 841352, Society of Automotive Engineers, Warrendale, PA, 1984.
6. Westbrook, S.R., in *Significance of Tests for Petroleum Products,* 7th edn., edited by S.J. Rand, ASTM International, West Conshohocken, PA, 2003, pp. 63–81.
7. Nadkarni, R.A.K., in *Guide to ASTM Test Methods for the Analysis of Petroleum Products and Lubricants*, American Society for Testing and Materials, West Conshohocken, PA, 2000.
8. Botros, M.G., Enhancing the Cold Flow Behavior of Diesel Fuels, in *Gasoline and Diesel Fuel: Performance and Additives*, SAE Special Publication SP-1302, Paper No. 972899, Society of Automotive Engineers, Warrendale, PA, 1997.
9. Denis, J., and J.-P. Durand, Modification of Wax Crystallization in Petroleum Products, *Rev. Inst. Fr. Pétrole 46:* 637–649 (1991).
10. Beiny, D.H.M., J.W. Mullin, and K. Lewtas, Crystallization of *n*-Dotriacontane from Hydrocarbon Solution with Polymeric Additives, *J. Crystal Growth 102:* 801–806 (1990).
11. Brown, G.I., and G.P. Gaskill, Enhanced Diesel Fuel Low Temperature Operability: Additive Developments, *Erdöle Kohle-Erdgas Petrochem. 43:* 196–204 (1990).
12. Coley, T.R., Diesel Fuel Additives Influencing Flow and Storage Properties, in *Critical Reports on Applied Chemistry, Vol. 25: Gasoline and Diesel Fuel Additives*, edited by K. Owen, Wiley and Sons, Chichester, 1989, pp. 105–123.
13. Chandler, J.E., and J.A. Zechman, Low Temperature Operability Limits of Late Model Heavy Duty Diesel Trucks and the Effect Operability Additives and Changes to the Fuel Delivery System Have on Low Temperature Performance, in *Gasoline and Diesel Fuel: Performance and Additives 2000*, SAE Special Publication SP-1563, Paper No. 2001-01-2883, Society of Automotive Engineers, Warrendale, PA, 2000.
14. Heraud, A., and B. Pouligny, How Does a "Cloud Point" Diesel Fuel Additive Work?, *J. Colloid Interface Sci. 153:* 378–391 (1992).
15. Brown, G.I., R.D. Tack, and J.E. Chandler, An Additive Solution to the Problem of Wax Settling in Diesel Fuels, in *Proceedings of the SAE International Fuels and Lubricants Meeting and Exposition*, Paper No. 881652, Society of Automotive Engineers, Warrendale, PA, 1988.
16. Anonymous, D 5773 Test Method for Cloud Point of Petroleum Products (Constant Cooling Rate Method), in *Annual Book of ASTM Standards*, Vol. 05.03, ASTM International, West Conshohocken, PA, 2003.
17. Machado, A.L.C., and E.F. Lucas, Influence of Ethylene-co-Vinyl Acetate Copolymers on the Flow Properties of Wax Synthetic Systems, *J. Appl. Polym. Sci. 85:* 1337–1348 (2002).
18. Zanier, A., Application of Modulated Temperature DSC to Distillate Fuels and Lubricating Greases, *J. Therm. Anal. 54:* 381–390 (1998).
19. Heino, E.L., Determination of Cloud Point for Petroleum Middle Distillates by Differential Scanning Calorimetry, *Thermochim. Acta 117:* 125–130 (1987).
20. Claudy, P., J.-M. Létoffé, B. Neff, and B. Damin, Diesel Fuels: Determination of Onset Crystallization Temperature, Pour Point and Filter Plugging Point by Differential Scanning Calorimetry. Correlation with Standard Test Methods, *Fuel 65:* 961–964 (1986).
21. Damin, B., A. Faure, J. Denis, B. Sillion, P. Claudy, and J.M. Létoffé, New Additives for *Diesel Fuels: Cloud-Point Depressants,* in *Diesel Fuels: Performance and*

Characteristics, SAE Special Publication SP-675, Paper No. 861527, Society of Automotive Engineers, Warrendale, PA, 1986.
22. Hipeaux, J.C., M. Born, J.P. Durand, P. Claudy, and J.M. Létoffé, Physico-Chemical Characterization of Base Stocks and Thermal Analysis by Differential Scanning Calorimetry and Thermomicroscopy at Low Temperature, *Thermochim. Acta 348:* 147–159 (2000).
23. Redelius, P., The Use of DSC in Predicting Low Temperature Behavior of Mineral Oil Products, *Thermochim. Acta 85:* 327–330 (1985).
24. Noel, F.,Thermal Analysis of Lubricating Oils, *Thermochim. Acta 4:* 377–392 (1972).
25. Lee, I., L.A. Johnson, and E.G. Hammond, Reducing the Crystallization Temperature of Biodiesel by Winterizing Methyl Soyate, *J. Am. Oil Chem. Soc. 73:* 631–636 (1996).
26. Lee, I., L.A. Johnson, and E.G. Hammond, Use of Branched-Chain Esters to Reduce the Crystallization Temperature of Biodiesel, *J. Am. Oil Chem. Soc. 72:* 1155–1160 (1995).
27. Wu, W.-H., T.A. Foglia, W.N. Marmer, R.O. Dunn, C.E. Goering, and T.E. Briggs, Low Temperature Property and Engine Performance Evaluation of Ethyl and Isopropyl Esters of Tallow and Grease, *J. Am. Oil Chem. Soc. 75:* 1173–1178 (1998).
28. Dunn, R.O., Thermal Analysis of Alternative Diesel Fuels from Vegetable Oils, *J. Am. Oil Chem. Soc. 76:* 109–115 (1999).
29. Rickeard, D.J., S.J. Cartwright, and J.E. Chandler, The Impact of Ambient Conditions, Fuel Characteristics and Fuel Additives on Fuel Consumption of Diesel Vehicles, in *Proceedings of the SAE International Fuels and Lubricants Meeting and Exposition*, Paper No. 912332, Society of Automotive Engineers, Warrendale, PA, 1991.
30. McMillan, M.L., and E.G. Barry, Fuel and Vehicle Effects on Low-Temperature Operation of Diesel Vehicles: The 1981 CRC Field Test, in *Proceedings of the SAE International Congress and Exposition*, Paper No. 830594, Society of Automotive Engineers, Warrendale, PA, 1983.
31. Manning, R.E., and M.R. Hoover, Flow Properties and Shear Stability, in *Fuels and Lubricants Handbook: Technology, Properties, Performance and Testing*, edited by G.E. Totten, S.R. Westbrook, and R.J. Shah, ASTM International, West Conshohocken, PA, 2003, pp. 833–878.
32. Anonymous, D 975 Specification for Diesel Fuel Oils, in *Annual Book of ASTM Standards*, Vol. 05.01, ASTM International, West Conshohocken, PA, 2003.
33. Dunn, R.O., and M.O. Bagby, Low-Temperature Properties of Triglyceride-Based Diesel Fuels: Transesterified Methyl Esters and Petroleum Middle Distillate/Ester Blends, *J. Am. Oil Chem. Soc. 72:* 895–904 (1995).
34. Foglia, T.A., L.A. Nelson, R.O. Dunn, and W.N. Marmer, Low-Temperature Properties of Alkyl Esters of Tallow and Grease, *J. Am. Oil Chem. Soc. 74:* 951–955 (1997).
35. Fukuda, H., A. Kondo, and H. Noda, Biodiesel Fuel Production by Transesterification of Oils, *J. Biosci. Bioeng. 92:* 405–416 (2001).
36. Kalligeros, S., F. Zannikos, S. Stournas, E. Lois, G. Anastopoulos, Ch. Teas, and F. Sakellaropoulos, An Investigation of Using Biodiesel/Marine Diesel Blends on the Performance of a Stationary Diesel Engine, *Biomass Bioenergy 24:* 141–149 (2003).
37. Lang, X., A.K. Dalai, N.N. Bakhshi, M.J. Reaney, and P.B. Hertz, Preparation and Characterization of Bio-Diesels from Various Bio-Oils, *Bioresour. Technol. 80:* 53–62 (2001).

38. Peterson, C.L., J.S. Taberski, J.C. Thompson, and C.L. Chase, The Effect of Biodiesel Feedstock on Regulated Emissions in Chassis Dynamometer Tests of a Pickup Truck, *Trans. ASAE 43:* 1371–1381 (2000).
39. Isiğigür, A., F. Karaosmanoğlu, H.A. Aksoy, F. Hamdallahpur, and Ö.L. Gülder, Performance and Emission Characteristics of a Diesel Engine Operating on Safflower Seed Oil Methyl Ester, *Appl. Biochem. Biotechnol. 45–46:* 93–102 (1994).
40. Peterson, C.L., R.A. Korus, P.G. Mora, and J.P. Madsen, Fumigation with Propane and Transesterification Effects on Injector Cooking with Vegetable Oils, *Trans. ASAE 30:* 28–35 (1987).
41. Dunn, R.O., Alternative Jet Fuels from Vegetable Oils, *Trans. ASAE 44:* 1751–1757 (2001).
42. Dunn, R.O., and M.O. Bagby, Low-Temperature Filterability Properties of Alternative Diesel Fuels from Vegetable Oils, in *Proceedings of the Third Liquid Fuel Conference: Liquid Fuel and Industrial Products from Renewable Resources*, edited by J.S. Cundiff, E.E. Gavett, C. Hansen, C. Peterson, M.A. Sanderson, H. Shapouri, and D.L. Van Dyke, American Society of Agricultural Engineers, St. Joseph, MI, 1996, pp. 95–103.
43. Dunn, R.O., M.W. Shockley, and M.O. Bagby, Improving the Low-Temperature Flow Properties of Alternative Diesel Fuels: Vegetable Oil-Derived Methyl Esters, *J. Am. Oil Chem. Soc. 73:* 1719–1728 (1996).
44. Avella, F., A. Galtieri, and A. Flumara, Characteristics and Utilization of Vegetable Oil Derivatives as Diesel Fuels, *Riv. Combust. 46:* 181–188 (1992).
45. Kaufman, K.R., and M. Ziejewski, Sunflower Methyl Esters for Direct Injection Diesel Engines, *Trans. ASAE 27:* 1626–1633 (1984).
46. Clark, S.J., M.D. Schrock, L.E. Wagner, and P.G. Piennaar, in *Final Report for Project 5980—Soybean Oil Esters as a Renewable Fuel for Diesel Engines*, Contract No. 59-2201-1-6-059-0, U.S. Department of Agriculture, Agricultural Research Service, Peoria, IL, 1983.
47. Srivastava, A., and R. Prasad, Rheological Behavior of Fatty Acid Methyl Esters, *Ind. J. Chem. Technol. 8:* 473–481 (2001).
48. Anonymous, D 97 Test Method for Pour Point of Petroleum Products, in *Annual Book of ASTM Standards*, Vol. 05.01, ASTM International, West Conshohocken, PA, 2003.
49. Dorado, M.P., E. Ballesteros, J.M. Arnal, J. Gómez, and F.J. López, Exhaust Emissions from a Diesel Engine Fueled with Transesterified Waste Olive Oil, *Fuel 82:* 1311–1315 (2003).
50. Anonymous, D 6371 Test Method for Cold Filter Plugging Point of Diesel and Heating Fuels, in *Annual Book of ASTM Standards*, Vol. 05.04, ASTM International, West Conshohocken, PA, 2003.
51. Anonymous, D 4539 Test Method for Filterability of Diesel Fuels by Low Temperature Flow Test (LTFT), in *Annual Book of ASTM Standards*, Vol. 05.02, ASTM International, West Conshohocken, PA, 2003.
52. Yu, L., I. Lee., E.G. Hammond, L.A. Johnson, and J.H. Van Gerpen, The Influence of Trace Components on the Melting Point of Methyl Soyate, *J. Am. Oil Chem. Soc. 75:* 1821–1824 (1998).
53. Dunn, R.O., A.W. Schwab, and M.O. Bagby, Physical Property and Phase Studies of Nonaqueous Triglyceride/Unsaturated Long Chain Fatty Alcohol/Methanol Systems, *J. Dispers. Sci. Technol. 13:* 77–93 (1992).

54. Goering, C.E., A.W. Schwab, M.J. Daugherty, E.H. Pryde, and A.J. Heakin, Fuel Properties of Eleven Vegetable Oils, *Trans. ASAE 25:* 1472–1483 (1982).
55. Desai, N.M., A.S. Sarma, and K.L. Mallik, Application of Performance Polymers in Petroleum Products: Studies on Viscosity Modifiers and Pour Point Depressants, *Polym. Sci. 2:* 706–712 (1991).
56. Holder, G.A., and J. Thorne, Inhibition of Crystallisation of Polymers, in *ACS Polymer Chemistry Division, Polymer Preprints 20:* 766–769 (1979).
57. Monkenbusch, M., D. Schneiders, D. Richter, L. Willner, W. Leube, L.J. Fetters, J.S. Huang, and M. Lin, Aggregation Behaviour of PE-PEP Copolymers and the Winterization of Diesel Fuel, *Physica B 276–278:* 941–943 (2000).
58. El-Gamal, I.M., T.T. Khidr, and F.M. Ghuiba, Nitrogen-Based Copolymers as Wax Dispersants for Paraffinic Gas Oils, *Fuel 77:* 375–385 (1998).
59. Davies, B.W., K. Lewtas, and A. Lombardi, PCT Int. Appl. WO 94 10,267 (1994).
60. Lal, K., U.S. Patent 5,338,471: Pour Point Depressants for Industrial Lubricants Containing Mixtures of Fatty Acid Esters and Vegetable Oils (1994).
61. Lal, K., D.M. Dishong, and J.G. Dietz, Eur. Patent Appl. EP 604,125: Pour Point Depressants for High Monounsaturated Vegetable Oils and for High Monounsaturated Vegetable Oils/Biodegradable Base and Fluid Mixtures (1994).
62. Böhmke, U., and H. Pennewiss, Eur. Patent Appl. EP 543,356 (1993).
63. Lewtas, K., and D. Block, PCT Int. Appl. WO 93 18,115 (1993).
64. Demmering, G., K. Schmid, F. Bongardt, and L. Wittich, Ger. Patent 4,040,317 (1992).
65. Bormann, K., A. Gerstmeyer, H. Franke, G. Stirnal, K.D. Wagner, B. Flemmig, K. Kosubeck, W. Fuchs, R. Voigt, J. Welker, U. Viehweger, and K. Wehner, Ger. (East) Patent DD 287,048 (1991).
66. Müller, M., H.P. Pennewiss, and D. Jenssen, Eur. Pat. Appl. EP 406,684 (1991).
67. Nylund, N.-O., and P. Aakko, Characterization of New Fuel Qualities, in *State of Alternative Fuel Technologies 2000*, SAE Special Publication SP-1545, Paper No. 2000-01-2009, Society of Automotive Engineers, Warrendale, PA, 2000.
68. Lawler, P.J., and P.S. Dimick, Crystallization and Polymorphism of Fats, in *Food Science and Technology Series No. 87, Food Lipids: Chemistry, Nutrition, and Biotechnology*, Marcel Dekker, New York, 1998, pp. 229–250.
69. Hernqvist, L., Crystal Structures of Fats and Fatty Acids, in *Crystallization and Polymorphism of Fats and Fatty Acids*, edited by N. Garti and Y. Sato, Marcel Dekker, New York, 1988, pp. 97–137.
70. Gunstone, F.D., in *An Introduction to the Chemistry and Biochemistry of Fatty Acids and Their Glycerides*, 2nd edn., Chapman & Hall, London, 1967, pp. 69–74.
71. Larson, K., and P.J. Quinn, in *The Lipid Handbook*, 2nd edn., edited by F.D. Gunstone, J.L. Harwood, and F.B. Padley, Chapman & Hall, London, 1994, pp. 401–430.
72. Bailey, A.E., in *Melting and Solidification of Fats*, Interscience, New York, 1950, pp. 1–73.
73. Knothe, G., R.O. Dunn, M.W. Shockley, and M.O. Bagby, Synthesis and Characterization of Some Long-Chain Diesters of Branched or Bulky Moieties, *J. Am. Oil Chem. Soc. 77:* 865–871 (2000).
74. Scherer, M., and J. Souchik, PCT Int. Appl. WO 0140334: Synthesis of Long-Chain Polymethacrylates by Atom Transfer Radical Polymerization for Manufacture of Lubricating Oil Additives (2001).

75. Scherer, M., J. Souchik, and J.M. Bollinger, PCT Int. Appl. WO 0140339: Block Copolymers of Long-Chain Alkyl Methacrylates and Acrylates as Lubricating Oil and Biodiesel Additives (2001).
76. Auschra, C., J. Vetter, U. Bohmke, and M. Neusius, PCT Int. Appl. WO 9927037: Methacrylate Copolymers as Low-Temperature Flow Improvers for Biodiesel Fuels and Biologically-Derived Fuel Oils (1999).
77. Noureddini, H., U.S. Patent No. 6,015,440: Process for Producing Biodiesel Fuel with Reduced Viscosity and a Cloud Point Below 32°F (2000).
78. Zhang, Y., and J.H. Van Gerpen, Combustion Analysis of Esters of Soybean Oil in a Diesel Engine, in *Performance of Alternative Fuels for SI and CI Engines,* SAE Special Publication SP-1160, SAE Paper No. 960765, Society of Automotive Engineers, Warrendale, PA, 1996, pp. 1–15.
79. Klopfenstein, W.E., Effects of Molecular Weights of Fatty Acid Esters on Cetane Numbers as Diesel Fuels, *J. Am. Oil Chem. Soc. 62:* 1029–1031 (1985).
80. Anonymous, D 613 Test Method for Cetane Number of Diesel Fuel Oil in *Annual Book of ASTM Standards*, Vol. 05.01, ASTM International, West Conshohocken, PA, 2003.
81. Anonymous, *Chem. Market Rep. 265:* 20–24 (2004).
82. Roussett, P., Modeling Crystallization Kinetics of Triacylglycerols, in *Physical Properties of Lipids*, edited by A.G. Marangoni and S.S. Narine, Marcel Dekker, New York, 2002, pp. 1–36.
83. Illingworth, D., Fractionation of Fats, in *Physical Properties of Lipids*, edited by A.G. Marangoni and S.S. Narine, Marcel Dekker, New York, 2002, pp. 411–447.
84. Kellens, M., and M. Hendrix, Fractionation, in *Introduction to Fats and Oils Technology*, 2nd edn., edited by R.D. O'Brien, W.E. Farr, and P.J. Wan, AOCS Press, Champaign, IL, 2000, pp. 194–207.
85. O'Brien, R.D., in *Fats and Oils: Formulating and Processing for Applications*, Technomic, London, UK, 1998, pp. 109–121.
86. Anderson, D., A Primer on Oils Processing Technology, in *Bailey's Industrial Oil and Fat Products, Vol. 4 (Edible Oil and Fat Products: Processing Technology)*, 5th edn., edited by Y.H. Hui, Wiley-Interscience, New York, 1996, pp. 31–45.
87. Krishnamurthy, R., and M. Kellens, Fractionation and Crystallization, in *Bailey's Industrial Oil and Fat Products, Vol. 4 (Edible Oil and Fat Products: Processing Technology)*, 5th edn., edited by Y.H. Hui, Wiley-Interscience, New York, 1996, pp. 301–338.
88. Duff, H.G.,Winterizing, in *Introduction to Fats and Oils Chemistry*, edited by P.J. Wan, American Oil Chemists' Society, Champaign, IL, 1991, pp. 105–113.
89. Brown, J.B., and D.K. Kolb, Applications of Low Temperature Crystallization in the Separation of the Fatty Acids and Their Compounds, in *Progress in the Chemistry of Fats and Other Lipids*, Vol. 3, edited by R.T. Holman, W.O. Lundberg, and T. Malkin, Pergamon Press, New York, 1955, pp. 58–80.
90. Bailey, A.E., *Melting and Solidification of Fats*, Interscience, New York, 1950, pp. 328–346.
91. Rajah, K.K., Fractionation of Fat, in *Separation Processes in the Food and Biotechnology Industries: Principles and Applications*, edited by A.S. Grandison and M.J. Lewis, Technomic, Lancaster, UK, 1996, pp. 207–242.
92. Anonymous, Cc 6–25 Cloud Point Test, in *Official Methods and Recommended Practices of the AOCS*, American Oil Chemists' Society, Champaign, IL, 1997.

93. Anonymous, Cc 11–53 Cold (Stability) Test, in *Official Methods and Recommended Practices of the AOCS*, American Oil Chemists' Society, Champaign, IL, 1997.
94. Dunn, R.O., M.W. Shockley, and M.O. Bagby, Winterized Methyl Esters from Soybean Oil: An Alternative Diesel Fuel with Improved Low-Temperature Flow Properties, in *State of Alternative Fuel Technologies*, SAE Special Publication No. SP-1274, SAE Paper No. 971682, Society of Automotive Engineers, Warrendale, PA, 1997, pp. 133–142.
95. González Gómez, M.E., R. Howard-Hildige, J.J. Leahy, and B. Rice, Winterization of Waste Cooking Oil Methyl Ester to Improve Cold Flow Temperature Fuel Properties, *Fuel 81:* 33–39 (2002).
96. Formo, M.W., in *Bailey's Industrial Oil and Fat Products*, Vol. 1, 4th edn., John Wiley & Sons, New York, 1979, p. 214.
97. Hanna, M.A., Y. Ali, S.L. Cuppett, and D. Zheng, Crystallization Characteristics of Methyl Tallowate and Its Blends with Ethanol and Diesel Fuel, *J. Am. Oil Chem. Soc. 73:* 759–763 (1996).
98. Dunn, R.O., Effect of Winterization on Fuel Properties of Methyl Soyate, in *Proceedings of the Commercialization of Biodiesel: Producing a Quality Fuel (1997)*, edited by C.L. Peterson, University of Idaho, Moscow, ID, 1998, pp. 164–186.
99. Harrington, K.J., Chemical and Physical Properties of Vegetable Oil Esters and Their Effect on Diesel Fuel Performance, *Biomass 9:* 1–17 (1986).
100. Knothe, G., M.O. Bagby, and T.W. Ryan, III, Cetane Numbers of Fatty Compounds: Influence of Compound Structure and of Various Potential Cetane Improvers, in *State of Alternative Fuel Technologies*, SAE Special Publication SP-1274, SAE Paper No. 971681, Society of Automotive Engineers, Warrendale, PA, 1997, pp. 127–132.
101. Knothe, G., M.O. Bagby, and T.W. Ryan, III, The Influence of Various Oxygenated Compounds on the Cetane Numbers of Fatty Acids and Esters, in *Proceedings of the Third Liquid Fuels Conference: Liquid Fuels and Industrial Products from Renewable Resources*, edited by J.S. Cundiff, E.E. Gavett, C. Hansen, C. Peterson, M.A. Sanderson, H. Shapouri, and D.L. VanDyne, American Society of Agricultural Engineers, St. Joseph, MI, 1996, pp. 54–66.
102. Anonymous, Part V, EPA, 40 CFR Parts 69, 80, and 86: Control of Air Pollution from New Motor Vehicles: Heavy-Duty Engine and Vehicle Standards and Highway Diesel Fuel Sulfur Control Requirements; Final Rule, *Fed. Reg. 66:* 5001–5193 (EPA 40 CFR Parts 69, 80, and 86) (2001).
103. Stournas, S., E. Lois, and A. Serdari, Effects of Fatty Acid Derivatives on the Ignition Quality and Cold Flow of Diesel Fuel, *J. Am. Oil Chem. Soc. 72:* 433–437 (1995).
104 Holder, G.A., and J. Winkler, Wax Crystallization from Diesel Fuels, *J. Inst. Pet. 51:* 228–252 (1965).
105. Toro-Vasquez, J.F., M. Briceño-Montelongo, E. Dibildox-Alvarado, M. Charó-Alonso, and J. Reyes-Hernández, Crystallization Kinetics of Palm Stearin in Blends with Sesame Seed Oil, *J. Am. Oil Chem. Soc. 77:* 297–310 (2000).
106. Suppes, G.J., T.J. Fox, K.R. Gerdes, H. Jin, M.L. Burkhart, and D.N. Koert, Cold Flow and Ignition Properties of Fischer-Tropsch Fuels, in SAE Technical Paper Series, Paper No. 2000-01-2014, Society of Automotive Engineers, Warrendale, PA, 2000.
107. Manning, R.E., and M.R. Hoover, Cold Flow Properties, in *Fuels and Lubricants Handbook: Technology, Properties, Performance and Testing*, edited by G.E. Totten, S.R. Westbrook, and R.J. Shah, ASTM International, West Conshohocken, PA, 2003, pp. 879–883.

108. Geyer, S.M., M.J. Jacobus, and S.S. Lestz, Comparison of Diesel Engine Performance and Emissions from Neat and Transesterified Vegetable Oils, *Trans. ASAE 27:* 375–381 (1984).
109. Masjuki, H., A.M. Zaki, and S.M. Sapuan, Methyl Ester of Palm Oil as an Alternative Diesel Fuel, in *Fuels for Automotive and Industrial Diesel Engines*, Proceedings of the 2nd Institute of Mechanical Engineers Seminar, Institute of Mechanical Engineers, London, 1993, pp. 129–137.
110. Neto da Silva, F., A.S. Prata, and J.R. Teixeira, Technical Feasibility Assessment of Oleic Sunflower Methyl Ester Utilization in Diesel Bus Engines, *Energy Conv. Manag. 44:* 2857–2878 (2003).

6.4

Oxidative Stability of Biodiesel

6.4.1

Literature Overview

Gerhard Knothe

Biodiesel is susceptible to oxidation upon exposure to air. The oxidation process ultimately affects fuel quality. Accordingly, the oxidative stability of biodiesel has been the subject of considerable research (1–20). An oxidative stability specification was included in the European biodiesel standards EN 14213 and EN 14214 (see Appendix B). The method to be used for assessing oxidative stability utilizes a Rancimat apparatus. This method is very similar to the Oil Stability Index (OSI) method (21). The following chapter on the "Stability of Biodiesel" (BIOSTAB) project in Europe details the development of the oxidative stability parameter in the European biodiesel standards using the Rancimat test [see also (13,15,17)]. This chapter provides a brief overview of results reported in the literature on oxidative stability.

Biodiesel is also potentially subject to hydrolytic degradation, caused by the presence of water. This is largely a housekeeping issue although the presence of substances such as mono- and diglycerides (intermediates in the transesterification reaction) or glycerol which can emulsify water, can play a major role (4).

The reason for autoxidation is the presence of double bonds in the chains of many fatty compounds. The autoxidation of unsaturated fatty compounds proceeds at different rates depending on the number and position of the double bonds (22). The CH_2 positions allylic to the double bonds in the fatty acid chains are those susceptible to oxidation. The *bis*-allylic positions in common polyunsaturated fatty acids (PUFA) such as linoleic acid (double bonds at $\Delta 9$ and $\Delta 12$, giving one *bis*-allylic position at C-11) and linolenic acid (double bonds at $\Delta 9$, $\Delta 12$, and $\Delta 15$ giving two *bis*-allylic positions at C-11 and C-14) are even more prone to autoxidation than allylic positions. Relative rates of oxidation given in the literature [(22) and references therein] are 1 for oleates (methyl, ethyl esters), 41 for linoleates, and 98 for linolenates. This is essential because most biodiesel fuels contain significant amounts of esters of oleic, linoleic, or linolenic acids, which influence the oxidative stability of the fuels. The species formed during the oxidation process cause the fuel to eventually deteriorate. Small amounts of more highly unsaturated fatty compounds had a disproportionately strong effect on oxidative stability using the OSI method (16).

Initially, hydroperoxides are formed during oxidation, with aldehydes, acids, and other oxygenates constituting oxidation products further along the reaction chain (22). However, the double bonds may also be prone to polymerization-type reactions so that higher-molecular-weight products, leading to an increase in viscosity, can occur. This may lead to the formation of insoluble species, which can clog fuel lines and pumps. One study (13) reports that polymers formed during storage under controlled conditions are soluble in biodiesel due to its polar nature and are insoluble only when mixing the biodiesel with petrodiesel.

The issue of oxidative stability affects biodiesel primarily during extended storage. The influence of parameters, such as the presence of air, heat, light traces of metal, antioxidants, peroxides as well as the nature of the storage container, was investigated in most of the aforementioned studies. Summarizing the findings from these studies, the presence of air, light, or the presence of metals as well as elevated temperatures facilitate oxidation. Studies performed with the automated OSI method confirmed the catalyzing effect of metals on oxidation, with copper showing the strongest catalyzing effect; however, the influence of the compound structure of the fatty esters, especially unsaturation, was even greater (16). Numerous other methods, including wet-chemical ones such as acid value (AV), peroxide value (PV), and pressurized differential scanning calorimetry (P-DSC), have been applied in oxidation studies of biodiesel.

Long-term storage tests on biodiesel were conducted. Viscosity, PV, AV, and density increased in biodiesel stored for 2 yr, and heat of combustion decreased (6). Viscosity and AV, which can be strongly correlated (11), changed dramatically over 1 yr with changes in the Rancimat induction period depending on the feedstock (15); however, even in storage tests for 90 d, significant increases in viscosity, PV, free fatty acid, anisidine value (AnV), and ultraviolet absorption were found (2). Biodiesel from different sources stored for 170–200 d at 20–22°C did not exceed viscosity and AV specifications but induction time decreased, with exposure to light and air having the most effect (12).

The PV is less suitable for monitoring oxidation because it tends to increase and then decrease upon further oxidation due to the formation of secondary oxidation products (9,11,15). When the PV reached a plateau of ~350 meq/kg ester during biodiesel [soybean oil methyl esters (SME)] oxidation, AV and viscosity continued to increase monotonically (9). In addition to viscosity, the AV has good potential as a parameter for monitoring biodiesel quality during storage (14). P-DSC can be used for determining the oxidative stability of biodiesel with and without antioxidants (10).

Stability tests developed for petrodiesel fuels were reportedly not suitable for biodiesel or biodiesel blends with petrodiesel (8,11), although appropriate modification may render them more useful (8). However, another study (15) states that the petrodiesel method ASTM D4625 [Standard Test Method for Distillate Fuel Storage Stability at 43°C (110°F)] is suitable but relatively slow.

Vegetable oils usually contain naturally occurring antioxidants such as tocopherols. Therefore, unrefined vegetable oils that still contain their natural levels of antioxidants usually have improved oxidative stability compared with refined oils (1) but do not meet other fuel requirements. Natural antioxidants were also deliberately added to biodiesel to investigate their antioxidant behavior. In addition to natural antioxidants, a variety of synthetic antioxidants exists. Many of them are substituted phenols such as butylated hydroxytoluene (BHT; 2,6-di-*tert*-butyl-4-methylphenol), butylated hydroxyanisole [BHA;(3)-*t*-butyl-4-hydroxyanisole] *tert*-butylhydroquinone (TBHQ; 2-*tert*-butylhydroquinone), pyrogallol (1,2,3-trihydroxybenzene), and propyl gallate (3,4,5-trihydroxybenzoic acid propyl ester). These synthetic antioxidants were also investigated for their effect on biodiesel.

Different synthetic antioxidants have different effects on biodiesel, depending on the feedstock (18,19) without affecting properties such as viscosity, cold-filter plugging point, density, and others. In another study, different antioxidants studied by the AOM method had little or no effect (7). TBHQ and α-tocopherol retarded SME oxidation (14). A high-performance liquid chromatography method for detecting antioxidants in biodiesel was also developed (20).

A European standard (pr EN 14112) was established for potential inclusion of an oxidative stability parameter in the European biodiesel standard EN 14214. The biodiesel standard EN 14214 calls for determining oxidative stability at 110°C with a minimum induction time of 6 h by the Rancimat method. The Rancimat method is nearly identical to the OSI method, which is an AOCS method. The ASTM standard D6751 currently does not include any specification of this kind.

Another parameter that was originally included in some biodiesel standards for addressing the issue of oxidative stability is the iodine value (IV). The IV is a measure of total unsaturation of a fatty material measured in g iodine/100 g of sample when formally adding iodine to the double bonds. The IV of a vegetable oil or animal fat is almost identical to that of the corresponding methyl esters (see tables in Appendix B). However, the IV of alkyl esters decreases with higher alcohols.

The idea behind the use of IV is that it would indicate the propensity of an oil or fat to oxidize, but it may also indicate the propensity of the oil or fat to polymerize and form engine deposits. Thus, an IV of 120 was specified in EN 14214 and 130 in EN 14213. This would largely exclude vegetable oils such as soybean and sunflower as biodiesel feedstock.

However, the IV of a mixture of fatty compounds, as found in oils and fats, does not take into consideratin that an infinite number of fatty acid profiles can yield the same IV (23). Different fatty acid structures can also give the same IV (23). Other, new structural indices are likely more suitable than the IV (23). Engine performance tests with a mixture of vegetable oils of different IV did not yield results that would have justified a low IV (24,25). No relation between the IV and oxidative stability was observed in another investigation on biodiesel with a wide range of IV (4).

Thus the IV was not included in biodiesel standards in the United States and Australia; it is limited to 140 in the provisional South African standard (which would permit sunflower and soybean oils), and the provisional Brazilian standards require only that it be noted (see Appendix B).

References

1. Du Plessis, L.M., Plant Oils as Diesel Fuel Extenders: Stability Tests and Specifications on Different Grades of Sunflower Seed and Soyabean Oils, *CHEMSA 8:* 150–154 (1982).
2. Du Plessis, L.M., J.B.M. de Villiers, and W.H. van der Walt, Stability Studies on Methyl and Ethyl Fatty Acid Esters of Sunflowerseed Oil, *J. Am. Oil Chem. Soc. 62:* 748–752 (1985).
3. Bondioli, P., A. Gasparoli, A. Lanzani, E. Fedeli, S. Veronese, and M. Sala, Storage Stability of Biodiesel, *J. Am. Oil Chem. Soc. 72:* 699–702 (1995).
4. Bondioli, P., and L. Folegatti, Evaluating the Oxidation Stability of Biodiesel. An Experimental Contribution, *Riv. Ital. Sostanze Grasse 73:* 349–353 (1996).
5. Simkovsky, N.M., and A. Ecker, Influence of Light and Contents of Tocopherol on the Oxidative Stability of Fatty Acid Methyl Esters, [Einfluß von Licht und Tocopherolgehalt auf die Oxidationsstabilität von Fettsäuremethylestern.] *Fett/Lipid 100:* 534–538 (1998).
6. Thompson, J.C., C.L. Peterson, D.L. Reece, and S.M. Beck, Two-Year Storage Study with Methyl and Ethyl Esters of Rapeseed, *Trans. ASAE 41:* 931–939 (1998).
7. Simkovsky, N.M., and A. Ecker, Effect of Antioxidants on the Oxidative Stability of Rapeseed Oil Methyl Esters, *Erdoel Erdgas Kohle 115:* 317–318 (1999).
8. Stavinoha, L., and S. Howell, Potential Analytical Methods for Stability Testing of Biodiesel and Biodiesel Blends, SAE Technical Paper Series 1999-01-3520, SAE, Warrendale, PA, 1999.
9. Canakci, M., A. Monyem, and J. Van Gerpen, Accelerated Oxidation Processes in Biodiesel, *Trans. ASAE 42:* 1565–1572 (1999).
10. Dunn, R.O., Analysis of Oxidative Stability of Methyl Soyate by Pressurized-Differential Scanning Calorimetry, *Trans. ASAE 43:* 1203–1208 (2000).
11. Monyem, A., M. Canakci, and J.H. Van Gerpen, Investigation of Biodiesel Thermal Stability Under Simulated In-Use Conditions, *Appl. Eng. Agric. 16:* 373–378 (2000).
12. Mittelbach, M., and S. Gangl, Long Storage Stability of Biodiesel Made from Rapeseed and Used Frying Oil, *J. Am. Oil Chem. Soc. 78:* 573–577 (2001).
13. Bondioli, P., A. Gasparoli, L. Della Bella, and S. Tagliabue, Evaluation of Biodiesel Storage Stability Using Reference Methods, *Eur. J. Lipid Sci. Technol. 104:* 777–784 (2002).
14. Dunn, R.O., Effect of Oxidation Under Accelerated Conditions on Fuel Properties of Methyl Soyate (Biodiesel), *J. Am. Oil Chem. Soc. 79:* 915–920 (2002).
15. Bondioli, P., A. Gasparoli, L. Della Bella, S. Tagliabue, and G. Toso, Biodiesel Stability Under Commercial Storage Conditions over One Year, *Eur. J. Lipid Sci. Technol. 105:* 735–741 (2003).
16. Knothe, G., and R.O. Dunn, Dependence of Oil Stability Index of Fatty Compounds on Their Structure and Concentration and Presence of Metals, *J. Am. Oil Chem. Soc. 80:* 1021–1026 (2003).
17. Lacoste, F., and L. Lagardere, Quality Parameters Evolution During Biodiesel Oxidation Using Rancimat Test, *Eur. J. Lipid Sci. Technol. 105:* 149–155 (2003).
18. Mittelbach, M., and S. Schober, The Influence of Antioxidants on the Oxidation Stability of Biodiesel, *J. Am. Oil Chem. Soc. 80:* 817–823 (2003).
19. Schober, S., and M. Mittelbach, The Impact of Antioxidants on Biodiesel Oxidation Stability, *Eur. J. Lipid Sci. Technol. 106:* 382–389 (2004).
20. Tagliabue, S., A. Gasparoli, L. Della Bella, and P. Bondioli, Quali-Quantitative Determination of Synthetic Antioxidants in Biodiesel, *Riv. Ital. Sostanze Grasse 80:* 37–40 (2004).

21. AOCS Official Method Cd 12b-92, Oil Stability Index (OSI), AOCS, Champaign, IL, 1999.
22. Frankel, E.N., Lipid Oxidation, The Oily Press, Dundee, Scotland, 1998.
23. Knothe, G., Structure Indices in FA Chemistry. How Relevant Is the Iodine Value? *J. Am. Oil Chem. Soc. 79:* 847–854 (2002).
24. Prankl, H., M. Wörgetter, and J. Rathbauer, Technical Performance of Vegetable Oil Methyl Esters with a High Iodine Number, *Proceedings of the 4th Biomass Conference of the Americas*, Oakland, CA, 1999, pp. 805–810.
25. Prankl, H., and M. Wörgetter, Influence of the Iodine Number of Biodiesel to the Engine Performance, *Proceedings of the 3rd Liquid Fuel Conference, Liquid Fuels and Industrial Products from Renewable Resources*, edited by J.S. Cundiff. E.E. Gavett, C. Hansen, C. Peterson, M.A. Sanderson, H. Shapouri, and D.L. VanDyne, ASAE, Warrendale, PA, 1996, pp. 191–196.

6.4.2
Stability of Biodiesel

Heinrich Prankl

Introduction

In 1997, the European Commission mandated the European Committee for Standardization (CEN) to develop standards for biodiesel used as a fuel for diesel engines and as heating fuel as well as the necessary standards for the analytical methods (1). During the drafting process, the lack of knowledge regarding biodiesel stability became apparent. This was considered to be an important issue for which detailed research was required. Between 2001 and 2003, the European Commission funded the project "Stability of Biodiesel" (BIOSTAB; http://www.biostab.info; more details are available through Ref. 2) which was carried out to clarify relevant questions about stability determination methods, storage, and stabilization of biodiesel fuel. Both European standards for biodiesel (EN14213, FAME as heating fuel, and EN14214, FAME as automotive diesel fuel) have been available since July 2003.

Objectives

The objective of the project was to establish criteria and the corresponding analytical methods for determining the stability of biodiesel. The detailed aims were as follows: (i) to develop appropriate methods for the determination of stability under realistic conditions; (ii) to understand the influence of storage conditions on the quality of biodiesel; (iii) to define a minimum level of natural and/or synthetic antioxidants; and (iv) to determine the effects of fuel stability on the use of biodiesel as automotive diesel fuel and as heating fuel.

Project Partners

Nine partners from industry, science, and research were involved in the project. Seven were members of one or more working groups during the biodiesel standardization process. A very experienced biodiesel research consortium consisting of the Bundesanstalt für Landtechnik (BLT, Austria; project coordination), Institut des Corps Gras (ITERG, Pessac, France), Stazione Sperimentale Oli e Grassi (SSOG, Milan, Italy), Institute of Chemistry of the University of Graz (Austria), Graz University of Technology (TUG, Austria), OMV AG (Vienna, Austria), TEAGASC (Oak Park Research Centre, Carlow, Ireland), NOVAOL (Paris, France), and OLC-Ölmühle Leer Connemann (OLC, Germany) was assembled.

Project Workplan

The project was divided into four thematic work areas. For each area, a leader was responsible for coordination among the partners.

Determination Methods (Leader: ITERG, France)

The objective was to evaluate and develop accurate methods for the determination of oxidation, storage, and thermal stability. In the area of oxidation stability, the Rancimat test (EN 14112) had already been chosen for the biodiesel standards. The relation between the induction period provided by this test and other quality parameters required clarification. Due to a lack of knowledge, no test method had been chosen for thermal stability and storage stability. One of the main goals was to select and develop a method for each item considering criteria such as reflecting real conditions, correlating with quality parameters of biodiesel, precision, and cost.

Storage Tests (Leader: SSOG, Italy)

Previous research demonstrated that storage conditions (e.g., temperature, light, atmosphere, presence of prooxidant metals) have a strong effect on storage behavior. The nature of the feedstock might also have a considerable influence on the final result. The main task was to carry out a systematic study of the changes in biodiesel samples obtained from different feedstocks and prepared by different production technologies, during a long-term storage experiment under real-world conditions.

Antioxidants (Leader: Institute of Chemistry, University Graz, Austria)

Natural antioxidants, such as tocopherols and carotenoids, delay the oxidation of vegetable oils. The antioxidant effect depends on the type and amount of antioxidant present. Antioxidants are also present in biodiesel derived from vegetable oils, and the amounts will depend on the vegetable oil used and on the process technology. Consequently, it may be necessary to add synthetic and natural antioxidants to the biodiesel to improve oxidative stability. The objective of this task was to evaluate the stabilizing effects of available natural and synthetic antioxidants in commercial biodiesel. Over 20 synthetic and natural antioxidants were evaluated, and optimum usage levels were determined.

Utilization of Biodiesel

Biodiesel is used both as automotive diesel fuel and as heating fuel. Bench and field tests were carried out on vehicles and injection systems, as well as on heating systems to establish a connection between laboratory test methods and effects during use.

Biodiesel as Automotive Diesel Fuel (Leader: University of Technology, Graz, Austria). The test program comprised bench tests with three different injection systems and test fuels with low-, standard-, and high-stability, long-term tests with two diesel engines, a fleet test using biodiesel with low stability, and a fleet test with blends of fossil diesel fuel and biodiesel with low stability.

Biodiesel as Heating Fuel (Leader: OMV, Austria). The objective of this task was to investigate the effects of the fuel stability when used in heating systems. The effects of fuel stability during the application and the operation parameters of the residential heating systems using blended fuels were studied. The test program comprised bench tests (emissions, operation, long-term testing) with heating systems and field tests with eight heating systems.

Results

Determination Methods

Oxidative Stability. Seven biodiesel samples (methyl esters of rapeseed oil, sunflower oil, used frying oil, and tallow) were evaluated with the Rancimat test (EN 14112; Fig. 1). Determination of quality parameters was carried out on aliquot

Fig. 1. Rancimat for determining the oxidation stability.

samples every 0.5 h. At the end of the Rancimat induction period (RIP), the samples did not meet fatty acid methyl ester (FAME), or oil and fat specifications such as viscosity, acid value (AV), ester content, or peroxide value (PV). The main conclusion was that the induction period determined by conductivity correlates well with the degradation of quality parameters by the Rancimat test (3).

Storage Stability. At the beginning of the project, two test methods were evaluated, namely, ASTM D4625 (storage at 43°C for 24 wk) and an IP48/IP306-like method at 90°C with air flow above the surface of the sample. For each method, seven quality parameters were defined. Because it was difficult to correlate ASTM D 4625 and the results of an accelerated method initially proposed (accelerated IP48/IP306-like method at 90°C), it was decided to use the Rancimat apparatus with special modifications for storage stability evaluation. A stream of purified air (10 L/h) was passed over the surface of 3 g of sample heated to 80°C for 24 h. Then the PV, ester content, and polymer content were measured. The modified Rancimat test is suitable for use in terms of repeatability, significance, and ease of handling. The PV shows the best correlation with ASTM D 4625 (storage at 43°C for 24 wk). Using this method, "poor stability" and "good stability" samples can be distinguished (4).

Thermal Stability. Initially, it was decided to keep the aging conditions of ASTM D 6468 (150°C, 180 or 90 min) because they were considered to be reasonably close to real-world conditions. Seven quality parameters to be evaluated before and after the aging test were defined. However, the variation of quality parameters (AV, Rancimat, ester content) after the aging test was too low to be measured correctly. Finally, it was decided to use the Rancimat apparatus with a procedure especially modified for thermal stability evaluation. The sample (8 g) was aged for 6 h at 200°C in open tubes with air exposure. After aging and cooling, the polymer content was determined by high-performance liquid chromatography (HPLC). The modified Rancimat test is suitable for use in terms of repeatability and ease of handling.

Storage Tests

A systematic study of the chemical and physical changes in 11 different samples of biodiesel was carried out between July 2001 and October 2002. The results allow an evaluation of the effects of different feedstocks and technologies as well as the use of selected antioxidants on the chemical properties of aged biodiesel samples. It was not possible to observe strong changes in 15 monitored characteristics. All samples met the specification limits even at the end of the storage period, with the exception of RIP; PV changes differed, depending on samples. For samples initially not too oxidized, the PV increase was slow. For samples initially oxidized, the PV first increased and then decreased due to the formation of secondary oxidation products.

The most important changes were recorded in oxidative stability as shown by the Rancimat test. This means that aging of biodiesel occurs independently of the monitored parameters, making biodiesel less stable with time. The Rancimat induction period decreases with time. The rate depends on the quality of the sample and on storage conditions. The Rancimat takes a picture of the actual situation, but it is impossible to predict the RIP value after long-term storage. There are aging processes that cannot be observed by analyzing the parameters reported in EN 14213 and EN 14214; thus, a method for storage stability prediction is necessary.

Proper additization increases the induction period greatly and ensures that the sample meets the specification for oxidative stability for at least 6 mon. Super-additization procedures leading to an induction period >20 h have no meaning and might have a negative effect on other parameters (e.g., Conradson carbon residue). The need for correct storage and logistic solutions to avoid the contact of biodiesel with air during its complete life cycle has been pointed out.

Antioxidants

On the basis of the findings of an extensive literature survey, 20 natural and synthetic antioxidants were selected, all of which are commercially available at an affordable price. The stabilizing effect of the selected antioxidants was evaluated with biodiesel prepared from four different raw materials: RME (rapeseed oil methyl ester), SME (sunflower oil methyl ester), UFOME (used frying oil methyl ester), and TME (tallow methyl ester). Both distilled and undistilled biodiesel samples were evaluated, and Rancimat induction times were used to indicate oxidation stabilities. Antioxidants with a good stabilizing effect were tested at different concentrations to determine the optimum antioxidant levels. The effect of natural antioxidants on the oxidative stability of biodiesel was also evaluated. The relatively high stability of RME, which cannot be attributed solely to the natural antioxidants detected, is being investigated further.

Generally, the limit for the proposed oxidative stability parameter could be achieved by adding antioxidants to all different types of biodiesel. Within the variety of antioxidants, synthetic products were more effective than natural antioxidants. The efficiency and the required amount of the different antioxidants strongly depend on the feedstock and technology used for biodiesel production (5). Under the given conditions, no significant negative influence of antioxidants on fuel behavior was observed. The influence of antioxidants on engine performance was not investigated within the project. However, to minimize possible negative effects, it is recommended that antioxidants be used at very low concentrations. The present report does not include any recommendations for the use of specific antioxidants. Long-term engine tests must be conducted to study the influence of synthetic antioxidants on engine performance.

Tocopherols (α-, δ-, and γ-) delay the oxidation of SME, RME, waste cooking oil methyl ester (WCOME), and TME in some cases by more than a factor of 10 compared with methyl esters without tocopherols. γ-Tocopherol was the most and α-tocopherol the least effective of the three natural antioxidants. Their antioxidant effect increased with concentration up to an optimum level. Above the optimum level, the increase in antioxidant effect with concentration was relatively small. The stabilizing effect of tocopherols also depended on the composition of the methyl ester. The order of effectiveness was as follows: TME > WCOME > RME > SME.

Oxidation of unsaturated fatty compounds begins with the build-up of peroxides. Irreversible oxidation, indicated by viscosity increase, starts only after the peroxides reach a certain level. Tocopherols stabilize the unsaturated fatty compound esters by reducing the rate of peroxide formation, thereby extending the time required to reach the peroxide level at which viscosity starts to increase.

The carotenoids, astaxanthin and retinoic acid, had no detectable effect on the stability of SME. Similarly, β-carotene added to camelina oil methyl ester (CME) along with some α-tocopherol, to give the same maximum absorbance at 448 nm as RME, had no stabilizing effect on the methyl ester. However, a carotenoid, at much higher level than β-carotene, was detected in RME, but it was not present in less stable methyl esters such as CME and SME. The detected carotenoid did not extend the period of stability of RME, but it changed its oxidation pattern by reducing both the rate of peroxide formation and of viscosity increase during oxidation.

Utilization of Biodiesel

Biodiesel as Automotive Diesel Fuel. As bench tests, long-term tests were carried out with three different modern injection systems (heavy-duty common rail, passenger car distribution pump, passenger car common rail) on the test bench. Three fuel qualities, i.e., RME with low (induction period 1.8–3.5 h according to EN 14112), standard (induction period 6 h), and high stability (induction period 14–18 h), were used.

Wear and sedimentation were normal for the runtime of each injection system. All enumerated effects were more salient in those parts that were operated in test runs with fuels with lower oxidative stability. Fatty deposits were detected only in those parts of the system that were operated under very strong conditions (RME low oxidation stability, no change of the fuel during the complete test run at the injection system test bed). In all other systems, no critical sedimentations were detected. After the test runs, the functionality was given for each system tested.

Two long-term, real-world engine tests were carried out on the test bench fueled by biodiesel with low (induction period 3.5 h according to EN 14112) and high stability (induction period 20 h). The direct injection diesel engines were equipped with a modern common rail injection system. The duration of each test was 500 h. The measured differences in emissions and power between diesel and RME operation at the beginning of each test were considered normal for biodiesel use. The power loss and difference in the injection amount after 250 h of runtime were higher than expected. Engine inspection revealed normal abrasion for a 500-h runtime and no significant differences due to diesel operation.

For the fleet test with pure biodiesel, four passenger cars were operated with a low-stability biodiesel from July 2001 to November 2002. The test fuel was preaged by a special treatment with temperature and air. The oxidation stability determined by the induction period (Rancimat, 110°C) could be reduced from 7 h to <2 h. Two cars were equipped with a distributor injection pump; the other cars had a unit injector system. The cars were used in typical operation, mainly on the highway. Distance and fuel consumption were recorded in a log book. Some temperatures (engine oil, fuel filter, fuel tank, ambient) were recorded automatically during the entire test period. The total driving distance ranged from 21,000 to 60,000 km/car.

Initial and final tests were conducted at the roller test bed, comprising performance and exhaust emission analyses (CO, HC + NO_x, particles). The differences in emissions and performance could not be assigned to the biodiesel operation (in combination with the results of the injection system check after the test run). Before and after the field test, all limited emissions were below the EURO 3 level (EURO 3 limit vehicle group 2, 1305 kg < reference weight < 1760 kg).

After the tests, the injection systems were inspected by the manufacturer. All systems were functioning normally. Swelling of elastomers was found in a distributor injection pump, which can lead to leakage particularly when using petrodiesel fuel. RME deposits were found on several parts of a distributor injection pump. Traces of oxidation and notable corrosion were found on some parts of the unit injectors. One fuel filter blocked and some fuel filters had to be changed during the winter. It should be noted that the results of the field test were obtained from four individual cars. Thus, a general conclusion about performance with low-stability biodiesel cannot be drawn. This would require an extensive fleet of cars and total coverage of all field influences.

For the vehicle fleet test with a diesel fuel/FAME blend, a 19-mon test run with four diesel vehicles (light- and heavy-duty passenger cars) was carried out. The performance of a diesel fuel (EN 590) blended with 5% UFOME was evaluated. Parameters controlled regularly during the test were fuel quality, cleanness of the fuel storage and supply system, and its operability, vehicle exhaust emissions, engine lubricant performance, drivability of the vehicles in warm and cold conditions, cleanness and wear of the vehicles' fuel systems and fuel injection equipment.

After ~66,000 km driven in daily traffic, there were no significant complaints about the operability of the vehicles registered. No excessive wear or deposit build-up occurred during this period compared with the operation with diesel fuel, which was known from previous fleet tests. The engine lubricant gave a performance similar to that experienced with pure diesel fuel (regular drain interval 15,000 km), and there was no need for premature change of the lubricant. A moderate increase in exhaust emissions was measured during the test interval.

The main components of the fuel storage and delivery system did not suffer from deterioration. The quality of the fuel in terms of most specification parameters was generally constant. Microbial contamination did not occur, probably due to proper cleaning and especially drying of the logistic system at the start of the test. The oxidation stability in the bottom layers of the storage vessels increased dramatically and significantly exceeded the limit of EN 590 (25 g/m^3; EN ISO 12205).

Biodiesel as Heating Fuel. In general, the bench tests in the 1-h heating cycle test gave two different results depending on the operating mode of the heating system, i.e., stationary or nonstationary conditions. Under stationary conditions, the emissions were as low as expected and all units met the general standards for heating units. However, during the starting procedure for the heating unit (semi-warm start, nonstationary condition), concentrations of the hydrocarbons and the carbon monoxide in the flue gas were higher than some seconds after the start. These irregularities depended in large part on the technology of the heating units. In addition, those heating systems that have high emissions during the starting process have worse CO and C_xH_y emissions with an increased percentage of FAME in the blend.

In the long-term test of three different heating units with different fuels, one fact can be seen clearly. The blends made from FAME products having critical stability (aged artificially, stored for 1 y, or distilled and stored for 1.5 y) caused problems in the oil feed rate counter. In the field test, no significant problems have been reported to date. But it must be noted that the conditions for the test were as perfect as possible.

Issues to be resolved include the following: (i) microbiological attack of the blends by bacteria or microbes because of the fast biological decomposition of FAME; (ii) the storage stability of the blends, which is influenced by the materials of the tank, plastics, softening agents; (iii) higher foaming behavior during filling, which leads to tank runover and problems with correct measurement of volume at the filling procedure; (iv) decomposition of the Euromarker (European tax marker) or other additives; (v) swelling effect of the used plastic material, i.e., sealing, pipes, parts in the flow counter, pumps, and nozzles.

Summary

The method for determining oxidative stability (EN 14112-Rancimat test) correlates well with the development of quality parameters. Biodiesel shows high resistance against temperature. A method for detecting thermal stability using the Rancimat equipment by evaluation of the polymer content was proposed. A method for the determination of the storage stability using the Rancimat equipment was also proposed. Samples with poor and good stability can be clearly separated.

Biodiesel can be stored under normal storage conditions without a dramatic change in quality parameters for a period of 1 y. Most changes affect oxidative stability and PV, depending on the quality and the storage conditions. The right additization ensures that the samples continue to meet the specifications after a storage period of 1 y. Nevertheless, proper storage and fuel transport/logistics are absolutely necessary. Contact with air, water, and sunlight must be avoided.

The limit for oxidative stability can be reached by appropriate addition of antioxidants with all different types of biodiesel. Synthetic antioxidants are more effective than natural ones. The efficiency and the amounts of the different antioxidants required depend strongly on the feedstock and biodiesel production technology. No significant negative influence of antioxidants on fuel performance has been observed to date. The influence of additives on engine performance was not investigated within the project. To minimize possible negative effects, however, it is recommended that antioxidants be used at very low concentrations. α-, δ- and γ-Tocopherol significantly delay the oxidation of unsaturated fatty compounds, with γ-tocopherol as the most effective.

The results of bench tests with injection systems showed that functionality was present with all injection systems after long-term tests with different fuel stabilities. Wear and sedimentation were normal for the runtime. Deposits similar to fat were found only in parts being operated under very strong conditions. No effects related to fuel stability were observed in long-term engine tests.

The injection systems of four vehicles from a field test fueled with low-stability biodiesel were acceptable after the test. However, traces of oxidation could be found on some parts of the unit injector. Fuel deposits were found on several parts of a unit injector. Although no problems in relation to the low stability of the fuel occurred, it is recommended to ensure that the minimum stability required by the standards is met to avoid problems during application under very sharp conditions.

The field test using a low-stability biodiesel in a 5% blend with fossil diesel did not have negative effects in terms of wear, deposits, engine lubricant, fuel storage, and delivery, microbiological contamination, or fuel quality. Critical points that remain include oxidative stability (EN ISO 12205 as prescribed by EN 590) and the absolute requirement to clean and keep clean the storage infrastructure (vessels, pipelines) of biodiesel.

When using biodiesel as a heating fuel, no differences existed in short-term bench tests with five heating units under stationary conditions. In contrast, the HC- and CO-emissions during the starting procedure were higher than with pure fossil heating fuel. The results depend significantly on the technology of the heating units. In long-term tests of three units with different fuels, problems occurred in the oil feed rate counter when using fuels with very low stability.

To date, no significant problems have been reported from a field test with eight heating units during two heating seasons. Fossil heating fuel containing 5% FAME (RME, UFOME, and antioxidants) was used. Open questions are focused on microbiological attack, the storage stability of blends, as well as foaming and material compatibility of blends.

Acknowledgments

The paper is based on the results of the project "Stability of Biodiesel" carried out in the 5th framework program of the European Commission. Many thanks to the project partners for providing the results and for the excellent teamwork.

References

1. M/245 Mandate to CEN for the Elaboration and Adoption of Standards Concerning Minimum Requirement Specification Including Test Methods for Fatty Acid Methyl Ester (FAME) as Fuel for Diesel Engines and for Heating, European Commission, 29 January 1997.
2. Stability of Biodiesel Used as a Fuel for Diesel Engines and Heating Systems. Presentation of the BIOSTAB Project Results, Graz, July 3, 2003, published by BLT Wieselburg, Graz, Austria, 2003, (http://www.biostab.info).
3. Lacoste, F., and L. Lagardere, Quality Parameters Evolution During Biodiesel Oxidation Using Rancimat Test, *Eur. J. Lipid Sci. Technol. 105:* 149–155 (2003).
4. Bondioli, P., A. Gasparoli, L. Della Bella, and S. Tagliabue, Evaluation of Biodiesel Storage Stability Using Reference Methods, *Eur. J. Lipid Sci. Technol. 104:* 777–784 (2002).
5. Mittelbach, M., and S. Schober, The Influence of Antioxidants on the Oxidation Stability of Biodiesel, *J. Am. Oil Chem. Soc. 80:* 817–823 (2003).

6.5

Biodiesel Lubricity

Leon Schumacher

Introduction

The need to reduce the exhaust emissions of diesel engines has driven the development of new diesel engine technology. These innovations have focused on the development of the following: (i) diesel-fuel injection technology, (ii) exhaust after-treatment technology, and (iii) diesel-fuel that has been refined to higher standards. The diesel fuel injection technology of a modern diesel engine operates at higher pressures than its counterparts (1). This new technology has led to a demand for better lubrication from the diesel fuel that has traditionally lubricated the fuel injection system of the diesel engine.

Before October 1993, the diesel fuel that was sold in the United States had a sulfur level of ~5000 ppm. In 1993, the Environmental Protection Agency (EPA) mandated that all diesel fuel sold in the United States contain ≤500 ppm sulfur. The petroleum refineries, largely due to special hydrotreating of the diesel fuel, produced a cleaner diesel fuel that met this requirement. On June 1, 2006, the EPA will again lower the level of sulfur in petroleum diesel fuel. The new standard will be 15 ppm or less. This reduction in sulfur is projected to reduce diesel-engine exhaust emissions by as much as 90% compared with the 500 ppm low-sulfur diesel-fuel era. The reduction in engine exhaust emissions is projected for new diesel engines that are equipped with diesel-engine exhaust catalytic converters.

Research demonstrated that catalytic converters last longer, aromatic hydrocarbon emissions are lower, and oxides of nitrogen emissions are lower when cleaner fuels are burned in diesel engines. Unfortunately, the hydrotreating that was used to reduce the sulfur produced a fuel that sometimes failed to provide adequate lubrication for the fuel injection system of the diesel engine (1–4).

Lubricity analysis using the scuffing load ball on cylinder lubricity evaluator (SL-BOCLE) and high-frequency reciprocating rig (HFRR) test procedures indicated that the new 15 ppm low-sulfur diesel fuel will have lower lubricity than the 500 ppm diesel fuel (5). Engine manufacturers proved that a single tankful of diesel fuel with extremely low lubricity can cause the diesel-fuel injection pump to fail catastrophically.

Research conducted using blends of biodiesel mixed with petroleum diesel fuel revealed an increase in lubricity (6). HFRR test procedures using a 2% blend of biodiesel reduced the wear scar diameter by nearly 60% (from 513 to 200 μm).

Background Information Concerning Lubricity

Lubricity can be defined in many ways. "Lubricity is the ability of a liquid to provide hydrodynamic and/or boundary lubrication to prevent wear between moving parts" (7). It can also be defined as follows: "Lubricity is the ability to reduce friction between solid surfaces in relative motion" (7). Another definition (2) is the "quality that prevents wear when two moving metal parts come in contact with each other."

The production of a cleaner diesel fuel could in fact lower the lubricity of diesel fuel (8). These authors reported that the lubricating quality of diesel fuel dropped significantly in 1993 when the United States mandated the use of a diesel fuel that had ≤500 ppm sulfur. The petroleum industry expects the lubricity of petroleum diesel fuel to drop even lower when the limit of sulfur is lowered to 15 ppm in June of 2006.

Although the viscosity of diesel fuel was believed to be related to lubricity (9), many researchers suggested that the lubricity of the fuel is not provided by fuel viscosity (4,8). Researchers found that lubricity is provided by other components of the fuel such as "polycyclic aromatic types with sulfur, oxygen, and nitrogen content." Oxygen and nitrogen were shown to impart natural lubricity in diesel fuel (1). It was reported (4) that oxygen definitely contributes to the natural lubricity of diesel fuel, but that nitrogen is a more active lubricity agent than oxygen. The authors determined that diesel fuels that were high in sulfur but low in nitrogen exhibited poor lubricity.

Some researchers stated that lowering sulfur or aromatics might not lower fuel lubricity. However, as early as 1991, hydrotreating was documented as lowering the lubricity of diesel fuel (5,10–12). It was noted that the special hydrotreating that was used to reduce the sulfur content of diesel fuel also lowered the lubricity of the diesel fuel (8). These authors further theorized that the components (oxygen and nitrogen) "may be rendered ineffective as a result of severe hydrotreatment to desulfurize the fuel."

It is important to note that some fuel injection system diesel engines rely entirely upon diesel fuel to lubricate the moving parts that operate with close tolerances under high temperatures and high pressure (2). Lubricity-related wear problems have already surfaced in Canada, California, and Texas when fleets elected to use low-sulfur fuels to reduce engine exhaust emissions (10). It was noted that rotary distributor injection pumps manufactured by several companies were most susceptible to boundary lubrication wear (2). It is important to note that failure of injection-system components was not limited to a single manufacturer. Several engines experienced problems with the Buna-N seals, which ultimately led to early failure of both the fuel injection system and engine components (11).

The ways to evaluate the lubricity of a fuel include the following: (i) vehicle test, (ii) fuel-injection test equipment bench test, and (iii) a laboratory test (7). The least expensive and most time-efficient of these tests is the laboratory lubricity test.

Fuel-injection equipment tests require 500–1000 h of closely monitored operations (1–3 mon). On road "vehicle tests" require a similar period of time (500–1000 h); however, the data may not be available for as long as 2 yr. The laboratory lubricity test provides a low-cost, accurate evaluation, in <1 wk.

The ASTM D 975 (13) standard specifications for diesel-fuel oils in the United States as of this writing do *not* include a specification for lubricity. Wielligh *et al.* (14) stated that there was a definite need for a diesel-fuel lubricity standard. The ASTM D 6078 (15) standard for lubricity was agreed upon by some engine manufacturers in Europe. These companies have selected test procedures to evaluate the lubricating quality of diesel fuel. For example, Cummins Engine Company determined that "3100 g or greater as measured with the U.S. Army SL-BOCLE (ASTM D 6078) test or wear scar diameter of 380 microns at 25°C as measured with the HFRR (ASTM D 6079) methods" are adequate lubricity values for modern diesel engines. Fuel with SL-BOCLE values >2800 g or an HFRR wear scar diameter that is <450 μm at 60°C, or <380 μm at 25°C, usually performs satisfactorily (7). According to LePera (2) ASTM D 975 will incorporate a lubricity standard by the year 2006, the next planned reduction in sulfur.

It was noted that several standards existed and but that the petroleum industry was divided concerning which was the best test procedure (12). The tests that are available to evaluate lubricity include the following: M-ROCLE (Munson roller on cylinder lubricity evaluator), SL-BOCLE, HFRR, and the SRV (optimal reciprocating rig).

The SRV test has a machine with a 10-mm steel ball sliding against a 25-mm diameter disc, in an off-center mode. The ball is loaded in increments that are adjusted, and the frequency and stroke of the sliding action can be changed. The friction between the ball and the disc results in torque being exerted on the disc, and the torque is measured. A computer calculates the friction coefficient based on the torque. The disc and ball are flooded by dripping the fuel onto the contracting surfaces (14).

The BOCLE and SL-BOCLE test devices press a steel ball bearing against a steel rotating-ring that is partially immersed in the lubricity fluid. Weight is applied until a "scuff" mark is seen on the rotating cylinder (15). More specifically, a 12.7-mm (0.5-in) diameter steel ball is placed on a rotating cylinder. A load is applied in grams. After each successful test, the ball is replaced with a new one, and more load is applied until a specific friction force is exceeded. Exceeding this friction force indicates that scuffing has occurred. The grams of force required to produce the scuff or scoring on the rotating ring are recorded according to ASTM D 6078.

The HFRR test is a computer-controlled reciprocating friction and wear test system. The HFRR test consists of a ball that is placed on a flat surface (16). The ball is then vibrated rapidly back and forth using a 1-mm stroke while a 200-g mass is applied. After 75 min, the flat spot that was worn in the steel ball is measured with a 100X microscope. The size of the spot is directly associated with the lubrication qualities of the fuel being tested.

Lubricity Test Procedures That Have ASTM and EuroNorm (EN) Recognition

Engine companies required a quick, dependable, cost-effective solution to predict fuel performance in a real injection pump. Two tests emerged, i.e., HFRR and SL-BOCLE. The SL-BOCLE was developed by modifying the existing instrument (BOCLE) that is used to measure the lubricity of jet fuel.

European engine manufacturers and fuel-injection pump manufacturers developed a round-robin program in an effort to determine which of these two test procedures was the most accurate. As noted earlier in the chapter, if the HFRR wear scar diameter is <450 μm, the fuel will usually perform satisfactorily. According to European engine manufacturers, the HFRR gave the best correlation with diesel-fuel injection pump durability. This test procedure was adopted as the Commission on European Communities (CEC) standard in 1996. The Europeans have amended EN 590 to include a lubricity standard. The HFRR test procedure was selected with a maximum wear scar diameter of 460 μm.

In the United States, the Engine Manufacturer's Association (EMA) guideline recommends the use of the SL-BOCLE test with a 3100-g minimum. The state of California recommends a 3000-g (SL-BOCLE) minimum. Investigations and additional discussion continue among engine manufacturers in the United States; these will lead ultimately to a specification. However, it is important to note that in the absence of a standard, each refiner has set its own threshold for diesel-fuel lubricity.

Because the data developed from these two test procedures are not exact, reports can be found that specify an HFRR of 500 or 550 μm and SL-BOCLE of 2800, 3000, or 3100, and even 3150 g. In short, comparing information taken using the SL-BOCLE and HFRR is not precise. Further, most supporting information suggests that the proposed 520-μm HFRR level is not a lower lubricity value compared with the 3100-g SL-BOCLE level. Some engine manufacturers suggest that the HFRR may be a better predictor of fuel lubricity for the engine.

According to the literature, the HFRR test method also is less operator-intensive than the SL-BOCLE test method. Because much of the variation noted when using the SL-BOCLE test procedure seems to be associated with operator differences/techniques, the HFRR may prevail as the test of choice. The adoption of the HFRR would ultimately allow engine and fuel system manufacturers to compare their test results more easily.

Analytical Variation of Lubricity Tests

With every analytical test procedure, the information obtained can sometimes vary from one laboratory to the next. Further, the information obtained can vary from one laboratory technician to the next in the same laboratory using the same analytical test equipment. Some companies report compensating for this variation of the SL-BOCLE test by allowing a ±300-g range from the target weight of 3100 g. The repeatability of the SL-BOCLE is ±900 g, and the reproducibility is ±1500 g (7). A similar effect, although with a smaller range, is noted for the HFRR test where the repeatability of the HFRR is ±0.8 and the reproducibility is ±0.136.

Effect of Using Biodiesel as a Lubricity Additive

A study evaluated the lubricity of virgin vegetable oil (17). This same study provided an overview of the lubricity of number one and number two 500 ppm low-sulfur diesel fuel (Table 1). In this table, a standard (3150 g) was noted (17) that was marginally different from the standards established by the ASTM for the SL-BOCLE (3100 g) and HFRR test procedures. The 500 ppm low-sulfur number one diesel fuel (kerosene) required a lubricity additive before use in a diesel engine.

When early research clearly supported the premise that biodiesel indeed had good lubricity, and that the tests conducted suggested that it was nearly two times more able to provide lubricity than petroleum diesel fuel, researchers set out to determine whether blends of the new low-sulfur diesel fuel (≤15 ppm) and biodiesel (1–2%) would provide adequate lubrication for the diesel fuel injection systems of the diesel engine. Blends of 1% biodiesel, 2% (and more), were prepared on a volumetric basis for lubricity testing. These blended fuels were analyzed by independent laboratories using ASTM SL-BOCLE test procedures. SL-BOCLE tests were conducted on the number one and number two diesel fuel (Tier 2 2004) and biodiesel. The results of these tests using ASTM D 6078 are reported in Table 2.

Several diesel-engine manufacturers indicated that an SL-BOCLE of 3100 g (Chevron reports 2800 g) provides adequate lubrication for a modern diesel-fuel injection system. The results in Table 2 indicate that a 1% replacement of number two diesel fuel with biodiesel will provide adequate lubrication for the injection system of a diesel engine. The increase in lubricity for the number one diesel fuel, when added to the level of 4%, fell short of the ASTM SL-BOCLE lubricity standard. Based on these data and information gathered subsequently (18), at least 5–6% biodiesel will have to be added to increase the lubricity of the new ultralow-sulfur number one diesel fuel above 3100 g.

TABLE 1
Lubricity Test Results for Low-Sulfur Diesel Fuel, Vegetable Oil, and Biodiesel Blends[a]

	F2 Commercial #2 with additives		F3 Kerosene (#1 diesel)		F4 Amoco #2 corrosion inhibitor/ no other additive	
Additive	SL-BOCLE	HFRR	SL-BOCLE	HFRR	SL-BOCLE	HFRR
None	4150	376	1250	675	4200	531
1% soybean oil	4150	365	3050	468	4550	303
1% methyl soyate	5200	251	3700	294	4775	233

[a]Engine manufacturer's standards: scuffing load ball on cylinder lubricity evaluator (SL-BOCLE) is >3150; high-frequency reciprocating rig (HFRR) is <450.

TABLE 2
SL-BOCLE Test Results for Ultralow-Sulfur Diesel Fuel (DF) and Biodiesel (BD) Blends[a]

Fuel	0% BD (100% DF)	0.5% BD	1% BD	2% BD	4% BD	12% BD	100% BD
Number 1	1250	N/A	2550	2880	2950	4200	5450
Number 2	2100	2600	3400	3500	N/A	N/A	5450

[a]SL-BOCLE, scuffing load ball on cylinder lubricity evaluator; N/A, not available.

Summary

The data available from engine manufacturers, ASTM, EN, CEC, and private companies suggest that the lubricity of the 15 ppm low-sulfur petroleum diesel fuel will be lower than the existing 500 ppm low-sulfur diesel fuel. Severe hydrotreating of the diesel fuel was used to remove the sulfur from the diesel fuel. The end result was a cleaner fuel, but also one that was poor in lubricity. Petroleum distributors are planning to use a lubricity additive to prevent premature failure of the diesel-fuel injection system when the new diesel fuel is mandated into use by the EPA on June 1, 2006.

The diesel-fuel injection system of a modern diesel engine requires better lubrication due to operating pressures that are higher than those used previously in diesel-fuel injection technology. Several lubricity test procedures were developed by the engine manufacturers together with the petroleum industry in an effort to ensure that the diesel-fuel injection system does not fail prematurely. Two of these test procedures emerged as bench lubricity evaluators, i.e., the SL-BOCLE and the HFRR procedures. Although several researchers contend that the SL-BOCLE correlates more closely with injection pump durability tests (19), the HFRR test procedure appears to be gaining in popularity because the EN has adopted this test procedure as a standard EN590.

The lubricity of petroleum diesel fuel was at one time believed to be related directly to the viscosity of the diesel fuel. Although viscosity and fuel temperature tend to be correlated with a high-lubricity diesel fuel, researchers determined that other compounds are responsible for the natural lubricity of diesel fuel. They also determined that removal of the sulfur did not lower the lubricity of the fuel; rather, the removal of oxygen and nitrogen during desulfurization resulted in a diesel fuel that was low in lubricity.

Lubricity research revealed that the lubricity of low-sulfur number one diesel fuel will be lower than that of number two diesel fuel. The lubricity of number two diesel fuel was noticeably lower than the acceptable level for diesel-fuel lubricity established by the EMA, EN, and the CEC. The addition of small quantities of biodiesel with number one and number two diesel fuel (20) significantly improved the lubricity of the diesel fuel.

Blending as little as 1–2% biodiesel with petroleum diesel fuel increased the lubricity to an acceptable level for the new ultralow-sulfur (15 ppm) number two diesel fuel. Because the new number one Tier 2 diesel fuel was not yet in production, the amount of biodiesel that will be required to raise the lubricity of number one diesel fuel to an acceptable level is unknown. However, based on lubricity research that was conducted using 15 ppm low-sulfur number one fuel (with a similar distillation curve and viscosity as the present-day 500 ppm number one diesel fuel), as much as 5–6% biodiesel may be required to raise the lubricity to a level that meets the lubricity guidelines set forth by the EMA, EN, and CEC.

References

1. Mitchell, K., Diesel Fuel Lubricity—Base Fuel Effects, SAE Technical Paper Series 2001-01-1928, 2001.
2. LePera, M., Low-Sulfur and Diesel Fuel Lubricity—The Continuing Saga, LePera and Associates, *Fuel Line Magazine 4:* 18–19 (2000).
3. Karonis, D., G. Anastopoulos, E. Lois, F. Stournas, F. Zannikos, and A. Serdari, Assessment of the Lubricity of Greek Road Diesel and the Effect of the Addition of Specific Types of Biodiesel, SAE Technical Paper Series 1999-01-1471, Warrendale, PA, 1999.
4. Barbour, R., D. Rickeard, and N. Elliott, Understanding Diesel Lubricity, SAE Technical Paper Series 2000-01-1918, Warrendale, PA, 2000.
5. Anonymous, Biodiesel, Indicators That the Biodiesel Industry Is Growing and Poised to Be a Significant Contributor to the U.S. Alternative Fuels Market, National Biodiesel Board, Jefferson City, MO, 2002.
6. Schumacher, L., J. Van Gerpen, and B. Adams, Diesel Fuel Injection Pump Durability Test with Low Level Biodiesel Blends, in *Proceedings of the 2003 American Society of Agricultural Engineers Annual Meeting*, Las Vegas, 2003.
7. Chevron U.S.A. Inc., *Diesel Fuel Technical Review* (FTR-2), San Francisco, 1998.
8. Keith, O., and T. Conley, *Automotive Fuels Reference Book*, 2nd edn., Society of Automotive Engineers, Warrendale, PA, 1995, pp. 487, 5195.
9. Lacey, P., and R. Mason, Fuel Lubricity: Statistical Analysis of Literature Data, SAE Technical Paper Series 2000-01-1917, 2000.
10. Anonymous, Low-Sulfur Diesel Fuel Requires Additives to Preserve Fuel Lubricity, Stanadyne White Paper on Diesel Fuel, Stanadyne Corporation, Windsor, CT, 2002.
11. Kidwell-Ross, R., Engine Damage from Low Sulfur Diesel Fuel, *American Sweeper* 17–21 (2001).
12. Munson, J., and P. Hetz, Seasonal Diesel Fuel and Fuel Additive Lubricity Survey Using the “Munson ROCLE” Bench Test, Saskatchewan Canola Development Commission, Saskatoon, Canada, 1999.
13. Anonymous, Standard specification for diesel fuel oils. ASTM D 975, ASTM, West Conshohocken, PA, 2002, pp. 1–19.
14. Wielligh, A., N. Burger, and T. Wilcocks, Diesel Engine Failure Due to Fuel with Insufficient Lubricity. Department of Mechanical and Aeronautical Engineering, University of Pretoria, Pretoria, Republic of South Africa, 2002.
15. Anonymous, Standard Test Method for Evaluating Lubricity of Diesel Fuels by Scuffing Load Ball-on-Cylinder Lubricity Evaluator (SLBOCLE), ASTM D 6078, ASTM, West Conshohocken, PA, 1999, pp. 1–9.
16. Anonymous, Standard Test Method for Evaluating Lubricity of Diesel Fuels by the High-Frequency Reciprocating Rig (HFRR), ASTM D 6079, ASTM, West Conshohocken, PA, 1999, pp. 1–4.

17. Van Gerpen, J., S. Soylu, and D. Chang, Evaluation of Lubricity of Soybean Oil-based Additives in Diesel Fuel, Iowa State University, Ames, IA, 1998.
18. Beach, M., and L. Schumacher, Lubricity of Biodiesel Blends, On Campus Undergraduate Research Internship Poster Session, University of Missouri, Columbia, MO, April 27, 2004.
19. Anonymous, Standard Specification for Biodiesel Fuel (B100) Blend Stock for Distillate Fuels, ASTM D 6751, ASTM, West Conshohocken, PA, 2002, pp. 1–6.
20. Anonymous, WK2571—Standard Specification for Diesel Fuel Oils, Revision of D975–04. ASTM International, ASTM, West Conshohocken, PA, 2004.

6.6

Biodiesel Fuels: Biodegradability, Biological and Chemical Oxygen Demand, and Toxicity

C.L. Peterson and Gregory Möller

Introduction

This chapter summarizes the University of Idaho research related to biodegradability, biochemical oxygen demand (BOD_5), chemical oxygen demand (COD), and toxicity of biodiesel (1–5). These studies were conducted in the mid-1990s using neat oils and biodiesel from a variety of feedstocks including soy, canola, rapeseed, and others. Both methyl and ethyl esters were included in most studies. Phillips 2-D low-sulfur diesel fuel was used as a comparison petrodiesel in all of the studies. In some studies, blends of the 2-D reference fuel with the test vegetable oil-based fuels were included as noted in each section. The fuel nomenclature is as follows: Phillips 2-D low-sulfur diesel control fuel, 2-D (petrodiesel); 100% rapeseed methyl ester, RME; 100% rapeseed ethyl ester, REE; 50% RME/50% 2-D, 50RME; 50% REE/50% 2-D, 50REE; 20% RME/80% 2-D, 20RME; 20% REE/80% 2-D, 20REE.

Biodegradability

The biodegradability of various biodiesel fuels in the aquatic and soil environments was examined by the CO_2 evolution method, gas chromatography (GC) analysis, and seed germination. The fuels examined included neat rapeseed oil (NR), neat soybean oil (NS), the methyl- and ethyl esters of rapeseed oil and soybean oil, and Phillips 2-D reference petroleum diesel. Blends of biodiesel/petrodiesel at different volumetric ratios, including 80/20, 50/50, 20/80, were also examined in the aquatic phase.

There are many test methods for accessing the biodegradability of an organic compound. Among them, the CO_2 evolution test (shaker flask system) and GC analysis are most common and were employed as the major method for the aquatic and soil experiments, respectively. One important difference between them is that CO_2 evolution measures ultimate degradation (mineralization) in which a substance is broken down to the final products, CO_2 and water, whereas GC analysis measures only primary degradation in which the substance is not necessarily transformed to the end products. Finally, because revegetation of soils contaminated by fuel spills is often a desirable goal, seed germination was employed to evaluate the toxicity of biodiesel on plants in the soil system.

The CO_2 evolution method employed in this work followed the Environment Protection Agency (EPA) standard method 560/6–82–003 using a shake flask system for determining biodegradability of chemical substances (4). The GC method involved extraction of the samples with a solvent and injection of a portion of the extract into a gas chromatograph.

A 500-mL Erlenmeyer flask system was used in the soil flask tests. Dry soil (30 g) was placed in the flask and thoroughly amended with the required amount of a test substance (10,000 mg/L) weighed by an accurate balance and stock solution. Deionized water was also added if necessary to bring the soil to a 30% moisture level. Each flask was sealed and incubated at room temperature.

At each time interval, 2 g of the soil (dry weight) was removed for extraction and GC analysis. The 2-g soil sample was placed in a 24-mL vial with a Teflon cap. The soil was mixed with the same volume of anhydrous sodium sulfate to absorb moisture in the sample. Then, 1 mL of internal standard (the same as those in the aquatic system) was added to the vial to determine the extraction efficiency and serve as a quantitative standard. Immediately after the addition of the internal standard, 9 mL of extraction solvent (the same as those in the aquatic system) was added; 2 mL of the extraction was then placed in a bath sonicator for 30 min with <10 vials at one time. The extract was transferred into a vial, sealed, and kept in a refrigerator (4°C) before GC analysis.

Seed germination involved four 32 cm × 6 cm (diameter × height) plates holding 2.0 kg soil (dry weight) that was contaminated with one of four different test substances including biodiesel REE, RME, NR, and 2-D at an average concentration of ~50,000 mg/L. A plate with no substrate was used as the control. Seeds (n = 100) of Legacy Alfalfa were seeded in each of the fuel-contaminated plates and the control at d 1 and wk 1, 3, and 6. The plates were covered by a thin plastic film (with small holes), kept in a greenhouse to maintain a favorable temperature for microorganisms and plants, and watered periodically to maintain required moisture.

Results

The average cumulative percentage theoretical CO_2 evolution from six biodiesel fuels (NR, NS, REE, RME, SEE, SME and 2-D) in 28 d is summarized in Figure 1. [All six samples in duplicate experiments were averaged and an arithmetic mean, SD, and relative standard deviation (RSD)% were calculated.] The maximum percentage of CO_2 evolution from REE, RME, SEE, SME was between 84 and 89%, the same as that of dextrose. Statistical analysis indicated that there was no difference in their biodegradability. The maximum percentage of CO_2 evolution from NR and NS was 78 and 76%, respectively, slightly lower than their modified products. This result might be explained by their higher viscosity. CO_2 evolution from 2-D was only 18.2% (averaged from several experiments).

The results demonstrate that all the biodiesel fuels are "readily biodegradable." Moreover, co-metabolism was observed in the biodegradation of the biodiesel diesel

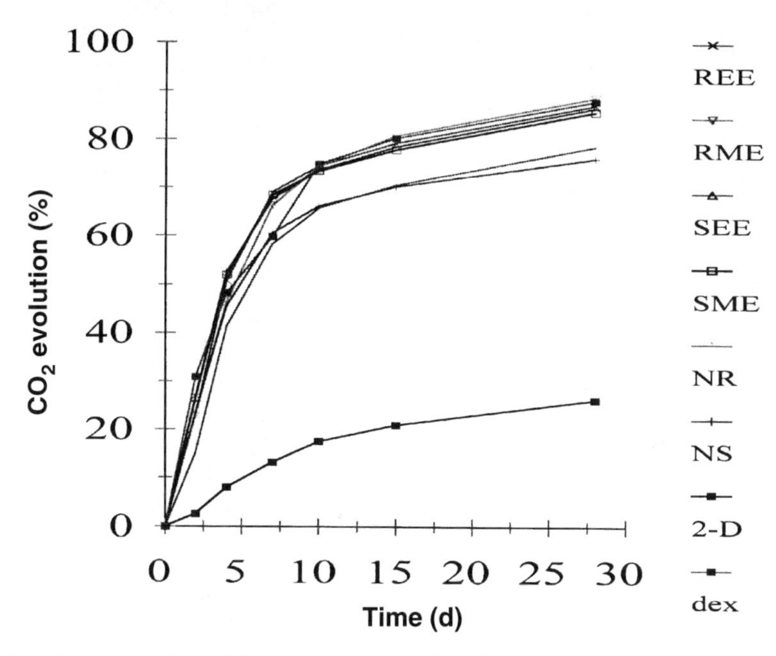

Fig. 1. Biodegradability of biodiesel compared with dextrose and 2-D reference fuel measured by the shake flask test.

blend in the aquatic phase, i.e., in the presence of REE, the degradation rate and extent of petroleum diesel increased to twice that of petroleum diesel alone.

The GC analysis showed much faster degradation for REE and 2-D. The disappearance of REE and 2-D on d 1 reached 64 and 27%, respectively. On d 2, all of the fatty acids in REE were undetectable, whereas only 48% of 2-D had disappeared. However, the ratios of the percentage of primary degradation to ultimate degradation for REE and 2-D were quite different, 1.2 vs. 2.7. The lower ratio for REE indicates that most of biodiesel was transformed into the end products, and the higher ratio for diesel implies that most of the diesel, about two-thirds of the 48% primary degradation, was changed to an intermediate product. Determining the nature of these intermediates warrants further study.

Blends. The percentage of CO_2 evolution from the REE/2-D blends (Fig. 2) increased linearly with the increase of REE concentration in the blend. The higher the volume of REE in the blend, the higher the percentage of CO_2 evolution. The relation can be described by a linear equation $Y = 0.629X + 20.16$ with $R^2 = 0.992$ (95% confidence limit), where X is the percentage concentration of REE in the blend and Y is the cumulative percentage of CO_2 evolved in 28 d.

Again, GC analysis showed a much faster and higher primary degradation in the REE/2-D 50/50 blend, 64 and 96% on d 1 and 2, respectively. In addition, co-

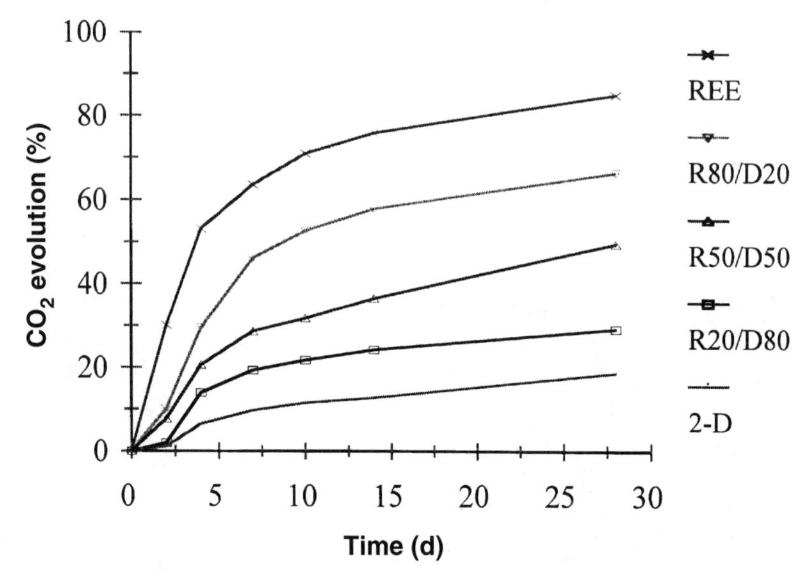

Fig. 2. Biodegradability of blends of biodiesel and 2-D reference fuel measured by the shake flask test.

metabolism was observed. The 2-D in the blend was degraded twice as fast as the 2-D alone, 63 vs. 27% on d 1. This suggests that in the presence of REE, microorganisms use the fatty acids as an energy source to promote the degradation of 2-D.

Soil Flask Results. The average substrate disappearance vs. time for five different biodiesel fuels and diesel at the initial concentration of 10,000 mg/L is summarized in Figure 3 (three samples for each substance were averaged). In 28 d, the percentage substance disappearance for five biodiesel fuels reached 83–95%, with an average of 88%; for diesel, it was 52%.

The results of the seed germination tests are shown in Table 1. Biodiesel fuels REE-, RME-, and NR-contaminated soils had lower seed germination rates on d 1 and at wk 1 of seeding compared with wk 3 and 4 because fungus grew rapidly across the plate by 2 wk after contamination. There were ~20 seeds that germinated but died underground. However, after wk 3, when most of the biodiesel was degraded and fungus began to disappear visually, the seed germination rates in biodiesel-spilled plates increased. After wk 6, the germination rates in all three biodiesel fuel-spilled plates reached 92–98%. Among the biodiesel fuels, NR had the highest germination rate, ~87%.

Seeds in the 2-D-spilled plate germinated at least 7 d later than those in biodiesel-spilled plates in the first seeding. Moreover, fungus growth was not observed in the diesel-spilled plate until wk 4. This is likely the reason why seed germination in the petrodiesel-spilled plate dropped dramatically after wk 4. These

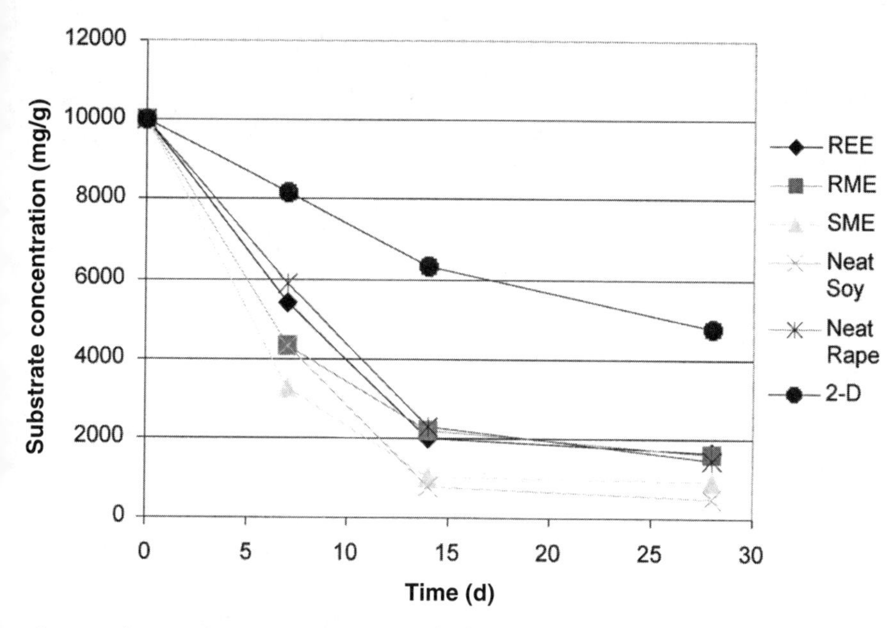

Fig. 3. Substrate disappearance vs. time for five biodiesel fuels and diesel at an initial concentration of 10,000 mg/L in soil.

results demonstrate that biodegradation can restore a biodiesel fuel-contaminated soil in 4–6 wk to such a degree that it can support plant germination.

Biodegradability Conclusions

1. All of the biodiesel fuels were "readily biodegradable" in the aquatic and soil environments. During a 28-d period, the average CO_2 evolution for all of the biodiesel reached 84% in the aqueous system, and the average substance disappearance was 88% in the soil environment.

TABLE 1
Seed Germination in the Fuel-Contaminated Soils Seeded[a]

	Seed germination (%)				
Time of seeding (d)	Control	Raw rapeseed oil	RME	REE	Diesel
1	100	84.0	60.5	51.9	19.8
7	100	76.1	55.4	73.9	62.0
21	100	91.1	82.2	83.3	87.8
42	100	95.4	92.0	97.7	19.5
Mean	100	86.6	72.5	76.7	47.3

[a]RME, rapeseed methyl ester; REE, rapeseed ethyl ester.

2. From the result of CO_2 evolution, the increase in REE concentration in the blends caused a linear increase in the percentage of ultimate biodegradation of the blends. According to the results of GC analysis, co-metabolism was observed in the primary biodegradation of the 50REE blend. The biodiesel in the blend appeared to promote and increase the extent of biodegradation of petrodiesel by up to 100%.

3. Biodegradation can restore a biodiesel fuel spill-contaminated soil in 4–6 wk to such a degree that it can support plant germination. However, the seed germination test showed that biodiesel-contaminated soil did have an effect on plant growth for the first 3 wk due to the rapid growth of microorganisms during the period of fuel degradation.

BOD_5 and COD

BOD_5 is a measure of the dissolved oxygen consumed during the biochemical oxidation of organic matter present in a substance. In the current study, BOD_5 was used as a relative measure of the amount of organic matter subject to the microbially mediated oxidative processes present in biodiesel fuel. This may also serve as a relative measure of biodegradability. COD is a measure of the amount of oxygen required to chemically oxidize organic matter in a sample. COD values were used in the study as an independent measure of the total oxidizable organic matter present in the fuels.

BOD_5 was determined using EPA Method 405.1. Dissolved oxygen is measured initially and after incubation. The BOD is computed from the difference between initial and final dissolved oxygen (DO). Replicate analyses were performed in triplicate (the method specifies that duplicate samples be used; thus, n = 6). Reference samples of glucose/glutamic acid solution, and a commercially available WasteWatR™ demand reference were also tested in duplicate. Fuels were tested at their appropriate water accommodated fraction (WAF: the highest concentration at which the test substance is maintained in the aqueous phase of the solution); WAF values were converted to pure substance for statistical comparison. COD was determined using EPA Method 410.1. Fuels were tested at their appropriate WAF, and these values were converted to pure substance for statistical comparison (1).

Results. BOD_5 values for REE (1.7×10^6 mg/L), RME (1.5×10^6 mg/L), SME (1.7×10^6 mg/L), NR (1.7×10^6 mg/L), and soybean oil (1.6×10^6 mg/L) were significantly higher than the 2-D reference fuel (0.4×10^6 mg/L) (Fig. 4). COD values were similar for all fuels tested (Fig. 5). Results indicate that biodiesel fuel substances contain significantly more microbially biodegradable organic matter than does 2-D reference fuel.

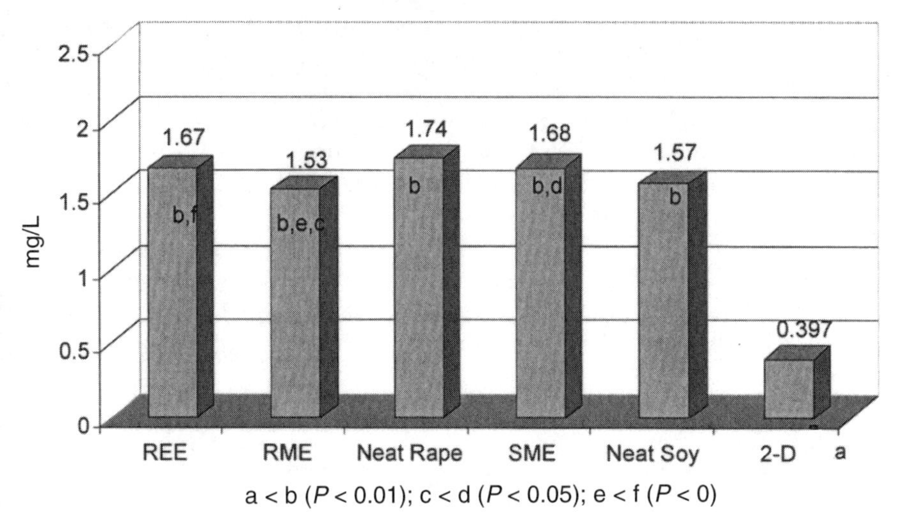

Fig. 4. Mean biochemical oxygen demand (BOD_5) values (n = 6).

BOD_5 and COD Conclusions

No significant difference was expected or observed between COD values for the 2-D control substance and any of the test substances. Due to the chemical nature of the test, a total measure of all oxidizable organic matter is given. This is in contrast to the BOD_5 test which more appropriately limits oxidative activity to microbial populations. The significant differences ($P < 0.01$ and $P < 0.05$, respectively) between REE and SME and between REE and RME may reflect slight differences

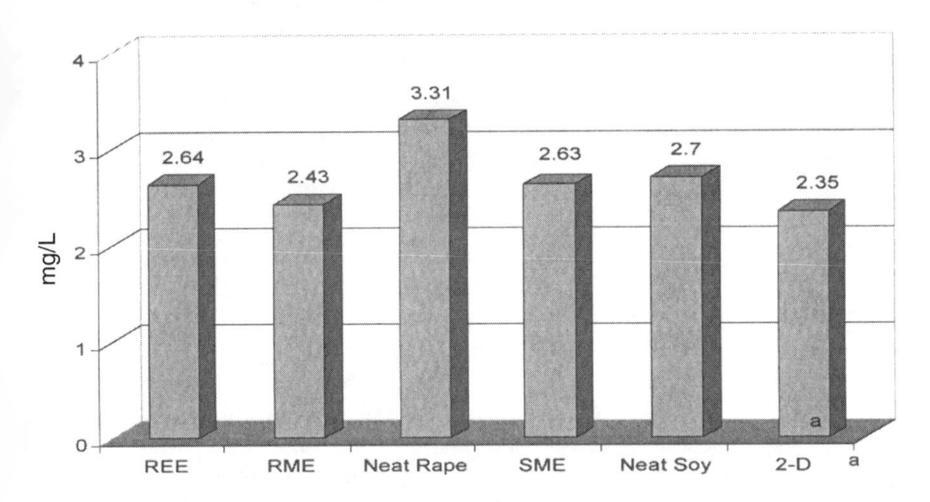

Fig. 5. Mean chemical oxygen demand (COD) values (n = 3).

in the amount of organic matter oxidized by microbial processes. Although the difference is significant, the magnitude of the differences is <10%. Therefore, the biodegradability of these substances may be considered to be similar. The significantly lower BOD_5 value for the 2-D control substance and the large magnitude of the difference (average 122% difference) may be attributed to various factors. The WAF value for 2-D was also noted at 3.8 mg/L, much lower than those of the test substances; this could occur because less of the substance was present in solution and available for microbial oxidative processes. The significantly lower ($P << 0.01$) BOD_5 values indicate the presence of a much smaller amount of microbial biodegradable organic matter in the Phillips 2-D diesel fuel. The lower BOD_5 value may also reflect the microbial toxicity of the diesel fuel or its components.

Toxicity

This section reports on acute oral and acute dermal toxicity tests and static acute aquatic toxicity tests with RME and REE and blends of each with diesel fuel. The acute oral toxicity tests were conducted with albino rats and the acute dermal toxicity tests were conducted with albino rabbits. Acute aquatic toxicity tests were performed with *Daphnia magna* and rainbow trout. Each of these studies was under contract with the University of Idaho. A separate set of studies with *D. magna* and juvenile rainbow trout was conducted by the University of Idaho.

The 50% lethal dose (LD_{50}; the point at which 50% have died and 50% are still alive determined by interpolation) values for each of the substances tested were >5000 mg/kg when administered once orally to rats and >2000 mg/kg when administered once for 24 h to the clipped, intact skin of male and female albino rabbits. The LD_{50} values for acute aquatic toxicity with *D. magna* in mg/L were 3.7 for table salt, 1.43 for 2-D, 23 for RME, 99 for REE, and 332 for methyl soyate. Duplicate tests with rainbow trout were run with 10 organisms/replicate. The median lethal concentration $(LC)_{50}$ values were not reported because of failure to kill a sufficient number of fish at the concentrations tested, even with petrodiesel. The 20 and 50% blends produced scattered losses of fish, but none of the tests had <85% survival at any concentration after 96 h.

The toxicology protocol was designed and conducted in compliance with the EPA Guidelines for Registering Pesticides in the U.S. (Pesticide Assessment Guidelines, Subdivision F, Hazard Evaluation: Human and Domestic Animals, Section 81-1) and the Toxic Substances Control Act (TSCA) Health Effects Test Guidelines, 40 CFR 798.1175. The studies were conducted in compliance with EPA Good Laboratory Practice Regulations (40 CFR Parts 160 and 792) and the Standard Operating Procedures of WIL Research Laboratories. The study was conducted and inspected in accordance with the Good Laboratory Practice Regulations, the Standard Operation Procedures of the contractor.

Procedures for aquatic toxicity testing are outlined in 40 CFR part 797.1300 (Daphnid acute toxicity test) and part 797.1400 (fish acute toxicity test), and ASTM

E 729-88. These procedures include, with the LC_{50}, an EC_{50} (median effective concentration), and an IC_{50} (inhibition concentration). All static tests for this study were performed according to Reference 6.

Acute Oral Toxicity Studies

One group of five male and five female rats was administered single doses at a level of 5000 mg/kg for each test substance. The rats were observed for mortality at ~1.0, 3.0, and 4.0 h postdose on d 0 and twice daily (morning and afternoon) thereafter for 14 d. The rats were observed for clinical observations at ~1.0, 3.0, and 4.0 h postdose on d 0 and once daily thereafter for 14 d. Body weights were obtained and recorded on study d –1, 0 (initiation), 7, and 14 (termination). Upon termination, all rats were killed by carbon dioxide asphyxiation. The major organ systems of the cranial, thoracic, and abdominal cavities were examined in all rats.

Acute Dermal Toxicity Studies

One group consisting of five male and five female albino rabbits was administered a single dose (24-h, semioccluded exposure) of each test substance at a dose level of 2000 mg/kg. The rabbits were observed for mortality at ~1.0, 3.0 and 4.0 h post-dose on d 0 and twice daily (morning and afternoon) thereafter for 14 d. The rabbits were observed for clinical observations at ~1.0, 3.0, and 4.0 h postdose on d 0 and once daily thereafter for 14 d. The application sites were examined for erythema, edema, and other dermal findings beginning ~30–60 min after bandage removal and daily thereafter for 13 d. The rabbits were clipped to facilitate dermal observations on study d 3, 7, 10, and 14. Body weights were obtained and recorded on study d 0 (initiation), 7, and 14 (termination). Upon termination, the rabbits were killed by intravenous injection of sodium pentobarbital. The major organ systems of the cranial, thoracic, and abdominal cavities were examined in all rabbits.

Results of the Acute Oral Toxicity Studies

100% RME and 100% REE. There were no deaths, remarkable body weight changes, or test material-related gross necropsy findings during the study. Single instances of wet yellow urogenital staining were noted for two female rats at d 1 for RME and three female rats at d 1 for REE. There were no other clinical findings. There were two individual clinical observations reported for RME and three for REE. All rats appeared normal on or before d 2 and throughout the remainder of the study. The LD_{50} of 100% RME and 100% REE was >5000 mg/kg when administered once orally *via* gastric intubation to food-deprived male and female albino rats.

50% RME/50% 2-D. There were no deaths, remarkable body weight changes, or test material-related gross necropsy findings during the study. Wet and/or dried yellow urogenital and/or ventral abdominal staining was noted for all rats. For 50% RME/50% 2-D, one rat each had clear ocular discharge or dried red material around the nose. For 50% RME/50% 2-D, single rats had clear ocular discharge, hypoactivity, hair loss on the dorsal head or dried red material around the nose or forelimb(s). There were no other clinical findings. There were 18 individual clinical observations reported for 50% RME/50% 2-D and 30 for 50% REE/50% 2-D. With the exception of one rat having hair loss on the dorsal head with the 50% RME/50% D-2 blend, all rats appeared normal by d 3 or earlier and throughout the remainder of the study. The LD_{50} of both blends was >5000 mg/kg when administered once orally *via* gastric intubation to food-deprived male and female albino rats.

20% RME/80% 2-D. There were no deaths, remarkable body weight changes, or gross necropsy findings during the study. Wet and/or dried yellow urogenital and/or ventral abdominal staining was noted for all rats. With the 20% RME/80% 2-D blend, two male rats each had clear wet matting around the mouth and clear ocular discharge; with the 20% REE/80% 2-D blend, one rat each had clear wet matting around the mouth, clear ocular discharge, and desquamation on the urogenital area. For 20% RME/80% 2-D, findings documented for one rat each included soft stool and hair loss on the hindlimb(s) and ventral abdominal area; for 20% REE/80% 2-D, two rats each had soft stool, dried red material around the nose, and hair loss on the hindlimb(s) and/or ventral abdominal area. There were no other clinical findings. There were 48 individual clinical observations reported for the RME blend and 66 for the REE blend. With the exception of hair loss on the hindlimb(s) and ventral abdominal area for one rat, all rats appeared normal on or before d 8 and throughout the remainder of the study. The LD_{50} of both blends was >5000 mg/kg when administered once orally *via* gastric intubation to food-deprived male and female albino rats.

100% 2-D. There were no deaths, remarkable body weight changes, or test material-related gross necropsy findings during the study. Wet and/or dried yellow urogenital and/or ventral abdominal staining was noted for all rats. Additional findings included various hair loss, clear wet matting around the mouth and/or urogenital area, soft stool, and hypoactivity. There were no other clinical findings. There were 105 individual clinical observations reported. With the exception of hair loss on the hindlimb(s) and/or urogenital area for four rats, all rats appeared normal on or before d 9 and throughout the remainder of the study. The LD_{50} of 100% 2-D was >5000 mg/kg when administered once orally *via* gastric intubation to food-deprived male and female albino rats.

Results of the Acute Dermal Toxicity Studies

100% RME. There were no deaths, test material-related clinical findings, body weight changes, or gross necropsy findings during the study. The test material induced very slight to slight erythema in all rabbits and very slight to slight edema in seven rabbits. All sites had desquamation. Fissuring was noted for one rabbit for 100% RME. There were no other dermal findings. All edema subsided completely by d 12 for 100% RME and d 10 for 100% REE. With the exception of two rabbits that had very slight erythema with 100% RME, all dermal irritation subsided completely on or before d 14. There were 8 individual clinical observations reported for 100% RME and 5 for 100% REE. There were 102 and 90 very slight erythema occurrences for RME and REE, respectively, as well as 8 and 10 slight, 32 and 55 very slight edema, 7 and 2 slight edema, and 70 and 62 desquamation occurrences for RME and REE, respectively; there was one fissuring occurrence for RME during the 14-d study. The LD_{50} of both 100% RME and 100% REE was >2000 mg/kg when administered once for 24 h to the clipped, intact skin of male and female albino rabbits. In addition, the 2000 mg/kg dose level was a no-observable effect level (NOEL) for systemic toxicity under the conditions of this study.

100% 2-D. There were no deaths, test material-related clinical findings, body weight changes, or gross necropsy findings during the study. The test material induced moderate erythema, slight to moderate edema, and desquamation in all rabbits. Two rabbits had fissuring. There were no other dermal findings. All erythema and edema subsided completely by d 14 or earlier. Desquamation persisted to d 14 in seven rabbits. There were no individual clinical observations reported. There were 42 very slight, 27 slight, and 31 moderate occurrences of erythema; 22 very slight, 28 slight, and 6 moderate occurrences of edema; three occurrences of fissuring and 120 desquamation occurrences during the 14-d study. The LD_{50} of 100% 2-D was >2000 mg/kg when administered once for 24 h to the clipped, intact skin of male and female albino rabbits. In addition, the 2000 mg/kg dose level was at a NOEL for systemic toxicity under the conditions of this study.

Acute Aquatic Toxicity

D. magna. The organisms were obtained from the contractor's in-house cultures and were <24 h old before initiation of the test. All organisms tested were fed and maintained during culturing, acclimation, and testing as prescribed by the EPA. The test organisms appeared vigorous and in good condition before testing. The *D. magna* were placed below the test surface at test initiation due to the nonsoluble nature of the sample.

Juvenile Rainbow Trout. The fish used in the first round of tests were 22 days old and 32 ± 2 mm in length. The rainbow trout were acclimated to test conditions (dilution water and temperature) for 10 d before test initiation. The rainbow trout used in the second round of tests were 24 days old and 28 ± 1 mm in length. The rainbow trout were acclimated to test conditions (dilution water and temperature) for 12 d before test initiation. All of the test organisms appeared vigorous and in good condition before testing.

Test Concentrations

D. magna. The concentrations tested in the definitive test on REE were 33, 167, 833, 4170, and 20800 mg/L of sample and dilution water for the control. The concentrations tested in the definitive test on RME were 67, 333, 1330, 6670, and 26700 mg/L of sample and dilution water for the control. The concentrations tested in the definitive test on D2 were 6.7, 13.3, 33.3, 66.7, and 1333 mg/L of sample and dilution water for control. The concentrations tested in the definitive test on methyl soyate were 13.3, 33.3, 66.7, and 6667 mg/L of sample and dilution water for control. The fuel mixture concentrations were run in quadruplicate with five organisms/replicate. Additional concentrations of 1.43 and 3.33 mg/L were set up for D2 with 10 organisms in one chamber. The fuel was stirred into the water before the *D. magna* were introduced into the chamber. There was a sheen of fuel on the top of each chamber.

Rainbow Trout. The concentrations tested for round 1 in the definitive test on D2, 20RME, 20REE, and REE were 100, 300, 600, 1200, and 2400 mg/L with dilution water for control. The concentrations tested for round 2 in the definitive test on RME and 50REE were 100, 500, 750, 1000, and 7500 mg/L, and the 50RME sample was tested at 100, 500, 600, and 7500 mg/L due to a shortage of the sample.

The rainbow trout bioassays were run in 5-gallon glass aquaria, with a volume of 5 L water. The samples were assayed in duplicate with 10 organisms/replicate. The photoperiod was 16 h light:8 h dark. The temperature range was 12 ± 1°C. Loading of test organisms was 0.53 g wet fish weight/L in round one, and 0.26 g wet fish/L in round two. Mortality was measured by lack of response to tactile stimulation and lack of respiratory movement. The fuel was stirred into the water before the rainbow trout were introduced into the chamber. There was a sheen of fuel on the top of each chamber.

Results of the Acute Aquatic Toxicity

D. magna. Some of the mortality seen in the tests may have been caused by the physical nature of the test substances. The raw data sheets noted when the *D. magna* were trapped on the oil sheen at the surface of the test containers. The LC_{50} for the REE sample was 99 mg/L. Results were summarized for the RME sample with a reported LC_{50} for RME of 23 mg/L as well as for methyl soyate with a reported LC_{50} of 332 mg/L.

The methyl and ethyl esters are not water-soluble and formed a sheen on the water surface. This sheen could be easily skimmed off, but the *D. magna* became captured in the sheen. At a concentration of 3.7 mg/L, 50% of the *D. magna* in common table salt had died. With petrodiesel, 50% had died at <1.43 mg/L and all were reported dead at this concentration. When this test was first completed, it was reported that the LC_{50} for diesel fuel was <6 mg/L because all of the *D. magna* had died. Four more concentrations <6 mg/L were tested and the petrodiesel fuel killed the *D. magna* at all concentrations. For RME, the LC_{50} was 23 mg/L, and at 26,700 mg/L, 30% were still alive. With REE, the LC_{50} was 99 mg/L and 20% were still alive at 20,800 mg/L. With methyl soyate, the LC_{50} was 332 mg/L; however, only 45% were alive at 667 mg/L. This difference between rapeseed esters and SME may be due to the high erucic acid content of the rapeseed. In the worst case, comparing the 23 mg/L for REE with the 1.4 mg/L for diesel fuel, the acute aquatic toxicity was 15 times less. What is even more significant is the 20 and 30% that were still alive at very high concentrations of biodiesel.

Rainbow Trout. The LC_{50} for D2 was not determined. These data compare cadmium chloride (CdCl), petrodiesel fuel, and the methyl and ethyl esters of rapeseed. The 50RME percentage survival summary results were identical to the 100% RME results. The 48-h LC_{50} value and Control Chart limits for the reference toxicant (CdCl) were at a concentration of 2.8 g/L for round one of the rainbow trout study and 4.6 g/L for round two. The results indicate that the test organisms were within their expected sensitivity range. Comments included in round one test data at 24 h indicated a general behavior of twitching, swimming on the sides, and skittering; at 48 h, their condition was the same as at 24 h. The trout in the 20 REE containers at 100 and 300 mg/L were swimming vertically; at 600 mg/L, the trout were on their sides at the bottom, and at 2400 mg/L they were barely moving at the bottom of the tank. The trout in the REE containers were not as active as in the other three test substances. The end condition of survivors was reported as being poor. The only comment in round two was at 48 h that the fish were dark and swimming vertically at concentrations as low as 500 mg/L in the 50RME and 50REE with the end condition of survivors reported as poor.

University of Idaho Static Nonrenewal and Flow-Through Acute Aquatic Toxicity Tests

D. magna (n = 20) were exposed to each of five concentrations of test/reference substance and a control for 48-h periods in static and flow-through environments as outlined in EPA TSCA Environmental Effects Test Guidelines at 40 CFR §797.1300 Daphnid Acute Aquatic Toxicity Test and additional guidelines in EPA Methods for Measuring the Acute Toxicity of Effluents and Receiving Waters to Freshwater and Marine Organisms, 4th edn. (EPA/600 4-90-027). Mortality data were collected at 24 and 48 h, and EC_{50} results were calculated using the EPA Probit Analysis Program.

Static Nonrenewal Toxicity Test. This is a system in which the test solution and test organisms are placed in the test chamber and kept there for the duration of the test without renewal of the test solution. The insoluble and glassware-coating nature of these test substances required the derivation of a WAF to minimize mortality occurring from suffocation of the daphnids. Tests were performed at or below the WAF levels, and pretest mixing for extended periods was necessary to avoid "floating" the test substance on the surface of the test chamber where mortality may be caused by other than toxic effects. The derived WAF was used as the highest concentration and was proportionally diluted to the other concentrations of the analysis. Test chambers were filled with appropriate volumes of dilution water, and the test chemical was introduced into each treatment chamber. The test started within 30 min after the test chemical was added and was uniformly distributed in static test chambers. At the initiation of the test, daphnids that had been cultured and acclimated in accordance with the test design were randomly placed into the test chambers. Daphnids in the test chambers were observed periodically during the test, the immobile daphnids removed, and the findings recorded. Dissolved oxygen, pH, temperature, concentration of test chemical, and water quality parameters were measured.

Flow-Through Tests. The test substances were initially mixed at the WAF for a minimum of 20 h in a 50-L holding tank under constant stirring in the flow-through test. The stirring action proceeded throughout the duration of the test. The test substance mixture at WAF was drawn from the mixing tank for each cycle. The amount withdrawn from the tank was replaced for each cycle, maintaining an equilibrium mixture at the WAF concentration.

Results of the University of Idaho Static and Flow-Through Tests

The 48-h LC_{50} values are presented in Figure 6 for static nonrenewal and in Figure 7 for the flow-through tests. The lowest static 48-h *D. magna* EC_{50}, indicative of

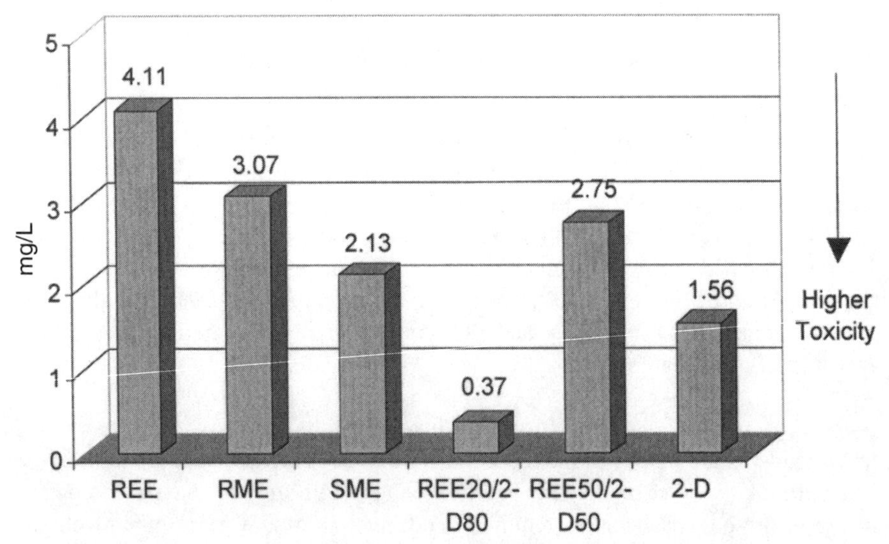

Fig. 6. Forty-eight-hour static, nonrenewal, *Daphnia magna* EC_{50}.

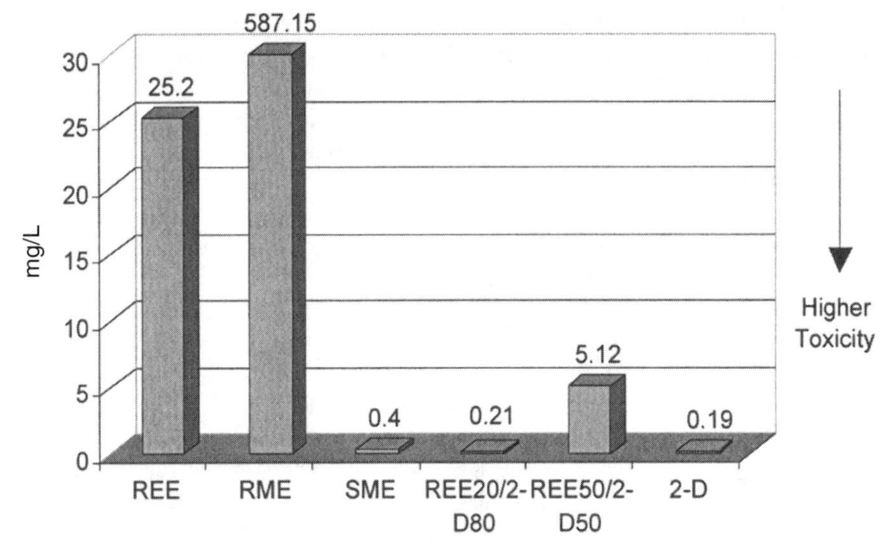

Fig. 7. Forty-eight-hour flow-through, *Daphnia magna* EC_{50}.

the highest toxicity, was 0.37 mg/L for the 20% REE/80% 2-D mixture. This was followed by the 2-D reference fuel at 1.56 mg/L and SME at 2.13 mg/L. The 50/50 mixture of REE and 2-D had a 48-h EC_{50} of 2.75 mg/L. The highest EC_{50} occurred with the REE and RME at 4.11 and 3.07 mg/L, respectively.

The lowest flow-through 48-h *D. magna* EC_{50}, indicative of highest toxicity, was 0.19 mg/L for the 2-D reference fuel. This was followed by the 20REE mixture at 0.21 mg/L and SME at 0.40 mg/L. The 50/50 mixture of REE and 2-D had an EC_{50} of 5.12 mg/L. The highest EC_{50} occurred with the RME and REE at 587 and 25.2 mg/L, respectively.

Toxicity Conclusions

The toxicity tests show that biodiesel is considerably less toxic than diesel fuel; however, one should continue to avoid ingesting biodiesel or allowing contact with skin. Although some adverse effects were noted in the tests with rats and rabbits, none died from either the biodiesel or the diesel fuel. The animals treated with diesel had more injurious clinical observations, but some effects were noted for both fuels.

The LD_{50} of each test substance was >5000 mg/kg (the limit dose) when administered once orally *via* gastric intubation to food-deprived male and female albino rats. The occurrences of clinical observations increased as the ratio of diesel fuel increased. The LD_{50} of 100% REE was >2000 mg/kg (the limit dose) when administered once for 24 h to the clipped, intact skin of male and female albino rabbits. In addition, the 2000 mg/kg dose level was found to be a NOEL for systemic toxicity under the conditions of this study for the three fuels tested. The 100% RME fuel was the least severe in the acute oral toxicity study, and the 100% REE was the least severe in the acute dermal toxicity study.

Biodiesel is not as toxic to *D. magna* as NaC1. Compared with the reference toxicant (NaCl), diesel fuel was 2.6 times more toxic, RME was 6.2 times less toxic, REE was 26 times less, and SME 89 times less toxic. Compared with number two diesel fuel, RME was 16 times less toxic, REE was 69 times less toxic, and SME was 237 times less toxic. An LC_{50} was not produced at or below the WAF when repeating the study using rainbow trout.

In the University of Idaho tests, the least toxic of the test substances in the flow-through test was REE, followed by RME, the 50/50 mixture of REE, and 2-D. The 2-D reference substance had the lowest EC_{50} value at 24 and 48 h. This was followed at both times by the 20/80 mixture of REE and 2-D, and the SME test substance.

In both the static and flow-through tests, the rapeseed-based fuels, REE and RME, displayed the highest EC_{50} values, signifying that they are less toxic than the other test substances. The EC_{50} values for the other vegetable oil-based fuel, SME, were lower than the rapeseed fuels (significance untested). It should be noted that the results of the static test for SME failed the χ^2 test for heterogeneity using a Probit model.

The REE and 2-D mixture results were as expected in that the 20/80 mixture had higher EC_{50} values than the 50/50 mixture for both the static and flow-through analyses. This agrees with the results of the other tests that indicate a higher toxicity (lower EC_{50}) for a higher percentage of 2-D in the mixture and a lower toxicity (higher EC_{50}) with an increasing percentage of REE.

References

1. Haws, R., Chemical Oxygen Demand, Biochemical Oxygen Demand, and Toxicity of Biodiesel, in *Proceedings of the Conference on Commercialization of Biodiesel: Environmental and Health Benefits*, University of Idaho, Moscow, ID, 1997.
2. Reece, D., and C. Peterson, Toxicity Studies with Biodiesel, in *Proceedings of the Conference on Commercialization of Biodiesel: Environmental and Health Benefits*, University of Idaho, Moscow, ID, 1997.
3. Zhang, X., C. Peterson, and D. Reece, Biodegradability of Biodiesel in the Aquatic Environment, in *Proceedings of the Conference on Commercialization of Biodiesel: Environmental and Health Benefits*, University of Idaho, Moscow, ID, 1997.
4. Zhang, X., Biodegradability of Biodiesel in the Aquatic and Soil Environments, Master's Thesis, Department of Biological and Agricultural Engineering, University of Idaho, Moscow, ID, 1996.
5. Zhang, X., C. Peterson, D. Reece, R. Haws, and G. Möller, Biodegradability of Biodiesel in the Aquatic Environment, *Trans. ASAE 41:* 1423–1430 (1998).
6. Weber, C. ed., *Methods for Measuring the Acute Toxicity of Effluents to Freshwater and Marine Organisms*, 1991, EPA/600/4-90/027; see http://www.epa.gov/waterscience/wet/atx.pdf.

6.7

Soybean Oil Composition for Biodiesel

Neal A. Bringe

Soybean Oil Composition

The composition of soybean oil can be modified to improve the usefulness of soybeans for food and fuel applications. Molecular marker, traditional breeding, and transgenic technologies enable seed companies to incorporate modified oil traits into high-yielding germplasm. It takes several years to deliver a modified oil composition to the marketplace; thus, it is prudent to select the right targets in the early stages of development.

Benefits sought by the biodiesel industry are improved oxidative stability and improved cold flow properties. These two properties are linked. In some situations, neat biodiesel has to be heated to ensure flow. The warm temperatures increase the rate of fatty acid oxidation. Thus, improvements in cold flow can reduce the stability target required to meet commercial needs.

The key fatty acids limiting the cold flow quality of biodiesel are palmitic (16:0) and stearic acids (18:0) as illustrated by the melting point of the fatty acid methyl esters (FAME) (Table A-1 in Appendix A). Polyunsaturated fatty acids (PUFA) improve cold flow properties but are most susceptible to oxidation. Thus one has to identify an optimum level of PUFA. Food processors' demand for PUFA must be considered. If a large segment of the food processing industry rejects the oil, the costs of segregating the grain will prohibit the practical use of the extracted oil for fuel purposes. Linoleic acid is a primary source of fried food flavor compounds such as 2,4 decadienal (1), and oleic acid is a source of fruity, waxy, and plastic tasting odors such as 2-decenal (2,3). The proportions of these fatty acids have to be selected to balance flavor and shelf-life objectives. Good potato chip flavor was obtained with an oil having 68% oleic and 20% linoleic acid (4). PUFA are essential in the diet and play a role in cardiovascular health particularly when they replace saturated fat (5,6). Thus, we hypothesized an optimized soybean oil composition for food and fuel use that retains ~24% PUFA (Table 1).

A synthesized biodiesel with the targeted composition was prepared from mixtures of pure FAME (>99% purity; modification from Table 1: 18:1 was 73.3% and the "other" category was 0%). The cold flow properties of the oil were compared with controls (Table 2). It was apparent from these data that the cold flow properties of the target biodiesel composition can be comparable to or better than those of petroleum diesel. Additional data from other biodiesel compositions

TABLE 1
Compositions of Typical Soybean Oil (Control), a Modified Composition, and a Target Soybean Oil Composition

	Control (%)	USDA[a] line (%)	Target (%)
18:1	21.8	31.5	71.3
18:2	53.1	52.7	21.4
18:3	8.0	4.5	2.2
16:0	11.8	5.2	2.1
18:0	4.6	4.1	1.0
Other	0.7	2.0	2.0

[a]United States Department of Agriculture.

were added to the data from this study, and an exponential relation was found between the saturated fat content of biodiesel and cloud point (Fig. 1). The effect of lowering the saturated fat level of biodiesel from 15 to 10% on cloud point was relatively minor compared with a change to a 3.5% saturated fat content.

The ignition quality of the synthetic biodiesel was tested using an Ignition Quality Tester (IQT) (see Chapter 6.1). The derived cetane number (CN) was 55.43 ± 0.4, the same as pure oleate methyl esters (10). The good CN of the synthetic biodiesel was attributed to the high-oleic fatty acid content (73%) of the biodiesel. When soybeans (United States Department of Agriculture line) were used to make biodiesel that had reduced saturated fat without large increases in oleic acid, the average derived CN of the biodiesel was 46.5 using the IQT. PUFA have low CN and do not compensate well for a reduction in saturated fatty acids in a new fuel composition. Palmitate and stearate had CN of ~75, whereas that for methyl linolenate was 33 (10). A CN of 46.5, if repeatable in engine tests, could be problematic, given the minimum CN of 47 in biodiesel specification D6751.

Biodiesel with low polyunsaturated fat levels, especially lower $C_{18:3}$ fatty acids, should also emit lower levels of nitrogen oxides. This expectation follows from linear correlations found between the level of biodiesel unsaturation (measured by the iodine value), the density of biodiesel, and nitrogen oxides emissions

TABLE 2
Cold Flow Properties of Fuels[a]

Sample	Cloud point (°C)	Pour point (°C)	Cold filter plugging point (°C)
Soy methyl esters[b]	2	–1	–2
Biodiesel, USDA line	–4	–6	–10
#2 Diesel	–11	–18	–17
Synthesized biodiesel	–18	–21	–21

[a]Using ASTM Methods (D2500, D97, IP3991D6371); USDA, United States Department of Agriculture.
[b]*Source:* Reference 7.

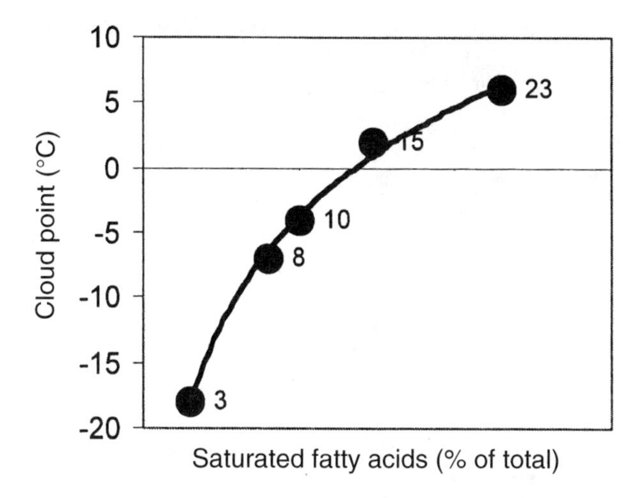

Fig. 1. Soy biodiesel cloud point and saturated fat content. Data for samples with 8, 15, and 23% (hydrogenated soy) saturated fat are from References 7, 8, and 9, respectively.

(11). Thus, reduced levels of fatty acid unsaturation should decrease the density of biodiesel and decrease nitrogen oxide emissions. Nitrogen oxide emissions from a 1991 DDC Series 60 engine may be predicted from the density of the biodiesel: $y = 46.959(\text{density}) - 36.388$, $R^2 = 0.9126$. The density of the synthetic biodiesel (0.8825 g/mL) gave a calculated nitrogen oxides emission of 5.05 g/(bhp·h), an improvement over soy biodiesel tested with the 1991 DDC Series 60 engine [5.25 g/(bhp·h)] (10). Full exhaust emissions testing is required to confirm any effect.

Conclusion

The oil of the target composition (Table 1) is suitable for use as a feedstock to produce biodiesel because biodiesel from the oil has improved cold flow, improved ignition quality (CN), improved oxidative stability, and presumably reduced nitrogen oxide emissions. The challenge is to create soybean oil near the target composition without sacrificing soybean yield. The composition also must be marketed successfully and tested in foods so that the soybeans are sought as an improved source of vegetable oil. These accomplishments will enable soybeans to be grown on a large percentage of the total acres and create value across the food and fuel chain.

References

1. Polorny, J., Flavor Chemistry of Deep Fat Frying in Oil, in *Flavor Chemistry of Lipid Foods*, edited by D.B. Min and T.H. Smouse, AOCS Press, Champaign, IL, 1989, pp. 113–155.
2. Neff, W.E., Odor Significance of Undesirable Degradation Compounds in Heated Triolein and Trilinolein, *J. Am. Oil Chem. Soc. 77:* 1303–1313 (2000).
3. Warner, K., W.E Neff, C. Byrdwell, and H.W. Gardner, Effect of Oleic and Linoleic Acids on the Production of Deep-Fried Odor in Heated Triolein and Trilinolein, *J. Agric. Food Chem. 49:* 899–905 (2001).

4. Warner, K., P. Orr, L. Parrott, and M. Glynn, Effects of Frying Oil Composition on Potato Chip Stability, *J. Am. Oil Chem. Soc. 71:* 1117–1121 (1994).
5. Hu, F.B., J.E. Manson, and W.C. Willet, Types of Dietary Fat and Risk of Coronary Heart Disease: A Critical Review, *J. Am. Coll. Nutr. 20:* 5–19 (2001).
6. Kris-Etherton, P., K. Hecker, D.S. Taylor, G. Zhao, S. Coval, and A. Binkoski, Dietary Macronutrients and Cardiovascular Risk, in *Nutrition in the Prevention and Treatment of Disease*, edited by A.M. Coulston, C.L. Rock, and E.R. Monsen, Academic Press, San Diego, 2001, pp. 279–302.
7. Lee, I., L.A. Johnson, and E.G. Hammond, Use of Branched-Chain Esters to Reduce the Crystallization Temperature of Biodiesel, *J. Am. Oil Chem. Soc. 72:* 1155–1160 (1995).
8. Peterson, C.L., J.S. Taberski, J.C. Thompson, and C.L. Chase, The Effect of Biodiesel Feedstock on Regulated Emissions in Chassis Dynamometer Tests of a Pickup Truck, *Trans. ASAE 43:* 1371–1381 (2000).
9. Kinast, J.A., Production of Biodiesels from Multiple Feedstocks and Properties of Biodiesels and Biodiesel/Diesel Blends. Report 1 in a Series of 6. NREL/SR-510–31460, 2003.
10. Bagby, M.O., B. Freedman, and A.W. Schwab, Seed Oils for Diesel Fuels: Sources and Properties, ASAE Paper No. 87-1583, American Society of Agricultural Engineers, St. Joseph, MI, 1987.
11. McCormick, R.L., M.S. Graboski, T.L. Alleman, and A.M. Herring, Impact of Biodiesel Source Material and Chemical Structure on Emissions of Criteria Pollutants from a Heavy-Duty Engine, *Environ. Sci. Technol. 35:* 1742–1747 (2001).

7

Exhaust Emissions

7.1

Effect of Biodiesel Fuel on Pollutant Emissions from Diesel Engines

Robert L. McCormick and Teresa L. Alleman

Introduction

In the United States, diesel engines are regulated for smoke opacity, total oxides of nitrogen (NO_x), total particulate matter <10 μm (PM-10 or PM), carbon monoxide (CO), and total hydrocarbon (THC) according to test procedures defined by the Environmental Protection Agency (EPA) in the Code of Federal Regulations. Because the magnitude of diesel emissions depends on fuel composition, emission certification testing is conducted with a "certification diesel fuel" that represents the U.S. national average. Other emissions from diesel engines such as aldehydes and polyaromatic hydrocarbons (PAH) may be regulated in the future in an attempt to control ambient levels of toxic substances in the air.

An important property of biodiesel is its ability to reduce total particulate emissions from an engine. Particulate emissions are defined by the EPA as condensed or solid material collected on an appropriate filter at a temperature ≤52°C. Particulate matter thus includes soot carbon, fuel and lubricating oil derivatives, and sulfuric acid aerosols. Particulate matter is often fractionated in terms of sulfate, soluble organic fraction (SOF), or volatile organic fraction (VOF), and carbon or soot (1). Biodiesel can affect soot and fuel-based SOF but not lubricating oil-based SOF.

Diesel engines are significant contributors of NO_x and PM to ambient air pollutant inventories (2). The quantity of CO and THC derived from diesel engines is generally small compared with emissions from light-duty gasoline vehicles. For this reason, the effect of biodiesel on PM and NO_x emissions is the primary concern of this review. The effect of biodiesel on emissions from 2-stroke engines was reviewed previously (3).

Heavy-Duty Engine Emissions

Heavy-duty engine emissions are regulated using an engine dynamometer test (4) with results reported in g/(bhp·h) [0.7457 g/(bhp·h) = 1 g/(kW·h)]. The EPA recently completed a review of published biodiesel emissions data for heavy-duty engines (5). The results for NO_x, PM, CO, and THC are summarized in Figure 1,

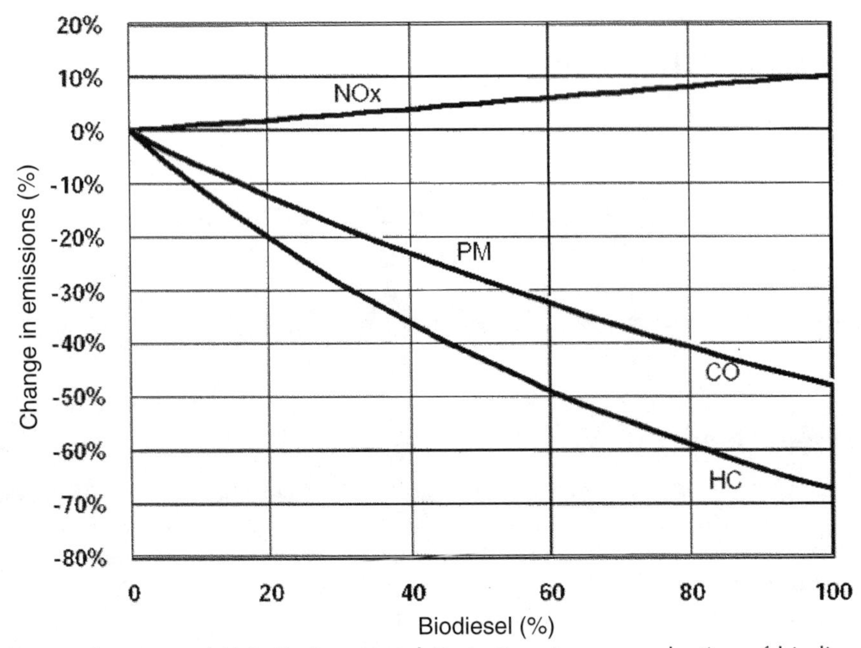

Fig. 1. Summary of U.S. Environmental Protection Agency evaluation of biodiesel effects on pollutant emissions for heavy-duty engines. NO_X, oxides of nitrogen; PM, particulate matter; CO, carbon monoxide; HC, hydrocarbon. *Source:* Reference 5.

taken from that report (5). It is clear that, on average, substantial reductions in emissions of PM, CO, and THC can be obtained by using biodiesel. However, the data also show an increase in NO_x emissions. Importantly, few studies of biodiesel emissions from newer engines (1998 and later) with more advanced technology have been published. A more detailed analysis of biodiesel emissions data indicates that the solid carbon fraction of the PM is reduced but the SOF may increase. The effect on total PM depends on the engine operating conditions. Under the conditions of the heavy-duty engine test, the solid carbon effect dominates such that PM emissions go down. Table 1 summarizes the average emission changes found by the EPA for B20 (a blend of 20% biodiesel with conventional diesel). Studies also showed significantly lower levels of emissions of specific toxic compounds for biodiesel and biodiesel blends, including aldehydes, PAH, and nitro-polyaromatic hydrocarbons (6–8).

The increase in NO_x may limit the market in areas that exceed ozone air quality standards. Considerable effort has therefore been devoted to understanding the environmental effect of the increase in NO_x. Recently, the air quality effect of 100% market penetration of B20 into on-road heavy-duty fleets in several major urban areas in the United States was examined (9). The study employed pollutant inventory and air quality models sanctioned by the EPA and the California Air Resources Board to model the effect of the increase in NO_x emissions on ground

TABLE 1
Average Heavy-Duty Emission Effect of 20% Biodiesel Relative to Average Conventional Diesel Fuel[a]

Air pollutant	B20 change (%)
Nitric oxide	+2.0
Particulate matter	–10.1
Carbon monoxide	–11.0
Hydrocarbon	–21.1

[a]*Source:* Reference 5.

level ozone concentrations. The air quality modeling indicated changes in ozone concentration of <1 ppb for all areas modeled. This suggests that the 2% NO_x increase does not have serious air quality implications.

A number of factors can cause biodiesel emissions to differ significantly from the average values as assessed by the EPA. For example, different fuel system designs and engine calibrations can result in measurably different emissions from biodiesel. Figure 2 shows how biodiesel emissions change with blend level in engines from two manufacturers and model years from 1987 to 1995. The results include four independent studies using different Detroit Diesel Corporation (DDC) Series 60 engines (10–13) and data for Cummins L10, N14, and B5.9 engines (6,14,15).

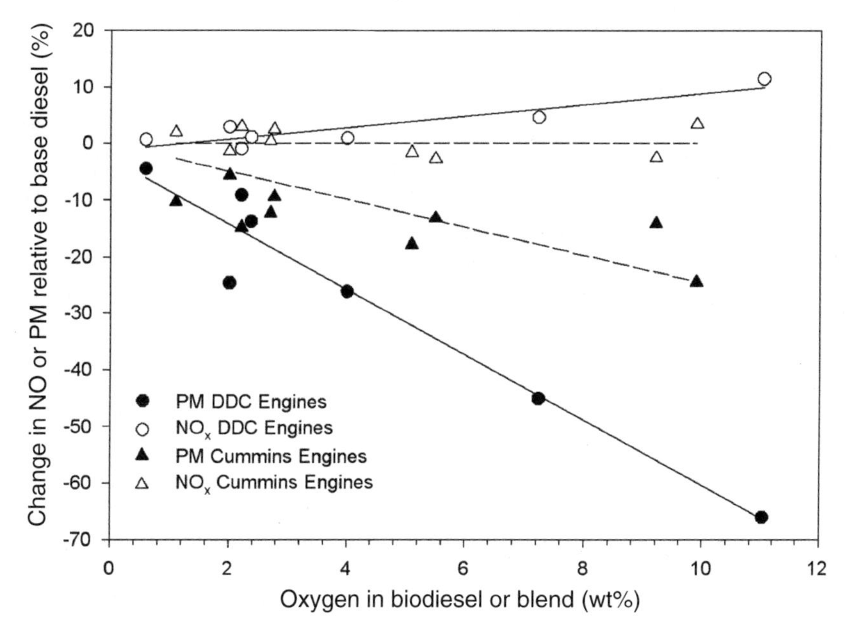

Fig. 2. Change in particulate matter (PM) and nitric oxide (NO_x) emissions for soy methyl ester as well as rapeseed methyl and ethyl ester blends in stock, 4-stroke engines. *Source:* Reference 3.

As Figure 2 indicates, a greater PM reduction can be obtained from the DDC engines, but at the cost of a larger increase in NO_x emissions. Differences between engines from the two manufacturers are probably related to the different approaches taken toward optimizing the NO_x/PM trade-off; in some cases, the type of fuel injection system also differs (i.e., unit injector vs. common rail system). For the DDC engines at 2 wt% oxygen (B20), the NO_x increase is ~3% and the PM reduction is ~15%. For the three Cummins engines, the average NO_x increase was only 0.4% and the slope of the regression line was <0.02. Therefore, on the basis of the regression and within the uncertainty of these data, NO_x emissions did not increase for the Cummins engines as a group. The PM emission reduction was ~12% for the Cummins engines operating on B20. Inspection of the data used to make this plot indicated that the L10 and N14 engines did produce a NO_x increase. The 1995 B5.9 engine actually exhibited a decrease in NO_x emissions when operating with soybean or rapeseed ester fuels.

Another factor that can affect NO_x emissions is biodiesel source material or formulation. Biodiesel fuels from several sources and having a wide range in iodine number were tested (16). Iodine number is a measure of degree of saturation, or the number of carbon-carbon double bonds in the biodiesel fatty acid chain. The results, summarized in Figure 3, indicate that the more saturated biodiesel exhibited lower NO_x emissions but essentially the same level of PM emissions. All biodiesel formulations exhibited the same brake-specific fuel consumption and engine thermal efficiency. A fuel with an iodine number of ~40 would be NO_x equivalent to certification diesel, but would exhibit a very high cloud point.

Heavy-Duty Vehicle Emissions

Only limited heavy-duty chassis emission tests have been conducted using biodiesel. Results from biodiesel testing on heavy-duty engines do not always correlate directly with test results on heavy-duty vehicles tested on chassis dynamometers. Several class 8 trucks using conventional diesel fuel and B35 (35% biodiesel) were tested over the 5-Peak cycle (17,18). The vehicles were powered by Cummins, Detroit Diesel, and Mack engines. Trends in NO_x emissions varied, with emissions increasing in some vehicles and decreasing in others. PM emissions were more consistent across the engine types, with PM decreasing by ~20% for the Detroit Diesel and Cummins engines, but showing no change or a slight increase for the Mack engines. A much greater body of information on emissions from in-use vehicles operating on biodiesel or biodiesel blends is required to assess fully the environmental effect of this fuel in the heavy-duty fleet.

Light-Duty Engine and Vehicle Emissions

Rapeseed methyl and ethyl ester fuels in 1994 and 1995 Cummins B5.9 engines in two Dodge pickup trucks were tested on a chassis dynamometer using the double arterial cycle and the chassis version of the heavy-duty transient test (19). The arterial cycle gave lower emissions on a g/mile basis compared with the heavy-duty chassis cycle,

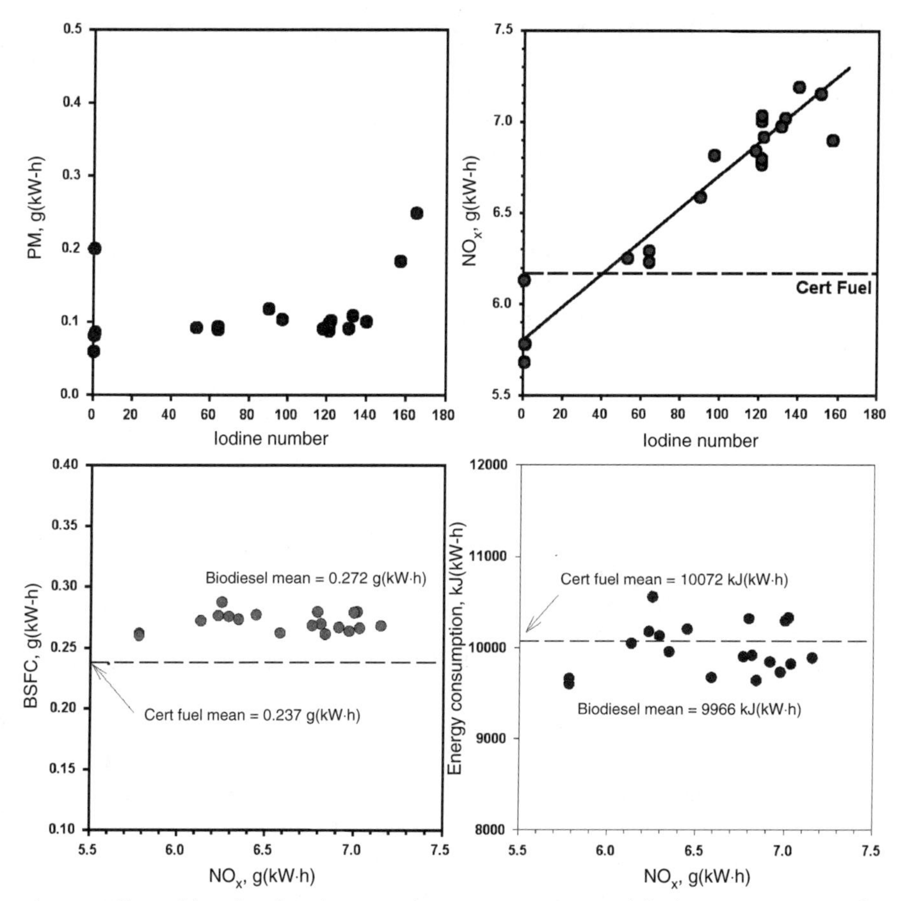

Fig. 3. Effect of biodiesel iodine number on emissions and fuel economy. Data for B100 fuels tested in a 1991 Detroit Diesel Corporation (DDC) Series 60 engine *via* the heavy-duty transient test. NO_x, oxides of nitrogen. *Source:* Reference 16.

but the relative effect of biodiesel on emissions was essentially the same. For B100 ethyl ester, NO_x decreased by 13% for each truck, and PM increased by 16 and 43% for the 1994 and 1995 trucks, respectively, compared with diesel. NO_x for the methyl ester was 3.7% higher than for the ethyl ester with the 1994 truck, whereas PM was 6% higher. Increasing PM emissions and only small changes in NO_x emissions were reported for biodiesel tested in similar large pickup trucks (20,21).

The European data on regulated emissions from rapeseed methyl ester (B100) were reviewed (22). Most data are for light-duty diesel passenger cars using various European multimode steady-state engine tests and the U.S. light-duty federal test procedure (FTP) driving cycle. Emissions are highly dependent on the driving cycle. NO_x increased for all cycles and engines discussed, typically by 10%. PM generally decreased, and the PM reduction was cycle and engine dependent, with the FTP giving the smallest decrease (0–20%), and the 13-mode steady-state test giving a 10–50% decrease. Data on toxic emissions from these vehicles were also reviewed (22), and it was found that emissions were highly dependent on the driving cycle. PAH emissions with biodiesel in direct injection engines ranged from 80 to 110% of those for conventional fuel for the FTP, and from near zero to 80% of diesel PAH emissions for various steady-state tests.

Results from the light-duty engine dynamometer testing appeared to follow the heavy-duty engine trends more closely, i.e., NO_x increased and PM decreased. In a more recent study (23), a Mercedes Benz OM904LA engine was tested using rapeseed methyl ester (B100) and conventional diesel. A small NO_x increase and PM decrease were observed for this steady-state testing. The PM emission reductions were attributed to a lower emission of soot compared with the other test fuels.

A Daimler Benz OM611 light-duty diesel engine was tested on ultralow sulfur diesel, conventional diesel, and a B20 blend (24). Neat soy methyl ester fuel was blended with the ultralow sulfur diesel fuel to produce the B20 blend. Results from the B20 blend were NO_x neutral compared with both reference diesel fuels. PM emission reductions were greater compared with the conventional diesel fuel than those for the ultralow sulfur diesel fuel (32 and 14%, respectively), due mainly to fuel property differences in the base diesel fuels.

Two types of rapeseed methyl ester and conventional diesel fuel were tested in a European light-duty diesel engine (25). Results of steady-state testing showed that NO_x emission increased and PM emission decreased. The authors attributed the NO_x emission increase to changes in the operation of the fuel injection pump. Biodiesel caused premature needle lift, effectively advancing injection timing. Analysis of the PM emission revealed that soot from the biodiesel had a higher percentage of SOF compared with diesel-derived PM; however, total soot decreased for the biodiesel fuels. Unregulated emission analysis revealed lower PAH emissions for biodiesel compared with conventional diesel fuel. For both conventional diesel and biodiesel, the total unregulated emissions were engine load dependent, with higher emissions observed at lower engine loads.

Because of the very small market share of light-duty diesel vehicles in the United States, very few light-duty emission studies have been performed. The results available suggest that for large pickup trucks and sport utility vehicles, the pollutant emission effects of biodiesel may differ from those observed for heavy-duty engines. Under light-loaded duty cycles such as those encountered with light-duty vehicles, pickup trucks, and sport utility vehicles, the effect of biodiesel on soot emissions can be smaller than that observed in heavy-duty engines. The increase in SOF caused by biodiesel dominates, and total PM may actually increase. The NO_x increase with biodiesel observed for heavy-duty engines may also be reversed with light loads. However the limited data for passenger car engines showed that NO_x and PM effects were similar to those observed for heavy-duty engines.

NO_x Reduction Strategies

The cause of the biodiesel NO_x increase, at least for unit injection systems, was shown to be related to a small shift in fuel injection timing caused by the different mechanical properties of biodiesel relative to conventional diesel (26,27). Because of the higher bulk modulus of compressibility (or speed of sound) of biodiesel, there is a more rapid transfer of the fuel pump pressure wave to the injector needle, resulting in earlier needle lift and producing a small advance in injection timing. Recently, this effect was examined in more detail (28). It was found that soy-derived B100 produces a 1° advance in injection timing but a nearly 4° advance in the start of combustion. The duration of fuel injection was also shorter for biodiesel.

Even before that work (26,27), timing changes were investigated in Cummins L10 (14) and Cummins N14 (15) engines. Retarding injection timing can reduce NO_x with a loss of some effectiveness for PM reduction and a loss of fuel economy. For example, in the N14 study performed by Ortech, the retardation in timing actually increased PM emissions using B20 to 4.1% above the base diesel level. Similar results were observed in the L10 study. In the United States and most other countries, changing injection timing constitutes tampering, or changing of the engine's emission control system, and would require recertification of the engine for emissions standard compliance.

Injection timing, injection pressure, and exhaust gas recirculation were investigated with various soy biodiesel blends and a conventional diesel using a 13-mode, steady-state test with a Navistar 7.3 L HEUI engine (29). It was found that NO_x increased with biodiesel under all conditions of speed and load. However, the rate of generation of PM and its form varied with conditions. Relatively low blending levels of 10–30% soy methyl ester are more responsive to engine parameter changes over the engine map than high blends (50 and 100%). For these lower biodiesel blend levels, it was possible to lower NO_x at fixed PM but not to reduce PM and NO_x simultaneously using engine timing and pressure changes. Any change made to pressure or timing for biodiesel blends worked equally well for conventional diesel. Thus, NO_x emissions increased and PM, CO, and THC emissions fell relative to diesel at any fixed configuration.

One approach to reducing biodiesel NO_x emissions to a level equal to that from conventional diesel involved either increasing the cetane number (CN) or decreasing aromatics. The effect of these properties on diesel emissions in general is revealed in several studies (30). It was found that B20 did respond to di-*t*-butyl peroxide (DTBP), a CN improver when tested in a 1991 DDC Series 60 engine (11). NO_x were lowered by 6.2% while preserving a 9.1% benefit in PM emissions from biodiesel. However, B20 produced no noticeable increase in NO_x for this engine. DTBP and 2-ethyl hexyl nitrate were examined in a similar engine (31) and the results are shown in Figure 4. This study confirms that cetane additives can reduce NO_x, at least for engines that do not have highly retarded injection timing. However, the relatively high additive blending levels required, in excess of 5000 ppm, may not be economical.

Another approach is to blend low aromatic or high-CN components, such as alkylate or Fischer-Tropsch (FT) diesel, into the biodiesel. Multimode steady-state testing

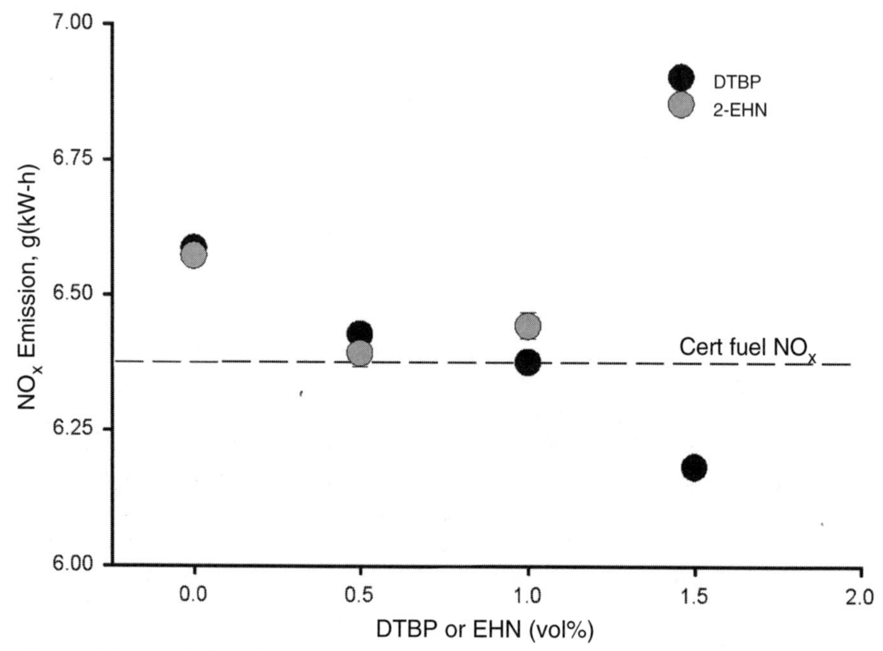

Fig. 4. Effect of fuel additives on nitric oxide (NO_x) emissions for soy B20, with di-*tert*-butyl peroxide (DTBP) and 2 ethyl hexyl nitrate (2-EHN) testing in a 1991 Detroit Diesel Corporation (DDC) Series 60 engine *via* the U.S. heavy-duty transient test. *Source:* Reference 31.

was conducted using a Cummins L10 engine (32) in which NO_x increased from 6.4 to 7.4 g/(bhp·h) when comparing conventional diesel and 100% soy methyl ester. Adding 20% heavy alkylate made a NO_x neutral fuel compared with the base diesel. *n*-Hexadecane was employed to reduce NO_x emissions (unpublished data), and similar effects were shown for FT-diesel (31). These studies demonstrate how diesel can be reformulated to include biodiesel but with a lower aromatic content to control emissions of NO_x and PM simultaneously.

Summary

There is broad consensus that biodiesel and biodiesel blends produce significant reductions in PM and an increase in NO_x for heavy-duty engines up to about the 1997 model year. Some strategies for mitigating the NO_x increase were demonstrated; however, the increase in NO_x emissions remains a potential hurdle for dramatic expansion of biodiesel use. For engines meeting the 1998 and 2004 heavy-duty emission standards, there appear to be few data available for performance during operation with biodiesel. Chassis dynamometer testing of heavy- or light-duty vehicles operating on biodiesel has been very limited. Full understanding of the effect of the pollutant emissions of this renewable fuel will require a much larger body of data from engines and vehicles of all sizes.

References

1. Coordinating Research Council, Chemical Methods for the Measurement of Unregulated Diesel Emissions—Carbonyls, Aldehydes, Particulate Characterization, Sulfates, PAH/NO_2PAH, CRC Report No. 551, Atlanta, 1987.
2. United States Environmental Protection Agency, National Air Quality and Emissions Trends Report, 2003 Special Studies Edition, EPA 454/R-03-005, 2003.
3. Graboski, M.S., and R.L. McCormick, Combustion of Fat and Vegetable Oil Derived Fuels in Diesel Engines, *Prog. Energy Combust. Sci. 24:* 125–164 (1997).
4. United States Code of Federal Regulations, Vol. 40, Part 86, Subpart N.
5. United States Environmental Protection Agency, A Comprehensive Analysis of Biodiesel Impacts on Exhaust Emissions, Draft Technical Report, EPA420-P-02-001, 2002.
6. Sharp, C.A., Emissions and Lubricity Evaluation of Rapeseed Derived Biodiesel Fuels, Final Report from SWRI to Montana Department of Environmental Quality, November 1996.
7. Sharp C.A., S.A. Howell, and J. Jobe, The Effect of Biodiesel Fuels on Transient Emissions from Modern Diesel Engines, Part I Regulated Emissions and Performance, SAE Technical Paper No. 2000-01-1967, 2000.
8. Sharp C.A., S.A. Howell, and J. Jobe, The Effect of Biodiesel Fuels on Transient Emissions from Modern Diesel Engines, Part II Unregulated Emissions and Chemical Characterization, SAE Technical Paper No. 2000-01-1968, 2000.
9. Morris, R.E., A.K. Polack, G.E. Mansell, C. Lindhjem, Y. Jia, and G. Wilson, Impact of Biodiesel Fuels on Air Quality and Human Health. Summary Report. National Renewable Energy Laboratory, NREL/SR-540-33793, Golden, CO, 2003.
10. Graboski, M.S., J.D. Ross, and R.L. McCormick, Transient Emissions from No. 2 Diesel and Biodiesel Blends in a DDC Series 60 Engine, SAE Technical Paper No. 961166 (1996).
11. Sharp, C.A., Transient Emissions Testing of Biodiesel and Other Additives in a DDC Series 60 Engine, Southwest Research Institute Report for National Biodiesel Board, December 1994.
12. McCormick, R.L., J.D. Ross, and M.S. Graboski, Effect of Several Oxygenates on Regulated Emissions from Heavy-Duty Diesel Engines, *Environ. Sci. Technol. 31:* 1144–1150 (1997).
13. Liotta, F., Jr., and D. Montalvo, The Effect of Oxygenated Fuels on Emission from a Modern Heavy-Duty Diesel Engine, SAE Technical Paper No. 932734, 1993.
14. Stotler, R., and D. Human, Transient Emission Evaluation of Biodiesel Fuel Blend in a 1987 Cummins L-10 and DDC 6V-92-TA, ETS Report No. ETS-95-128 to National Biodiesel Board, November 30, 1995.
15. Ortech Corporation, Operation of Cummins N14 Diesel on Biodiesel: Performance, Emissions and Durability, Final Report for Phase 1 to National Biodiesel Board, Report No. 95 E11-B004524, Mississauga, Canada, January 19, 1995.
16. McCormick, R.L., T.L. Alleman, M.S. Graboski, A.M. Herring, and K.S. Tyson, Impact of Biodiesel Source Material and Chemical Structure on Emissions of Criteria Pollutants from a Heavy-Duty Engine, *Environ. Sci. Technol. 35:* 1742–1747 (2001).

17. Clark, N.N., and D.W. Lyons, Class 8 Truck Emissions Testing: Effects of Test Cycles and Data on Biodiesel Operation, *Trans. ASAE 42:* 1211–1219 (1999).
18. Wang, W.G., D.W. Lyons, N.N. Clark, M. Gautam, and P.M. Norton, Emissions from Nine Heavy Trucks Fueled by Diesel and Biodiesel Blend Without Engine Modification, *Environ. Sci. Technol. 34:* 933–939 (2000).
19. Peterson, C.L., and D.L. Reece, Emissions Testing with Blends of Esters of Rapeseed Oil Fuel with and Without a Catalytic Converter, SAE Technical Paper No. 961114, 1996.
20. Durbin, T.D., J.R. Collins, J.M. Norbeck, and M.R. Smith, Effects of Biodiesel, Biodiesel Blends, and Synthetic Diesel on Emissions from Light Heavy-Duty Diesel Vehicles, *Environ. Sci. Technol. 34:* 349–355 (2000).
21. Durbin, T.D., and J.M. Norbeck, Effects of Biodiesel Blends and Arco EC-Diesel on Emissions from Light Heavy-Duty Diesel Vehicles, *Environ. Sci. Technol. 36:* 1686–1991 (2002).
22. Krahl, J., A. Munack, M. Bahadir, L. Schumacher, and N. Elser, Review: Utilization of Rapeseed Oil, Rapeseed Oil Methyl Ester of Diesel Fuel: Exhaust Gas Emissions and Estimation of Environmental Effects, SAE Technical Paper No. 962096, 1996.
23. Krahl, J., A. Munack, O. Schröder, H. Stein, J. Bünger, Influence of Biodiesel and Different Designed Diesel Fuels on the Exhaust Gas Emissions and Health Effects, SAE Technical Paper No. 2003-01-3199, 2003.
24. Sirman, M.B., E.C. Owens, and K.A.Whitney, Emissions Comparison of Alternative Fuels in an Advanced Automotive Diesel Engine, SAE Technical Paper No. 200-01-2048, 2000.
25. Cardone, M., M.V. Prati, V. Rocco, M. Seggiani, A. Senatore, and S. Vitolo, *Brassica carinata* as an Alternative Oil Crop for the Production of Biodiesel in Italy: Engine Performance and Regulated and Unregulated Exhaust Emissions, *Environ. Sci. Technol. 36:* 4656–4662 (2002).
26. Tat, M.E., and J.H. van Gerpen, Measurement of Biodiesel Speed of Sound and Its Impact on Injection Timing. National Renewable Energy Laboratory, NREL/SR-510-31462, Golden, CO, 2003.
27. Monyem, A., J.H. van Gerpen, and M. Canakci, The Effect of Timing and Oxidation on Emissions from Biodiesel-Fueled Engines, *Trans. ASAE 44:* 35–42 (2001).
28. Sybist, J.P., and A.L. Boehman, Behavior of a Diesel Injection System with Biodiesel Fuel, SAE Technical Paper No. 2003-01-1039, 2003.
29. FEV Engine Technology, Inc., Emissions and Performance Characteristics of the Navistar T444E DI Engine Fueled with Blends of Biodiesel and Low Sulfur Diesel, Phase 1, Final Report to National Biodiesel Board, December 6, 1994.
30. Lee, R., J. Pedley, and C. Hobbs, Fuel Quality Impact on Heavy Duty Diesel Emissions: A Literature Review, SAE Technical Paper No. 982649, 1998.
31. McCormick, R.L., J.R. Alvarez, M.S. Graboski, K.S. Tyson, and K. Vertin, Fuel Additive and Blending Approaches to Reducing NO_x Emissions from Biodiesel, SAE Technical Paper No. 2002-01-1658, 2002.
32. Marshall, W., Improved Control of NOX Emissions with Biodiesel Fuels, Final Report, DOE Contract DE-AC22-94PC91008, March 1994.

7.2

Influence of Biodiesel and Different Petrodiesel Fuels on Exhaust Emissions and Health Effects

Jürgen Krahl, Axel Munack, Olaf Schröder, Hendrik Stein, and Jürgen Bünger

Introduction

To evaluate the emissions of biodiesel [rapeseed oil methyl ester (RME)] vs. petrodiesel on a broad basis, four different fuels were investigated. In addition to the Swedish low-sulfur diesel fuel (DF) MK1, complying with the Swedish standard SS 15 54 35, and German biodiesel RME, complying with the German standard DIN 51606 (these standards were superseded by the European standard EN14214, see Table B-3 in Appendix B), the fuels examined were a petrodiesel fuel, complying with the European standard EN 590, and a low-sulfur DF with a high content of aromatics and a flatter boiling curve, complying with EN 590 and referred to as DF05. Technical data for the fuels, engine (DaimlerChrysler OM904LA) and the 13-mode ECE-R 49 test cycle (the running conditions correspond to the 13-mode cycle used in the United States; however, the weighting factors differ) were reported elsewhere (1). Because diesel engine particles likely pose a lung cancer hazard to humans (2), the determination of mutagenic potential of particulate matter was carried out to estimate possible carcinogenic health effects.

Results

The weighted sums of measured specific emission rates for the modes of the 13-mode test are summarized in the figures below. Detailed results for each of the modes can be found in the project report (3).

Carbon Monoxide (CO; Fig. 1). For all fuels, the emissions were clearly far below the legal limit of 4.0 g/kWh (Euro II) valid for the engine used. RME led to a considerable decrease in CO emissions. This could be due in part to the oxygen in the ester bonds, which allows more CO to be oxidized to CO_2.

Hydrocarbons (HC; Fig. 2). For HC, the emission rates were also far below the legal limit of 1.1 g/kWh. RME caused a significant decrease.

Nitrogen Oxides (NO_X; Fig. 3). The emission rates were below the legal limit of 7 g/kWh; however, they approached that limit. This demonstrates that NO_x and, as shown below, particulate matter (PM) are the critical components for diesel engines.

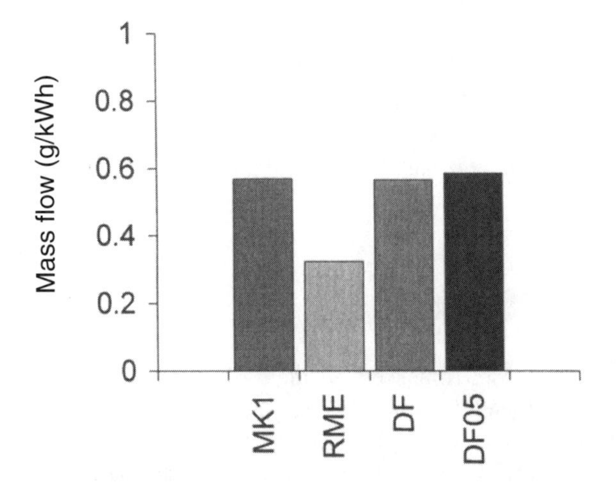

Fig. 1. Specific CO emission rates.

As reported earlier in many publications, RME leads to an increase in NO_x emissions if the engine management (timing and course of injection) remains unchanged. However, it is possible to optimize diesel engines on RME using software (4). A condition for the application of this strategy in practical use is a system for on-board fuel (blend) detection. Therefore, a biodiesel sensor was developed (5,6).

PM (Fig. 4). The legal limit of 0.15 g/kWh was met by all four fuels. The nonconventional fuels led to a reduction of 25 to nearly 40% compared with classical DF.

Particle Size Distribution (Fig. 5). Diesel engines are a major source of the emissions of fine particles (diameter <2.5 μm) and are main sources of ultrafine particles (diame-

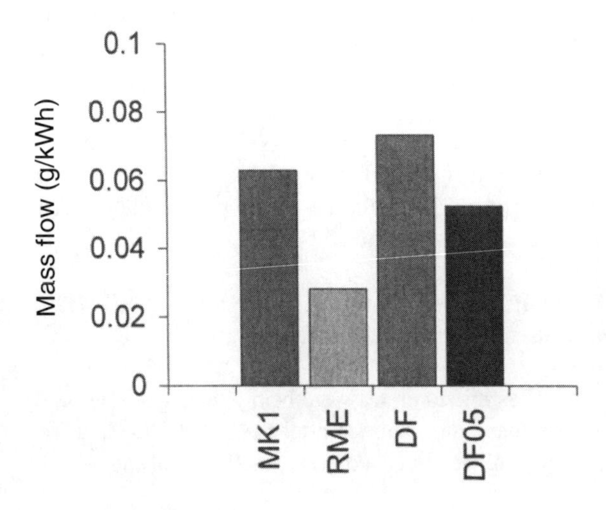

Fig. 2. Specific HC emission rates.

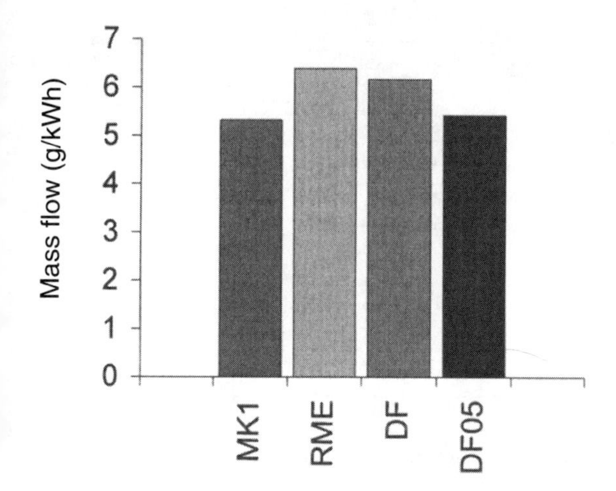

Fig. 3. Specific NO_x emission rates.

ter <0.1 μm). The ultrafine particles are regarded as being much more relevant toxicologically (7). Emissions from the use of diesel engines, as far as the particle numbers are concerned, occur mainly in the range of 10–300 nm. Therefore, this range was measured according to a literature procedure (8). The four fuels produced quite different emissions. RME led to more particles in the 10–40 nm range compared with DF and fewer particles with larger diameters. MK1 caused a reduction over the entire range measured, whereas DF05 yielded considerably higher numbers of particles. However, this must be different for higher diameters in the range not covered by the analyzer because the overall particle emissions were lower compared with DF.

Aromatic HC (Fig. 6). Aromatic compounds, determined according to a literature procedure (9), are observed mainly in idle and light-load modes. In the other

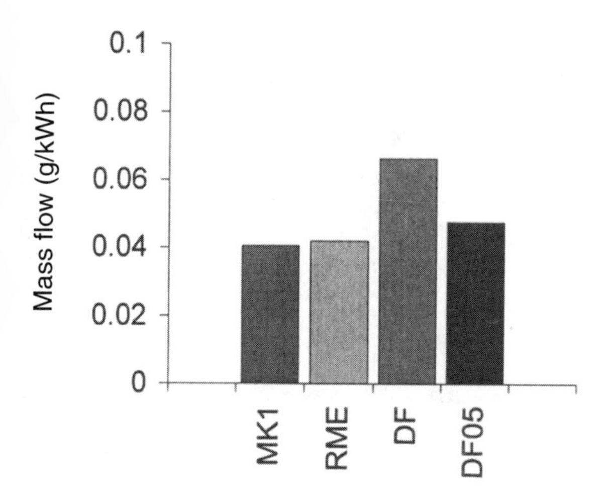

Fig. 4. Specific PM emission rates.

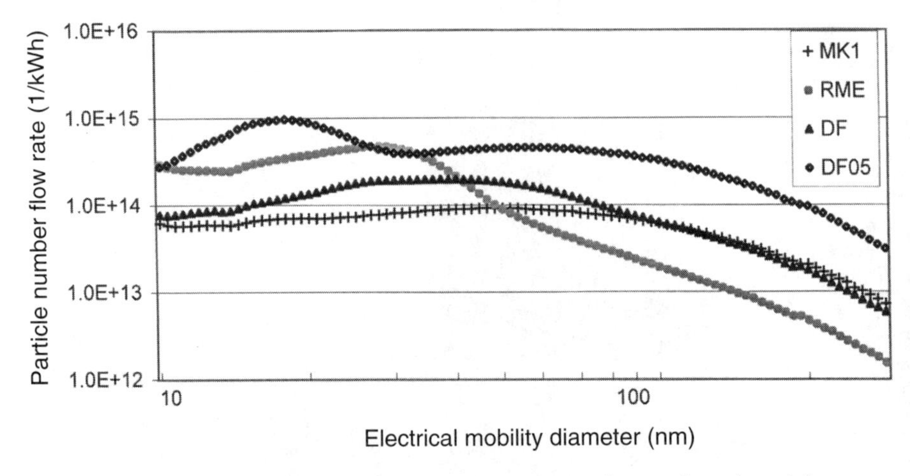

Fig. 5. Size distribution of particles with respect to the number of particles.

modes, the concentration in the exhaust emissions is <1 ppb, so that they cannot be distinguished from the background concentration. The results showed that RME led to a significant reduction in these emissions. As stated above, the very different combustion conditions are regarded as being the reason for this discrepancy.

Alkenes (Fig. 7). These species were again determined according to a literature procedure (9). Among the unsaturated HC, ethene, ethyne, and propene were the main exhaust emissions components. Similar to the aromatics, they were hardly detectable with the exceptions of idle and light-load modes. The "new" fuels, MK1

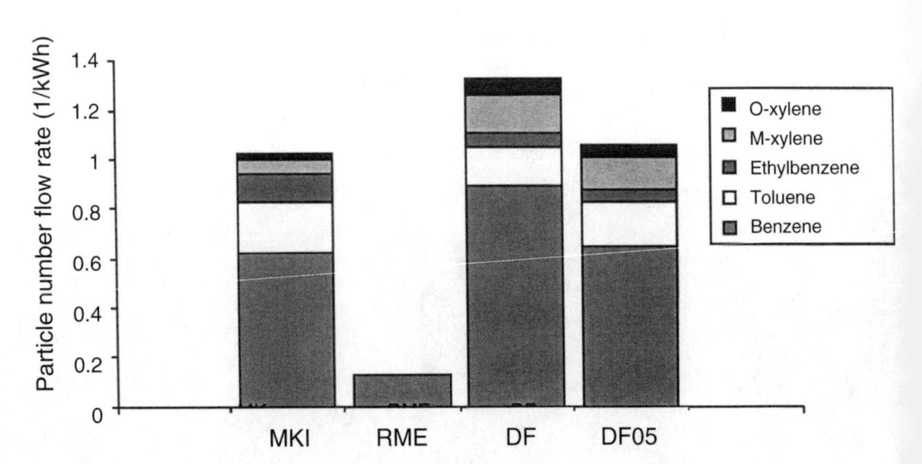

Fig. 6. Specific aromatic hydrocarbons emission rates.

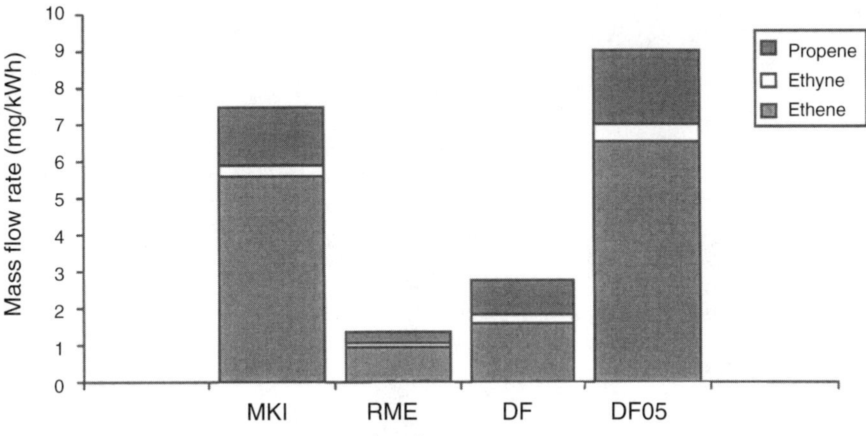

Fig. 7. Specific alkenes emission rates.

and DF05, had considerably higher emission rates; however, in this case, they were at a low level.

Aldehydes and Ketones (Fig. 8). Like alkenes, aldehydes and ketones, which were analyzed according to previous literature (10), contributed to summer smog formation. Aldehydes had a 30–50% share in the overall HC emissions. The results showed a reduction of 30% for RME and DF05 compared with DF, and a slight increase for MK1.

The results of the extraction of the PM produced by the investigated fuels are compared in Figure 9. RME, MK1, and, to a lesser extent, DF05, produced a considerably decreased particle mass compared with DF. This is probably due to the lower sulfur content of these fuels compared with DF as described in earlier stud-

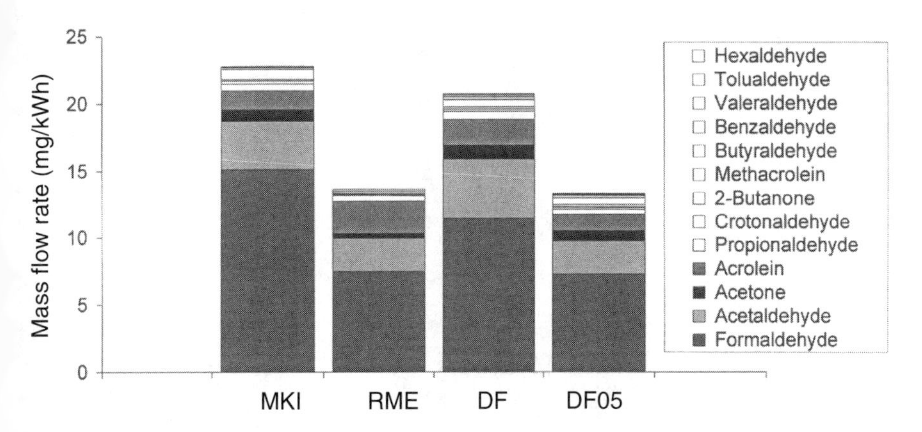

Fig. 8. Specific emission rates for aldehydes and ketones.

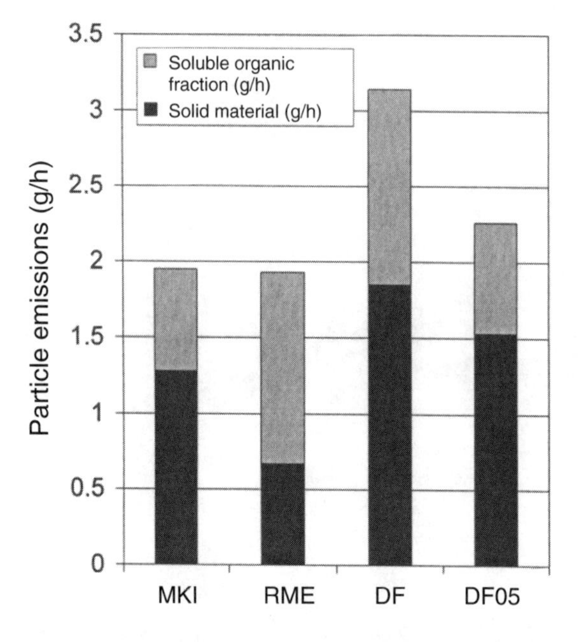

Fig. 9. Particle emissions stratified for solid and soluble fractions.

ies (11,12). The solid material (mainly soot/carbon) was lowest from RME, indicating a higher portion of unburned fuel in the soluble organic fraction of these extracts. In some load modes, RME produced almost no soot as noted in an earlier study (13).

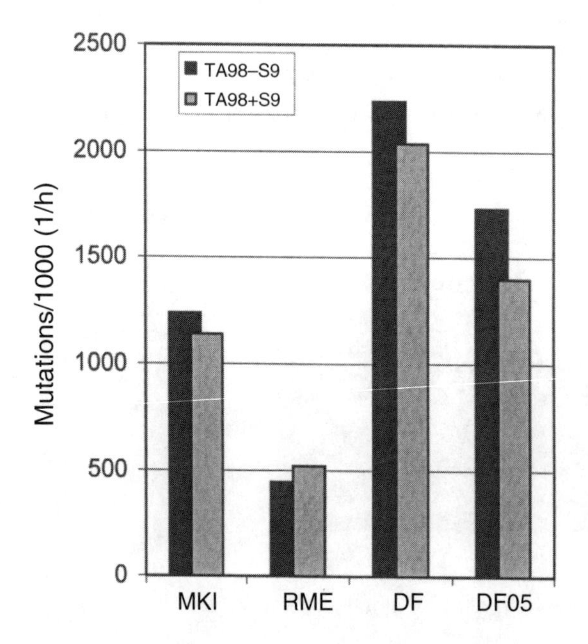

Fig. 10. Mutagenic effects of the particle extracts with (+S9) and without (–S9) metabolic activation.

Mutagenic properties were investigated with the *Salmonella typhimurium*/mammalian assay (14). The mutagenic effects of the particle extracts from the tested fuels varied greatly (Fig. 10). RME produced the lowest mutagenic effects. The mutagenicities of MK1-extracts were 2–3 times, of DF05 3–4 times, and of DF 4–5 times greater than those of RME. The results with (+S9) and without (–S9) metabolic activation by rat liver enzymes differed slightly.

The very small number of mutations for RME is ascribed to a lower content of polycyclic aromatic HC in the particle emissions of biodiesel fuels (12,15). Mutagenicity induced by MK1- and DF05-particle extracts was also generally lower than mutagenicity by DF-extracts. This effect is likely due to the low sulfur content of these fuels. There is a correlation between sulfur content of DF and mutagenic effects of its exhaust (12,13). Because RME, MK1, and DF05 contain almost no sulfur, a similar range of mutagenic effects could be expected from the use of these fuels compared with DF containing 41 ppm sulfur. However, mutagenicity of MK1- and DF05-particle extracts was stronger than the mutagenic effects of RME. This may be due to the aromatic compounds in MK1 and DF05 that are not found in RME. Aromatic compounds in DF were proven to increase the mutagenic effects of particle emission extracts (11,16).

Conclusions

Exhaust emissions from a modern diesel engine were measured using the following: (i) conventional DF according to DIN EN590; (ii) Swedish low-sulfur DF MK1; (iii) biodiesel (RME); and (iv) a new DF with lowered boiling characteristics, low-sulfur content, and a high level of aromatic compounds. The results for nonregulated emissions must be interpreted with great care because measurement errors are relatively high when analyzing gas components at very low concentrations. Biodiesel had both positive and negative effects on exhaust emissions. In addition, the mutagenicity of RME emissions was much lower than that of fossil fuels, indicating a reduced health risk from cancer.

References

1. Krahl, J., A. Munack, O. Schröder, H. Stein, and J. Bünger, Influence of Biodiesel and Different Designed Diesel Fuels on the Exhaust Gas Emissions and Health Effects, SAE Technical Paper Series 2003-01-3199 in Oxygenated and Alternative Fuels, and Combustion and Flow Diagnostics, 2003, pp. 243–251.
2. USEPA, U.S. Environmental Protection Agency, Office of Research and Development, National Center for Environmental Assessment, Health Assessment Document for Diesel Engine Exhaust, EPA/600/8-90/057F, Washington, 2002, pp. 1–669.
3. Munack, A., O. Schröder, H. Stein, J. Krahl, and J. Bünger, Systematische Untersuchungen der Emissionen aus der motorischen Verbrennung von RME, MK1 und DK, Final Report (in German), Institute for Technology and Biosystems Engineering, 2003, FAL, Braunschweig, Germany.

4. Tschöke, H., and G. Braungarten, Biodiesel und Partikelfilter, *Landbauforsch. Völkenrode 239:* 69–86 (2002).
5. Munack, A., J. Krahl, and H. Speckmann, A Fuel Sensor for Biodiesel, Fossil Diesel Fuel, and Their Blends, 2002 ASAE Annual Meeting/CIGR XVth World Congress, paper no. 02-6081, Chicago, 2002.
6. Munack, A., J. Krahl, and H. Speckmann, Biodieselsensorik, *Landbauforsch. Völkenrode, 239:* 87–92 (2002).
7. Wichmann, H.E., Dieselruß und andere Feinstäube—Umweltproblem Nr. 1? *Gefahrstoffe—Reinhaltung der Luft 62:* 1–2 (2002).
8. Bischof, O.F., and H.-G. Horn, Zwei Online-Messkonzepte zur physikalischen Charakterisierung ultrafeiner Partikel in Motorabgasen am Beispiel von Dieselemissionen, *MTZ Motortechnische Z. 60:* 226–232 (1999).
9. Krahl, J., K. Baum, U. Hackbarth, H.-E. Jeberien, A. Munack, C. Schütt, O. Schröder, N. Walter, J. Bünger, M.M. Müller, and A. Weigel, Gaseous Compounds, Ozone Precursors, Particle Number and Particle Size Distributions, and Mutagenic Effects Due to Biodiesel, *Trans. ASAE 44:* 179–191 (2001).
10. Krahl, J., G. Vellguth, and M. Bahadir, Bestimmung der Schadstoffemissionen von landwirtschaftlichen Schleppern beim Betrieb mit Rapsölmethylester im Vergleich zu Dieselkraftstoff, *Landbauforsch. Völkenrode 42:* 247–254 (1992).
11. Sjögren, M., H. Li, C. Banner, J. Rafter, R. Westerholm, and U. Rannug, Influence of Physical and Chemical Characteristics of Diesel Fuels and Exhaust Emissions on Biological Effects of Particle Extracts: A Multivariate Statistical Analysis of Ten Diesel Fuels, *Chem. Res. Toxicol. 9:* 197–207 (1996).
12. Bünger, J., M. Müller, J. Krahl, K. Baum, A. Weigel, E. Hallier, and T.G. Schulz, Mutagenicity of Diesel Exhaust Particles from Two Fossil and Two Plant Oil Fuels, *Mutagenesis 15:* 391–397 (2000).
13. Schröder, O., J. Krahl, A. Munack, and J. Bünger, Environmental and Health Effects Caused by the Use of Biodiesel, Society of Automotive Engineers, SAE Technical Paper Series No. 1999-01-3561, Warrendale, PA, 1999, pp. 1–11.
14. Ames, B.N., J. McCann, and E. Yamasaki, Methods for Detecting Carcinogens and Mutagens with the Salmonella/Mammalian-Microsome Mutagenicity Test, *Mutat. Res. 31:* 347–363 (1975).
15. Bagley, S.T., L.D. Gratz, J.H. Johnson, and J.F. McDonald, Effects of an Oxidation Catalytic Converter and a Biodiesel Fuel on the Chemical, Mutagenic, and Particle Size Characteristics of Emissions from a Diesel Engine, *Environ. Sci. Technol. 32:* 1183–1191 (1998).
16. Crebelli, R., L. Conti, B. Crochi, A. Carere, C. Bertoli, and N. del Giacomo, The Effect of Fuel Composition on the Mutagenicity of Diesel Engine Exhaust, *Mutat. Res. 346:* 167–172 (1995).

8
Current Status of the Biodiesel Industry

8.1
Current Status of Biodiesel in the United States

Steve Howell and Joe Jobe

Introduction

The mere fact that this book is being published is testament to the significant progress that has been made in the production and use of biodiesel in the United States and around the world. Many of our colleagues will be covering details of biodiesel production, emissions, quality, and by-products; therefore, we will not deal with these aspects in detail here. Instead, this chapter focuses on the background of biodiesel in the United States, the driving forces for biodiesel use, how these driving forces are manifesting themselves in the marketplace and in public opinion, and provides some insight into the future of biodiesel use as a fuel in the United States.

Why Biodiesel?

Throughout the early to late part of the 20th century, petroleum-based fuels were cheap and abundant. New oil fields were discovered throughout the world, and it seemed we would be able to rely forever on crude oil as a cheap, readily available source of energy. Throughout the 20th century, motorized transportation proliferated after the invention of the automobile and was fueled by society's ever-increasing desire for mobility. Now, there is almost one automobile for every home and in many cases two or more. The trend over the last 10 yr has been toward larger, gas-guzzling sport utility vehicles.

During this time, industry grew to meet the needs of an ever-growing population; the most efficient means of transporting industrial goods and services is through the use of diesel transportation whether by truck, rail, or ship. As the U.S. population and the industry to support it grew, pollution (air, water, and land) became more and more of a concern. It became difficult to obtain permits for building new petroleum refineries, and the cost of the refineries became staggering. U.S. refinery capacity is now stretched to the limit, which has resulted in a sustained increase in gasoline and diesel fuel imports. At this time, we are dependent on other countries for ~60% of our total petroleum needs.

During this time, agriculture was undergoing significant change. Improvements in breeding and planting techniques, soil conservation, pesticide management, and overall productivity made farming a highly competitive business. Over the past 30 yr, U.S. farmers dramatically increased yield, lowered costs, and reduced the environmental effects of farming. Although this efficiency resulted in lower food prices for consumers, it generally did not bring improved income opportunities for farmers. Many farmers are forced to take a second job outside farming or become involved in turning their crops into value-added products and commodities to earn a decent income. Turning crops into value-added products also provides high-paying manufacturing jobs, jobs that will most likely be in or near rural areas. Loss of manufacturing jobs to cheap overseas labor is also a growing concern in the United States.

Increased dependence on foreign oil, increased pressure to reduce pollution, the need for value-added products from agriculture, and the need to create manufacturing jobs in the United States—this is the background for the current marketplace in which biodiesel is starting to play an increasing role. Each of these important driving forces is manifesting itself in changes in our society and our public policies, and it is the combination of these forces that is driving the success of biodiesel.

Current Industry Status

The U.S. Department of Agriculture (USDA) and other researchers in the United States began investigating vegetable oils as a fuel source in the late 1970s and early 1980s after the Organization of Petroleum Exporting Countries (OPEC) oil crisis. The general conclusion at that time was that vegetable oils were too viscous to be used in modern diesel engines over long periods of time. Transesterifying the vegetable oil into its methyl ester yielded a fuel or blending component more similar to petrodiesel that could serve as a drop-in replacement in existing engines. This modified vegetable oil fuel was still uneconomical compared with conventional diesel and was viewed largely only as an emergency fuel; thus, research efforts dwindled. A small amount of technical research continued during the 1980s and early 1990s. The topic sprang up again in earnest in the early 1990s as Congress began investigating alternatives to imported petroleum fuels after the Desert Storm war. The subsequent passage of the Energy Policy Act of 1992 and the formation of the National Soydiesel Development Board by 11 soybean farmer-run Qualified State Soybean Boards were the beginning of the commercial biodiesel industry in the United States. The National Soydiesel Development Board embraced other biodiesel feedstocks in 1995, changing its name to the National Biodiesel Board (NBB) and focusing its efforts mainly on addressing the technical and regulatory needs to commercialize a new fuel in the United States.

Early efforts included development of ASTM specifications for biodiesel (see Appendix B), a significant amount of emissions testing in a variety of applications, and maintaining the legal status of biodiesel through registration with the Environment Protection Agency (EPA) and completion of the Tier 1 and Tier 2 health effects testing requirements of Section 211(b) of the Clean Air Act amendments of 1990. Biodiesel is the only alternative fuel to complete the Tier 1 and Tier 2 (1,2) testing and submit the data to the EPA at a cost of over $2.2MM. Since its inception, the NBB and private industry have invested >$50MM in biodiesel research, development, and promotion, the majority of which was derived from U.S. soybean farmers through the Soybean Check-off program. During this time, real-world field demonstrations were successfully conducted; interest from government and academia and investment began to increase as well as some investment by non-soy interests. With these data and efforts in place, and other driving forces discussed below, the industry has begun to see a significant increase in sales volumes over the past several years. B100 sales in the United States have increased to 25 million gal (~95 million L; 83,300 t)/yr in 2003, up from only 500,000 gal (1665 t) in 1999 (Fig. 1).

Although not currently tracked through existing government tracking mechanisms such as the Energy Information Administration, NBB estimates that most of the biodiesel used today is as a blend with petrodiesel. The NBB estimated in 2003 that ~79% of the biodiesel was used by state, federal, and government fleets in a 20% blend of biodiesel with 80% petrodiesel (B20). NBB estimates that 30% of the biodiesel used in 2003 was a blend of 2% biodiesel with 98% diesel fuel (B2); this was used mainly by farmers in the Midwest. The remainder of the biodiesel was used as a pure fuel by environmentally conscious individuals or entities, such as the City of Berkeley in California and owners of Volkswagen diesel automobiles.

The increase in fuel use and other beneficial policies have also given rise to a dramatic increase in biodiesel production companies and distributors. In 1995, there was one biodiesel production and distribution company in the United States. As of November 2003, >1100 distributors and >300 retail locations (Fig. 2) existed in the United States with the numbers growing every day. In 2004, 20 biodiesel plants were in operation with >15 announced or proposed (see Fig. 3).

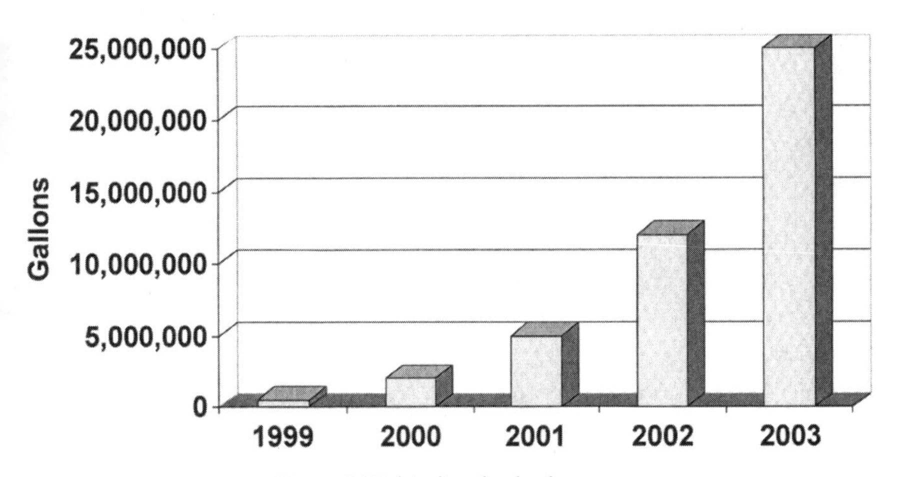

Fig. 1. U.S. biodiesel sales by year.

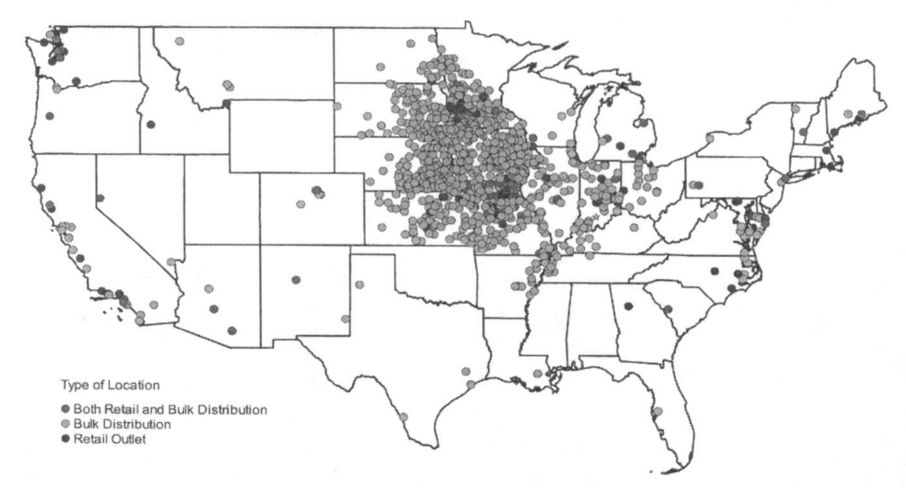

Fig. 2. Biodiesel retail locations and biodiesel bulk distribution locations (November 2003).

Continued growth in the biodiesel industry will depend on the driving forces for biodiesel use and the value that these driving forces take on in the marketplace. These factors are discussed in more detail below. Any one of them could cause a significant increase in biodiesel volume in the future.

Biodiesel Market Drivers–2004

To better understand the real reasons for the interest in biodiesel, the NBB commissioned a survey. The recent national internet survey among adult residents ≥18 yr old

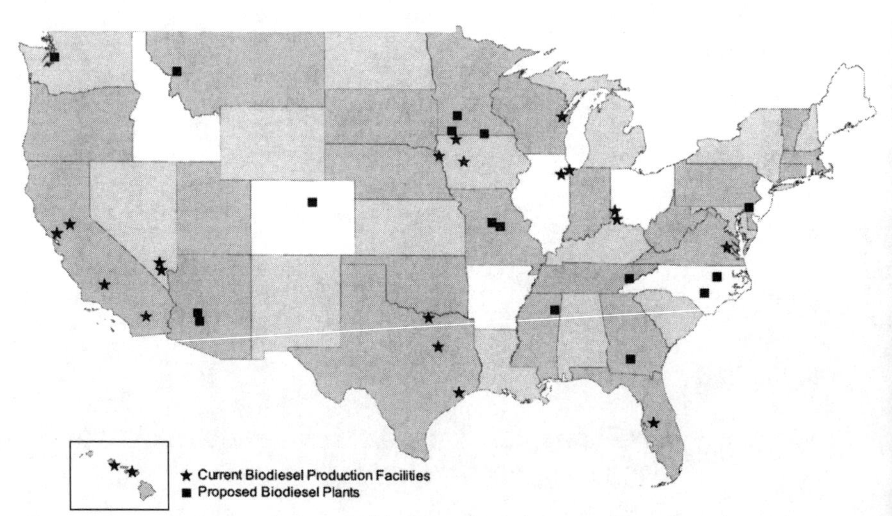

Fig. 3. Current and proposed biodiesel production plants (March 2004).

revealed that biodiesel has significant potential in the national marketplace. Some of the key findings from that survey (conducted online June 26–27, 2004, among a representative sample of 1042 U.S. consumers ≥18 yr old; potential sampling error was ±3% at the 95% confidence level) were that 27% of U.S. consumers have heard of biodiesel and that after hearing about the benefits and features of biodiesel, 77% would be likely to use biodiesel, 61% would pay at least $0.01–0.04 more per gallon (3.785 L) for biodiesel, whereas 39% said they would not pay more, 63% would consider purchasing a diesel car, 89% support a federal tax incentive to make biodiesel approximately the same price as regular diesel, and that the most important benefits of biodiesel to American consumers were the reduced dependency on foreign oil and potential health benefits, in that order.

The tragic events of terrorism of September 11, 2001, in New York and Washington, D.C., and the recent war in Iraq have clearly affected public opinion. A similar survey 2 yr ago ranked health benefits number one and environmental benefits number two with reduced dependence on foreign oil a distant third. The remainder of this chapter will focus on how these market drivers have made their way into public policy and on the consumer purchase preferences that have been driving increased biodiesel sales.

Reduced Dependence on Foreign Petroleum

The desire to reduce dependence on foreign energy sources manifested itself through a variety of public policies over the years that have served to increase biodiesel use in the United States. After the Desert Storm war in 1990, the Energy Policy Act (EPACT) of 1992 was enacted. The goal of this legislation was to derive 10% of the nation's energy from alternative fuels by 2000 and 30% by 2010 from a starting point in 1992 of <1%. Congress assigned much of the implementation of this legislation to the U.S. Department of Energy (DOE). Initial efforts focused largely on the use of compressed natural gas, liquefied natural gas, methanol, and 85% ethanol (E85). Alternative fuels were defined as those fuels that were substantially not petroleum and provided significant environmental and energy security benefits.

Most of the initial alternative fuels under consideration required a type of engine other than the gasoline or diesel engines currently in use; therefore, DOE chose to implement the legislation by requiring federal and state government fleets and alternative fuel companies (primarily utilities) to purchase an increasing level of light-duty vehicles (<8500 lbs = 3860 kg gross vehicle weigh) designed for the alternative fuel when purchasing new vehicles. Currently, >75% of most federal and state light-duty vehicle purchases and 90% of utility company light-duty vehicle purchases must be alternative fuel vehicles. There was no corresponding requirement for the purchase and use of the alternative fuel in the vehicle, however, and many bought vehicles that could run on either the alternative fuel or the conventional fuel, i.e., so-called flexible fueled vehicles. These vehicles met the AFV purchase requirements, but in the end, resulted in very little alternative fuel use because many fleets simply used conventional fuel, which was much more readily available.

Biodiesel can be used in an existing diesel vehicle with little or no modifications, but the rules gave credits only for the purchase of new vehicles rather than fuel use, and there are very few light-duty diesel vehicles in the United States. Biodiesel was essentially locked out of one of the most beneficial pieces of alternative fuel legislation. In 1998, the American Soybean Association and the NBB led a successful effort to convince Congress to modify EPACT to allow the use of fuel in existing heavy-duty vehicles to gain credits toward the light-duty AFV purchase requirements. The Energy Conservation Reauthorization Act of 1998 provided a mechanism whereby 450 gal of B100, used in any vehicle in a blend of B20 or higher, gains an EPACT fleet operator one AFV purchase credit. Fleets can use the biodiesel fuel use option to derive up to 50% of their credit requirements. B20, although still more expensive than conventional petrodiesel, proved to be the most cost-effective EPACT compliance option for many fleets. B20 use for EPACT and to meet Executive Order 13149 (described below) currently accounts for ~70% of total biodiesel sales.

Late in his administration, President Clinton signed Executive Order (EO) 13149. EO 13149 provides instructions to all federal agencies to reduce energy consumption in 2005 by 20% compared with 1999 levels. With EO 13149 requiring actual energy displacement, and the ECRA biodiesel options for EPACT, B20 became a very desirable option for meeting both of these requirements for federally controlled vehicles. It also provided a reason for use of B20 by other federally controlled diesel vehicles that are not covered by EPACT such as those of the armed forces. Today, B20 is being used by all branches of the armed services (Army, Navy, Air Force, Marines) at various locations throughout the United States. In fact, some military bases such as Port Hueneme, CA, are even putting in small biodiesel plants to make biodiesel from used cooking oil at the base.

Decreased Effects on Human Health

The use of biodiesel in existing diesel engines provides substantial reductions in unburned hydrocarbons (HC), carbon monoxide (CO), and particulate matter (PM), but slight increases in nitrogen oxides (NO_x). The average effects of biodiesel on today's engines were evaluated by the EPA (see Chapter 7.1)

Biodiesel emissions contain decreased levels of polycyclic aromatic hydrocarbons (PAH) and nitrated polycyclic aromatic hydrocarbons (nPAH), which were identified as potential cancer-causing compounds. In the NBB Tier 1 Health Effects testing performed by the Southwest Research Institute, PAH compounds were reduced by 75–85%, with the exception of benzo(a)anthracene, which was reduced by ~50%. Targeted nPAH compounds were also reduced dramatically with biodiesel, with 2-nitrofluorene and 1-nitropyrene reduced by 90%, and the rest of the nPAH compounds reduced to only trace levels. The effect of biodiesel blends on these compounds is believed to be mostly linear with concentration, but the data base with blends is smaller due to the extreme costs associated with testing of this nature.

The lower emissions from biodiesel, both regulated emission (HC, CO, PM) and unregulated (PAH, nPAH), are among the leading factors for the use of biodiesel and biodiesel blends by fleets and individuals alike. Many mechanics as well as riders are highly enthusiastic about the cleaner exhaust of B20 blends, i.e., how it does not burn their eyes or affect their breathing like conventional petrodiesel. Interestingly enough, the EPA curves show a phenomenon that has given rise to the desire to use biodiesel blends over neat biodiesel, or B100. The blends have a slightly larger per unit emissions decrease for CO, HC, and PM than does B100. Therefore, use of 100 gallons (378.5 L) of biodiesel as a B20 blend (500 gal ≅ 1892 L of total B20 blend) actually reduces emissions of HC, CO, and PM more overall than use of that same 100 gal as B100 alone and 400 gal (1514 L) of petrodiesel alone.

Although the lower emissions are important, and users cite this as the #2 reason for using biodiesel, it is very difficult to quantify how many customers use biodiesel solely for this purpose. Similarly, if biodiesel is used for other reasons (for example, EPACT or EO compliance), it is difficult to ascertain just how much value can be assigned to the emissions reduction of biodiesel. It is clear that fewer customers use biodiesel solely for its emissions reduction than when combined with other incentives.

New Engine Technology

One area in which the driving force related to human health and diesel emissions has manifested itself in legislation was the EPA regulations requiring increasingly lower exhaust emissions from diesel engines. Starting in 2007, the allowable levels of NO_x and PM exhaust emissions from new on-road diesel engines will be 90% less than in today's engines. This will make 2007 diesel engines much cleaner than the cleanest compressed natural gas engine available. Environmentalists have been heavily encouraging compressed natural gas over conventional diesel as a way to clean up urban pollution and have been instrumental in working on a federal and state basis (especially California) to put into place favorable legislation toward cleaner burning technologies. Sulfur in the fuel poisons the new catalysts and after-treatment technologies that are required to reduce the NO_x and PM levels from diesel engines. Therefore, the EPA mandated that all on-road diesel fuel must contain $\leq$15 ppm of sulfur beginning in 2006, the year before the implementation of the engine standards. The engine and sulfur rules will also go into effect for the off-road, locomotive, and marine fuels, with some starting in 2010 and the remainder going into effect over time so that eventually almost all the #1 or #2 diesel fuel in the United States will have $\leq$15 ppm sulfur.

Biodiesel from most U.S. feedstocks such as soybean oil already contains virtually no sulfur. In addition, removal of sulfur from petrodiesel removes other compounds that help to impart lubricity (see Chapter 6.5) important for the operation of most diesel fueling systems. This requires the use of a lubricity additive in virtually all diesel fuel. In addition to its lack of sulfur, biodiesel is an excellent component for restoring the lubricity of diesel fuel to safe levels, even at levels as low as ≤2%. The reduction in particulate matter, the presence of oxygen, and the relatively higher boiling point of biodiesel may also actually provide benefits beyond lubricity and sulfur reduction in new after-treatment catalyst and trap systems. Although these attributes are technical benefits of biodiesel with the new diesel fuel and after-treatment systems, they take on added value in the emissions arena because the EPA is seeking to reduce the health effects of diesel engine exhaust emissions through these new technologies.

Reduced Environmental Effect

Because the growing of plants that produce vegetable oil consumes CO_2, biodiesel has a closed carbon cycle, dramatically reducing CO_2. In addition, biodiesel is produced only from the oil contained in the seed; this normally ranges from a few percent (corn oil) to 20% (soybeans) to 40% (canola oil). A triglyceride oil such as soybean oil or animal fats is nature's way of storing energy so that oils and fats are already intrinsically high in energy content. Biodiesel therefore differs greatly from other fuels that may use the entire plant or seed for production. Starting with bare ground and counting all inputs for growing, harvesting, processing, and transportation, a DOE/USDA analysis found that biodiesel produced from soybean oil provides a 78% life cycle decrease in CO_2 emissions compared with petroleum diesel fuel and a 3.24:1 positive life cycle energy balance compared with petroleum diesel fuel.

Biodiesel is also biodegradable and nontoxic (see Chapter 6.6), so that spills are of much less concern than those of petrodiesel or crude oil. The value of biodiesel's CO_2 reductions, which are truly impressive at 78% reduction vs. petrodiesel, may be even greater in the future as the growing acceptance of global warming becomes a reality.

These environmental benefits are important to customers; however, as with emissions reductions, it is difficult to quantify the value of these benefits. Only a select few are using biodiesel for environmental benefits alone; this is occurring mainly in wetlands, national parks, or other environmentally sensitive areas.

Increased Economic Development

Increased use of biodiesel means a new use for fats and oils in addition to the need to build biodiesel production plants and staff these plants with new employees. Using biodiesel, therefore, has economic benefits for farmers, local communities, and the nation as a whole. Moreover, the spending associated with increasing investment in biodiesel production and higher agricultural output will stimulate aggregate demand, create new jobs, and generate additional household income.

Increased utilization of renewable biodiesel results in significant economic benefits for both the urban and rural sectors, as well as the balance of trade. A study completed in 2001 by the USDA found that an average annual increase in demand for soy-based biodiesel equivalent to 200 million gal would boost total crop cash receipts by $5.2 billion cumulatively by 2010, resulting in an average net farm income increase of $300 million/yr. The price for a bushel of soybeans would increase by an average of $0.17 annually during the 10-yr period.

Several states have also conducted independent macroeconomic studies, and these predict increased employment, increased economic activity, and a corresponding increase in state and local tax revenue, as well as other indirect and induced economic effects. In today's slumping rural economy, this factor alone is causing legislators to pass incentives encouraging biodiesel use and the building of production plants in their state. On a national basis, a reduction in imported oil also improves the national balance of trade, and this is also a factor encouraging beneficial legislation.

Legislation

The combination of all of these benefits, along with crude petroleum oil prices reaching almost $50/barrel (1 barrel; 42 gal; 159 L), has driven a recent upsurge in legislation benefiting biodiesel. Beneficial legislation has been introduced in >30 states over the past 2 yr, and the number is growing. The list is too long to include in this chapter, but they vary from mandates (Minnesota has mandated that most petrodiesel contain a minimum of 2% biodiesel beginning in the summer of 2006), to production plant incentives, to forgiving sales tax. For example, Illinois forgives a portion of the sales tax on biodiesel blends ≥10% and forgives it entirely for blends ≥11%. Current information is available at www.biodiesel.org.

The most critical piece of legislation is a proposed federal tax credit that is currently working its way through Congress. This landmark legislation would provide a tax credit of ~$0.01 for each 1% of biodiesel used in a blend. Depending on the prices for oils and fats as well as crude diesel fuel, this tax credit could make biodiesel blends very cost competitive with conventional petrodiesel. The industry sees passage of this tax credit as a number one priority and is hopeful that the bipartisan support it enjoys will allow it to be signed into law soon.

Future Growth Areas

Clearly, biodiesel has many benefits. How far it will progress—and which markets it will penetrate the fullest—is not known. Biodiesel will very likely see a significant increase as a low-blend component (B2 to B5) in future diesel fuel as a way to solve the lubricity problem with 15 ppm sulfur diesel fuel and as one avenue toward decreasing dependence on foreign oil, improving the environment, creating manufacturing jobs, and being an outlet for value-added agricultural products. We foresee increased penetration of B20 into state, federal, and government fleets, especially the military, as well as school buses and garbage truck fleets. This could easily change the current blend level use from ~70% B20 and 30% B2 to one approaching 70% B2 and 30% B20 with a significant growth in overall volume. There will still be a small number of users of B100, but that volume is expected to be low.

There are three other fuel-related applications, electrical generation, home heating oil and industrial boilers, and fuel cells, that may also see significant growth, depending on policies and other societal pressures. Biodiesel can be used as a blend or as a neat fuel for generation of electricity in diesel generator sets (both small and large), which could potentially be used to meet renewable electricity mandates in 12 states. Some have suggested placing a biodiesel generator set at the bottom of windmills at wind farms to create a totally renewable and reliable source of electric power. Biodiesel can also be used in gas turbine applications to create electricity, in home heating oil systems (predominantly in the Northeast), or as a fuel for industrial boilers anywhere #2 fuel oil is used. In these open-flame applications, biodiesel appears to reduce NO_x (due to its oxygen content), and some companies or municipalities may choose biodiesel use in these applications for the same societal reasons as for low blends in the transportation sector (foreign oil, health, environment, jobs) while also reducing NO_x.

Last, although the debate concerning whether fuel cells and electric vehicles will storm the country will no doubt continue for years, pure biodiesel makes an excellent high-density source that can be easily reformed into hydrogen for fuel cell applications. Biodiesel's high flash point, high biodegradability, and low toxicity may give it a competitive edge over methanol and natural gas as a fuel cell fuel. Extremely high full life cycle energy balance, full life cycle CO_2 reduction, and renewability may also tip the scales in favor of biodiesel.

Conclusions

From the first use of peanut oil in 1900 to today's stringent ASTM standards for vegetable oil methyl esters, i.e., biodiesel, the use of vegetable oils and animal fats as a source for diesel fuel applications has come a long way. The industry is currently enjoying exponential growth, and public policy will likely encourage further growth over time. The driving forces for biodiesel use are many and may become even more important as time goes on. Future technologies, such as ultraclean diesel engines or fuel cells, actually represent further opportunities for biodiesel—not threats. In our eyes, there is no longer a question of whether biodiesel will succeed; it is only a question of how large it will grow and how soon.

Acknowledgments

We thank the staff members of the NBB and MARC-IV for their diligent support in providing the statistics and data for this chapter and for their never-ending support of this American-made, cleaner burning fuel: biodiesel.

References

1. Lovelace Respiratory Research Institute (LRRI), Tier 2 Testing of Biodiesel Exhaust Emissions, Final Report Submitted to the National Biodiesel Board, Study Report No. FY98-056, May 2000.
2. Finch, G.L., C.H. Hobbs, L.F. Blair, E.B. Barr, F.F. Hahn, R.J. Jaramillo, J.E. Kubatko, T.H. March, R.K. White, J.R. Krone, M.G. Ménache, K.J. Nikula, J.L. Mauderly, J. Van Gerpen, M.D. Merceica, B. Zielinska, L. Stankowski, K. Burling, and S. Howell, Effects of Subchronic Inhalation Exposure of Rats to Emissions from a Diesel Engine Burning Soybean Oil-Derived Biodiesel Fuel, *Inhal. Toxicol. 14:* 1017–1018 (2002).

8.2

Current Status of Biodiesel in the European Union

Dieter Bockey

Introduction

The current status of biodiesel production in the European Union (EU) as described in this chapter represents political measures with the fundamental objective of achieving full utilization of renewable energy sources in a stepwise manner (see Table 1). With these policies, the EU is pursuing a strategy directed toward utilizing renewable energy for the following reasons: (i) combating climate change; (ii) reducing local environmental loads; (iii) creating jobs and income in an EU increasing to 25 member countries; and (iv) contributing toward a secure supply of energy.

To achieve these objectives, the following timeline was embodied in EU law:

- 1997—White Book COM (97)599: Energy for the Future: Renewable Sources of Energy.
- 2000—Green Paper COM (2000)769: Towards a European Strategy for the Security of Energy Supply.
- 2001—Directive 2001/77/EC on the Promotion of Electricity Produced from Renewable Energy Sources in the Internal Electricity Market.
- 2002—Directive 2002/91/EC on the Energy Performance of Buildings.
- 2003—Directive 2003/30/EC on the Promotion of the Use of Biofuels or Other Renewable Fuels for Transport.
- 2004—Directive 2003/96/EC Restructuring the Community Framework for the Taxation of Energy Products and Electricity.

In its white book, the EU and its member countries set the goal of increasing the production and use of renewable energy to a minimum of 12% of the total domestic energy consumption by the year 2010. The amounts of green power and biofuels should be 2.2 and 5.75%, respectively.

It should not be overlooked that these ambitious goals were not approved without reservations by representatives of commercial and political interests. The different conversions into national legislation by individual member countries reflect their different priorities regarding energy and climate policy. This is the result of intensive political discussions in the individual countries and at the EU level in which the variability provided in the EU directives is utilized in very different fashions relative to the individual climate protection responsibilities. The

TABLE 1
Current EU Targets and Feasibility: Will the White Paper Targets Be Achieved[a]

Type of energy	1995 Eurostat	2001 Eurostat	AGR 1995–2001	White paper targets 2010	AGR needed 2001–2010
1. Wind	2.5 GW	17.2 GW	37.9%	40 GW	9.8%
2. Hydro	87.1 GW	91.7 GW	0.9%	100 GW	1.0%
3. Photovoltaics	0.04 GWp	0.26 GWp	36.6%	3 GWp	31.2%
4. Biomass	44.8 Mtoe	56.5 Mtoe	3.6%	135 Mtoe	10.3%
5. Geothermal	2.72 Mtoe	3.43 Mtoe	3.9%	5.2 Mtoe	4.7%
6. Solar thermal	6.5 Mio m^2	11.4 Mio m^2	9.8%	100 Mio m^2	27.2%

[a]AGR, annual growth rate; EU, European Union.

strategies of the member countries for promoting renewable energy, therefore, differ greatly and can be exemplified by the different tax advantages for biofuels described below. Thus, a well-balanced consensus of all political and economical parties is required to achieve the strategic objectives without conflict and without distorting competition. It is already clear that the speed at which the directives will be applied varies greatly, and it is possible that the desired proportions of biofuels according to the EU directive may not be attained. The German example shows that technical problems concerning the commercialization of biodiesel and other biofuels (ethanol and sundiesel) can be solved only within existing economical and political conditions.

Directive Promoting Biofuels

On May 28, 2003, the European directive for promoting biofuels went into effect. The centerpiece of this directive is an action plan that prescribes minimum proportions of biofuels for each member country based on its share of the fuel market. The action plan calls for biofuel to comprise a 2% share of the overall fuel consumption in the EU beginning in 2005. This share is to increase stepwise to 5.75% by 2010 (see Table 2).

Depending on how total fuel consumption develops, ~14–16 MMT of biofuels should be produced in the EU by 2010. Biodiesel would account for ~7.5 MMT. In terms of absolute biofuel quantities, satisfying these goals is an enormous challenge for the individual member states. Therefore, it was not surprising that these goals, which originally were suggested as obligatory by the EU Commission, were the subject of an intensive discussion among the EU Parliament, Council of Ministers, and the EU Commission. The resulting compromise was that the goals are not obligatory; rather, they are indicated objectives. However, the Commission reserved the right to prescribe the goal for an individual member country if the country does not make any effort to meet an objective.

Furthermore, the realization of the directive promoting biodiesel makes the Commission responsible for reporting to the EU Parliament, and therefore from

TABLE 2
Biofuel Action Plan of the EU Commission[a]

Year/minimum share[b]	Gasoline consumption	Diesel consumption	Total
	(1000 metric tons)		
2005/2.00%	2341	2532	4873
2006/2.75%	3219	3482	6701
2007/3.50%	4096	4431	8527
2008/4.25%	4974	5381	10355
2009/5.00%	5852	6331	12183
2010/5.75%	6730	7280	14010

[a]*Source:* Reference 1. EU, European Union.
[b]Based on fuel consumption in 1998.

member countries to the Commission. Subjects to be reported on include the environmental efficiency of the individual biofuels, their contributions to preserving natural resources and climate protection, as well as energy independence. It must be emphasized here that the directive promoting biofuels contains an empowerment for the member countries to tailor the mineral oil tax preference based on comparative life-cycle assessments (LCA), depending on the individual biofuel. An intensive discussion on LCA has therefore begun in Germany and also on a European level.

The biofuels industry in the growing EU therefore must adjust to the increasing transparency of the sector from production of the feedstock to the value-added biofuel. Therefore, the Union for the Promotion of Oil and Protein Plants (UFOP) participated in financing and conducting an assessment of LCA based on globally available studies. This work was commissioned by the Technical Association for Combustion Engines (Fachvereinigung Verbrennungskraftmaschinen; FVV), an institution founded by the German automotive industry for promoting research on problems transcending those of individual companies. In 2003, the UFOP commissioned the Institute for Energy and Environmental Research to update the LCA of rapeseed oil methyl ester, opening up to political discussion the contribution of the biodiesel chain to climate protection and preservation of resources. Although the future tax frameworks will be determined largely by comparison of LCA, it must be noted critically that there is no concerted strategy at the EU level between the biodiesel or biofuels industry and their trade associations. On the other hand, this subject is of intense interest to the automotive industry. The underlying goal is to accelerate the strategic development of new fuels and engine technologies to achieve increasingly CO_2-neutral mobility, thus satisfying the self-imposed obligation of the European Association of the Automobile Industry to reduce the CO_2 emissions to 140 g/km by 2008.

Another important strengthening aspect for the biodiesel industry is the production of biodiesel as prescribed in the EU directive according to the standard EN

14214, which went into effect in November 2003. According to the directive promoting biodiesel, the member countries are obligated to monitor the biodiesel quality, i.e., the amount of biofuel permitted in gasoline (petrol) or diesel fuel at filling stations. The new European standard for conventional diesel fuel, EN 590, now permits the blending of a maximum of 5% biodiesel on the basis of the European standard. The directive promoting biodiesel calls for special labeling of the fuel at blend levels >5% biodiesel. The new member countries must meet these quality requirements after joining the EU on May 1, 2004.

Energy Tax Directive

After nearly 12 yr of intensive coordination between the member countries, the energy tax directive went into effect on October 31, 2003. The energy tax directive is the legal basis for national legislation and regulations concerning tax advantages for biofuels. Article 16 of this directive empowers the member countries to apply tax exemptions or reduced tax rates to biofuels. The energy tax directive specifically limits the tax exemption to the biomass portion of the biofuel. This limitation is significant for bio-ethanol used for producing ethyl *tert*-butylether (ETBE) but also for biodiesel, whose production involves the use of methanol of fossil origin. The energy tax directive requires that the tax reduction or exemption must be examined for overcompensation, taking the corresponding development of raw materials into account and, if needed, be changed. The tax exemption or reduction is valid for only 6 yr, although this period may be extended. Furthermore, the member countries have to report the tax reductions or exemptions to the EU Commission every 12 mon with the first report due December 31, 2004.

In summary, the promoting and energy tax directives of the EU grant the member countries significant variability for creating tax advantages for biofuels as a prerequisite for meeting the goals of the action plan, but also require extensive reporting, amounting to constant monitoring, from the cultivation of the raw material and production up to the final use of the biofuels in their corresponding markets.

Conversion into National Legislation

The possibilities for conversion into national legislation described above currently vary greatly among the EU member countries. In some member countries, tax advantages were created and are based on legislation. In other countries, a parliamentary discussion has begun and legislation is pending; in some countries, however, no government or parliamentary initiatives for creating a tax advantage for biofuels appear to exist.

In Germany, a change in the law regarding the tax advantage for biofuels went into effect on January 1, 2004. Formally, no tax exemption for biofuels was granted. Instead, a tax rate of 0 is introduced without limiting the quantity of biofuels receiving this advantage. The tax rate will be examined annually with the first examination by April 30, 2005, by means of a report from the government to the federal parliament. The government, i.e., the finance ministry, can reserve the right to carry out an adjustment based on this report. The question of overcompensation, in particular, will be examined but also whether the production of biofuels contributes to both energy security based on the feedstocks produced in Germany and to the quantity goal or whether the tax exemption leads to significant imports of biofuels. To satisfy the reporting requirement, the law on energy statistics was extended by the introduction of a reporting requirement for biofuel producers beginning in 2004. Thus, fuel producers for the most recent calendar year have to report retroactively to the German Federal Statistics Agency (Statistisches Bundesamt) the quantity of fuel produced, classified according to feedstock and marketing routes. Imports and exports are also covered. Below are listed the most important changes to the mineral oil tax act and the mineral oil tax implementing ordinance.

Amendments to the Mineral Oil Tax

Biofuel = Mineral Oil. Fatty acid methyl esters (biodiesel) appear in the list of mineral oils as further taxable items. In line with the energy tax directive, a further amendment to the law relating to the definition of biofuels specifies that biofuels are to be treated as mineral oils if the corresponding products (rapeseed oil, biodiesel, bioethanol, biogas) are destined to be used as fuel, although in chemical terms, they are not hydrocarbons like mineral oils. The exception to this is biomass to be used for heating purposes such as rapeseed oil/biodiesel used for the operation of oil heaters or block-type thermal power stations. In other words, the intended use determines how the product is classified for tax purposes.

The changes with respect to tax relief for biofuels for transport and heating purposes provide for preferential tax treatment to be granted up to December 31, 2009. This period may be extended in line with the energy tax directive. However, the tax relief is limited to the part of the biofuel that can be proven to consist of biomass. The specification states that, to the extent that they are manufactured by esterification, fatty acid methyl esters (biodiesel) are also to be regarded as biofuels, recognizing that the process uses methanol made from fossil raw materials but simultaneously yields an equivalent amount of biogenic glycerol. Although a full exemption of biofuels from the mineral oil tax does not require the approval of the EU Commission, German law makes it clear that tax relief must not lead to overcompensation. Thus, the level of the tax break can be adjusted to developments on the crude oil market and the prices of biomass and fuel. The amended legislation requires the situation to be monitored in a report to be submitted each year to the German federal parliament (Bundestag). This report is to include consideration of the effects on climate and environmental protection, the conservation of natural resources, the external costs of the various fuels, and progress in achieving the targets laid down in the EU directive in terms of the contribution of biofuels to energy supply. The first report is to be submitted to the Bundestag on March 31, 2005.

Notification Requirements. To qualify for tax relief, biofuel manufacturers and/or storage facility operators storing biofuel for transportation or heating purposes are required to register their activity with the appropriate customs office. This obligation applies to all biofuel manufacturers, including the operators of small-scale pressing plants (rapeseed oil as fuel). The customs office is required to confirm receipt of the notification and to inform the applicants that an application to set up a tax zone in their production facility (in which the amount of product for sale to customers is registered) must be submitted. It can be determined in individual cases whether it suffices to set up calibrated sampling stations because the risk of tax loss is low. It appears that red tape is to be minimized for the operators of small plants.

Changes to the Mineral Oil Tax Regulation

Proving the Nature of the Biomass. Granting tax relief is also linked to proof of the nature of the biomass in the biofuel in question. The regulation specifies that a general declaration by itself is not sufficient. An agreement has to be reached in individual cases between the applicant and the tax authorities. According to information from the federal finance ministry, in the case of plant oils or fatty acid methyl esters, the nature of the biomass is established by product analysis. Hence, records kept by the producer regarding the quantities manufactured from the respective plant-based raw materials can be regarded as sufficient proof. When biodiesel is blended with petrodiesel in a mineral oil storage facility, as a condition of tax relief, the operator must provide proof of both the nature and amount of the biomass in the biofuel added to the mixture. According to the finance ministry, a declaration from the manufacturer of the biodiesel suffices to establish the nature of the biomass.

Fuel Quality

Notwithstanding the tax possibilities involved in the manufacture of fuel/biofuel mixtures, the quality and/or quality assurance of the product is an important issue. The European standards for diesel fuel EN 590 and for gasoline (petrol) EN 228 permit a maximum addition of 5% biodiesel or bioethanol. Although other mixtures containing a high proportion of biofuels are possible in tax terms, these mixtures no longer conform to either of these fuel standards and therefore have to be classified separately in accordance with the EU directive on the promotion of the use of biofuels and the national fuel quality and classification ordinance (10th Federal Emission Protection Ordinance). This aspect is particularly relevant when diesel/biodiesel blends are used with respect to the end customer (product liability) because the original quality of the fuels in question is impossible to determine in a mixture.

Against this background, the exemption provided for in the amended implementing ordinance means that practically any mixture can be produced in the warehouse of the end customer without tax implications, providing the fuel mixture created is exclusively for the end customer's own use. The German Farmers' Association pressed for this regulation to promote the marketing of biodiesel and encourage businesses to convert. In practice, the regulation means that a second storage tank can be dispensed with. However, when using biodiesel/petrodiesel blends, it is still necessary to carry out appropriate measures such as cleaning the storage tank and changing over the fuel pumps to dispense fuel containing biodiesel. Nevertheless, it is important to emphasize that biofuels are mineral oils (see above) within the context of tax law, with the consequence that the mixing of biofuels with fossil fuels (except in the case of the end customer) outside the tax warehouse has tax consequences and will lead to the entire amount being subject to retrospective assessment.

Situation in the EU Member Countries

In France, the tax advantage for biodiesel is 33/100 L with a tax rate of 41.69/100 L for conventional diesel fuel. The government determines annually the quantity of biodiesel receiving this advantage. The amounts for 2003 and 2004 are 320,000 and 390,000 metric tons, respectively. However, the total production capacity for biodiesel is 470,000 metric tons. This surplus capacity results in fuel exports and supply pressure in other countries. With the objective of satisfying the minimum amount of biofuels in gasoline and diesel fuels according to the EU directive for promoting biofuels, the French cabinet recently decided to take specific measures for promoting biofuels within the framework of a planned energy orientation law. The details for realizing this basic decision are to be specified in the fall of 2004 within the framework of a "plan for developing biofuels." An improved promotion by taxes, in particular, should contribute to this plan.

The government of the United Kingdom set a tax advantage of 0.20/100 L for biodiesel without limiting quantity. This tax advantage does not suffice for a market breakthrough, according to the British Association for Biofuels. The developing production capacity is concentrated largely around the use of used frying oils and animal fats; however, these must be evaluated critically relative to satisfying the quality criteria set forth in EN 14214. The Italian government also determines an annual quantity of tax-exempt biodiesel. For 2003, this amount was 120,000 metric tons.

Like Germany, Sweden has exempted biofuels since the beginning of 2004 from the combined CO_2^- and energy tax, thereby setting the "mineral oil tax rate" to 0. The tax advantage is financed by tax increases on conventional diesel fuel and power for the manufacturing industry as well as households and the service sector. The tax exemption is 36/100 L. In Austria, too, biodiesel is not subject to the mineral oil tax, and there is no limiting quantity.

The Dutch cabinet announced it would do its utmost to ensure that biofuels become available beginning January 1, 2006, and to implement a subsidy scheme to achieve the maintained target value (2% of the energy content for gasoline and diesel fuel). In 2005, the government will publish the results of a study, which includes the financial aspects and preparations required. The policy will be evaluated in 2007 and will help the government determine whether the EU guideline's indicative target value (5.75% in 2010) can be achieved.

In June 2004, the EU Commission approved the subsidies of the Czech Republic for the promotion of biofuels. The measure is a reduction in the consumption tax of 95/1000 L if the biodiesel/petrodiesel blend contains at least 31% biodiesel (rapeseed methyl ester). The tax rate thus decreases from 306/ 1000 L to 211/1000 L. Until the end of 2006, the Czech government will provide a direct support of 257/1000 L for rapeseed methyl ester producers in the Czech Republic. This support is limited to 100,000 metric tons.

Currently, no satisfactory information is available regarding tax legislation in all EU member countries. By the end of 2004, the member countries must inform the EU Commission about the conversion of the directive into national legislation. The incomplete compilation underscores the still insufficient association structure at the European level. Biodiesel producers are challenged to join the European Biodiesel Board (EBB) and to supply the corresponding data for the necessary statistics as a way of safeguarding their interests. Strengthening trade associations is imperative so that the biodiesel industry can enter into the expected political discussions on national and EU levels in a unified fashion.

Capacity and Development of Production

With the background of the promotional and legislative framework described above, biodiesel capacity has increased in an unexpected fashion in the EU and especially in Germany. From 1996 to 2003 with more than 2.2 MMT capacity, biodiesel production capacity in the EU more than quadrupled. Biodiesel production capacity in the EU as of 2004 is depicted in Table 3. In Germany, 24 companies with a total capacity of 1.1 MMT t are producing biodiesel. An additional 0.5 Mio tons capacity is currently under construction. The capacity will rise at the by the end of 2006 to ~16 Mio metric tons/yr. Germany will be the first member state fulfilling the EU-promotion directive in the diesel market. The development of biodiesel sales in Germany since 1991 is shown in Table 4.

In Germany, 90% of the feedstock for biodiesel is rapeseed oil, whereas sunflower oil is also used in southern Europe. The decisive parameter for feedstock use is the minimal quality requirement of the standard EN 14214. Because biodiesel is marketed mainly as neat fuel in Germany, the automotive industry exerts significant pressure on the biodiesel industry to rigorously meet the requirements in the standard. This pressure will increase through the altered directive effective April 2004 requiring the labeling of fuel quality. Biodiesel is contained in this directive; therefore, like gasoline (petrol) or conventional diesel, as so-called "common commercial fuel," it is subject to the same unannounced controls by regulatory agencies. Serious deviations can lead to the temporary closure of the

TABLE 3
Biodiesel Production Capacity in Europe in 2004[a,b]

Country	Capacity (1000 metric tons/yr)
Germany	1097
France	520
Italy	370
Austria	120
Spain	70
Slovenia	70
Czech Republic	63
Denmark	30
Sweden	8
UK	5
Ireland	2
Belgium	0

[a]*Source:* Reference 2.
[b]Total capacity is 2,355,000 metric tons/yr.

biodiesel pump. The service station operator is liable for the product. In Germany, the "Working Group for Quality Management of Biodiesel" will extend the contract conditions concerning additization with an antioxidant to avoid liability claims from service stations directed toward biodiesel distributors and producers. For creating an "aging reserve" in light of the very different turnover at service stations or even self-use tanks, this quality assurance measure is absolutely necessary. This was shown by an extensive check conducted in cooperation with Daimler Chrysler AG and Volkswagen AG at 170 service stations. The AGQM regularly conducts field analyses and producer checks, quality is constantly monitored, and

TABLE 4
Development of Biodiesel Sales in Germany

Year	Sales (metric tons)
1991	200
1992	5,000
1993	10,000
1994	25,000
1995	45,000
1996	60,000
1997	100,000
1998	100,000
1999	130,000
2000	340,000
2001	450,000
2002	550,000
2003	700,000
2004	850,000

the results entered into updates of the quality assurance concept and the information sheets of the AGQM for transporting and storing biodiesel.

More intensive international cooperation and sharing of experience are also urgently needed in the area of quality assurance. A first step would be an extension of the round-robin tests conducted by biodiesel producers; to date, this is taking place only in Germany. Subsequent workshops for training laboratory personnel and quality assurance specialists were successful.

References

1. Directive 2003/30/E6 of the European Parliament and of the Council, May 8, 2003.
2. UFOP, European Biodiesel Board.

8.2.1

Biodiesel Quality Management: The AGQM Story

Jürgen Fischer

Introduction

AGQM is the abbreviation for **A**rbeits**G**emeinschaft **Q**ualitäts **M**anagement Biodiesel e.V., the Association for the Quality Management of Biodiesel. It may appear strange to have a special quality management association for biodiesel. Obviously, fuels must meet all quality requirements to ensure that a vehicle can be driven without any problems caused by the fuel. What is the purpose of such an association?

The German biodiesel market differs from those of other countries. In most countries (Czech Republic, France, Italy, Sweden, and the United States), biodiesel is sold as a blend with petrodiesel in varying concentrations. Germany is the only country in which biodiesel is sold as pure fuel, available at public filling stations and with a growing market share. In the beginning, only a few producers supplied the growing demand on an equilibrium level. Beginning in 1999, the situation changed completely, showing a demand of biodiesel exceeding the production capacity (Fig. 1). Decreasing prices for vegetable oils and increasing petroleum prices increased interest in the production of biodiesel, and many new producers appeared on the market.

The capacities of the production plants vary between 2000 and 150,000 MT/yr (Fig. 2). Investment costs for small units are fairly low, and subsidies by the gov-

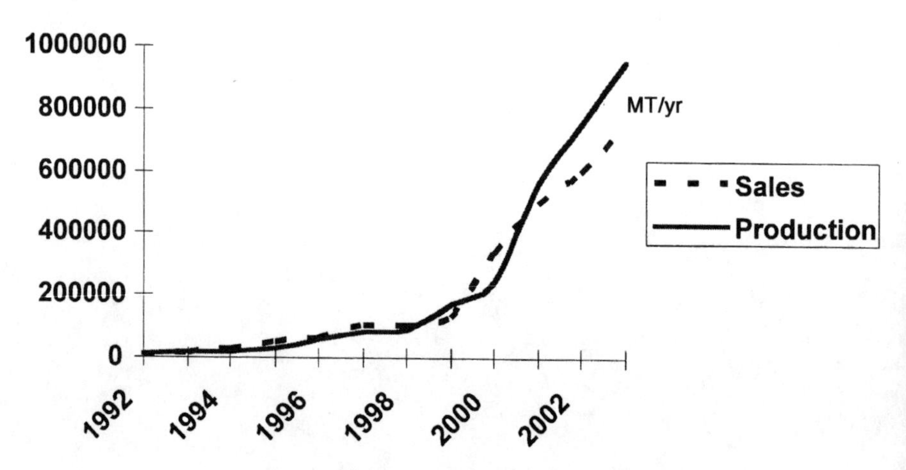

Fig. 1. The German biodiesel market.

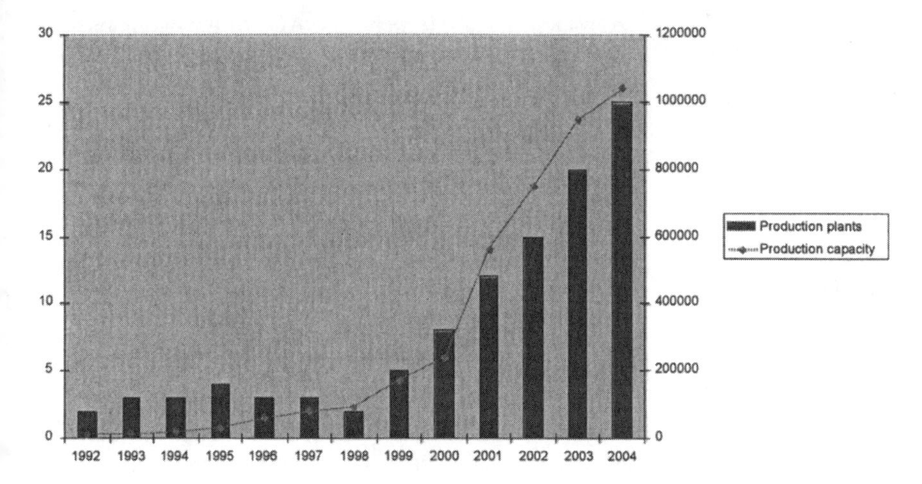

Fig. 2. Biodiesel production in Germany.

ernment made it very economical to invest in biodiesel plants. In addition to the ecological advantages, biodiesel prices dropped, and the public interest in biodiesel began to grow, which was also demonstrated by an increase in the number of filling stations to the current level (>1700).

In 1999, the German automobile industry, which by then had approved the use of biodiesel, started to complain about biodiesel quality. An increasing number of complaints, including filter blocking and damaged injection equipment, were related to the quality of biodiesel. The threat of losing the approval of the automobile industry if there was no improvement in quality made it necessary to act.

The structure of the German biodiesel industry is inhomogeneous. The plants are owned by the food industry, farmers, and investment groups, who lack the necessary technical background. From a chemical and technical standpoint, biodiesel production appears to be an easy process, leading to the opinion that it is really not necessary to invest a lot of money in quality control. In contrast to the petroleum industry, which has been familiar with fuel requirements for more than 100 yr, many of the investors ignored this issue. Many of the new production plants do not have a laboratory of their own because the investment for a laboratory equipped for analyzing biodiesel according to the standard would be ~ 1.5 million. Compared with the total costs of 15–20 million for a unit with a capacity of 100,000 MT, these costs are relatively high.

In December 1999, the AGQM was founded as an association of the German biodiesel industry, including biodiesel producers and trading companies, together with UFOP (Union zur Förderung von Öl- und Proteinpflanzen; Union for Promoting Oil and Protein Plants), a part of the German farmers' association. The original mandate of AGQM included ensuring the quality of biodiesel, setting up quality standards for this fuel, and offering help to the members in installing quality control systems for production and delivery. From the original 10 members, AGQM now comprises by far the largest part of the German biodiesel business with 16 producers, 29 trading companies, 11 subsidizing members (R&D, plant construction, additive industry), and UFOP. In a very short time, AGQM has become well-known in Germany. Many activities formerly carried out by single biodiesel producers have been taken over by AGQM, increasing the effectiveness of such measures. This section describes some of the most important activities of AGQM.

Standardization

One of the major concerns of the automotive industry was that biodiesel had no defined standard. After many discussions in 1995, the German preliminary standard DIN V 51606 was published. It was developed by a working group of automobile manufacturers, the German standardization board (DIN), and the biodiesel industry, consisting at that time of only one company (Oelmuehle Leer Connemann). This preliminary standard was followed by the draft standard E DIN 51606 in 1997. At the beginning of 1998, the European Commission mandated to the CEN (Comité Européen de Normalisation; European Committee for Standardization) the development of a European standard for biodiesel. Representation for the German biodiesel industry was taken over in 2000 by AGQM, making it possible to concentrate the activities of all German biodiesel producers and to benefit from the experience of this working group.

The AGQM members comprise a considerable part of the subgroup of the German standardization board, which is responsible for the conversion of the international standard to German rules. In the meantime, the experience of the AGQM members in the analysis of biodiesel made it possible to work on and improve the test methods. The AGQM coordinates the activities and, together with DIN, organizes round-robin tests and workshops to verify and solve analytical problems concerning biodiesel, i.e., the precision of test methods. The input of the AGQM members is extremely important for such activities.

Quality Management and Quality Control

Quality control is the main concern of the AGQM. All products sold by AGQM members to filling stations have to fulfill the requirements of the German biodiesel standard, i.e., some special AGQM requirements that were defined after discussions with the automotive industry. To ensure the quality and to have standard procedures for all members, the association installed a quality management system consisting of five steps that cover the whole life cycle of biodiesel from the raw material to the filling station:

1. Production: Producers are responsible for the choice of feedstock and product control. All documentation concerning production control, laboratory testing, and product quality must be available. The AGQM is auditing the producers

and is checking the production quality through independent auditors. Minimum laboratory equipment is recommended and analysis of the product before delivery is mandatory. In cooperation with independent laboratories, AGQM carries out round-robin tests to inspect and improve the performance of the laboratories. It is mandatory for AGQM members to participate in these tests. Producers must ensure proper transport conditions. Many of the problems occurring in the market are based on contamination during transport. Trucks and tank cars must to be checked properly before loading.

2. The storage of biodiesel differs greatly from that of the petrodiesel fuels. Traders must ensure that suitable materials are used, and that tank cleaning, housekeeping, and so on are acceptable. Separate logistics systems are necessary to avoid the mixing of petrodiesel or gasoline with biodiesel. Storage tanks are also under the control of AGQM.
3. In 2002, AGQM installed a licensing system for biodiesel filling stations. Stations having signed a contract with AGQM are permitted to use the AGQM logo (Fig. 3) for advertising, and they receive full support in all quality questions. To be allowed to use the symbol, they have to sell biodiesel according to the AGQM standard. The quality at these filling stations is checked by independent laboratories.
4. AGQM also offers service for biodiesel customers. The association publishes brochures, information about approval by automobile manufacturers, lists of filling stations, as well as reports available on its homepage. The AGQM office in Berlin is also a contact for customers in case of problems with its members.
5. Examination of real and supposed problems is an important part of the quality management system; this is done by technical working groups, consisting of experts from the member companies. Related to this are discussions with customers, garages, car manufacturers, and original equipment manufacturers.

Quality surveys

Field surveys of fuel quality are a common way of obtaining an overview on standard qualities. In Germany, these surveys are initiated by several interested groups, namely, government authorities, because fuel quality is regulated by law, the petroleum industry, the automotive industry, and automotive associations.

To date, biodiesel has not been subject to regulations by the authorities, and the German petroleum industry has not been interested in the quality of a niche product. For the biodiesel industry, these surveys are of great interest for obtaining information on the behavior of biodiesel in filling stations, storage, and transport conditions and to be able to take measures to avoid complaints by customers. In 1997, UFOP started with the first quality survey in Germany. When AGQM adopted this task in 2000, the number of filling stations had more than doubled and spread all over Germany (Fig. 4).

Fig. 3. The AGQM logo.

The AGQM quality survey consists of a main check at the beginning of the winter period and up to four smaller regional surveys spread over the year. In Europe, winter-grade fuel, biodiesel as well as petrodiesel fuel, has to be available at the pump from November 15 on. Because complaints about winter operability are very common, it is important to fulfill this requirement. Biodiesel producers who are members of AGQM have to deliver winter-grade biodiesel beginning October 15, so that every filling station will have this grade available when winter begins.

At this point, 100 filling stations all over Germany are tested. Samples are taken by an independent laboratory and several quality parameters are checked,

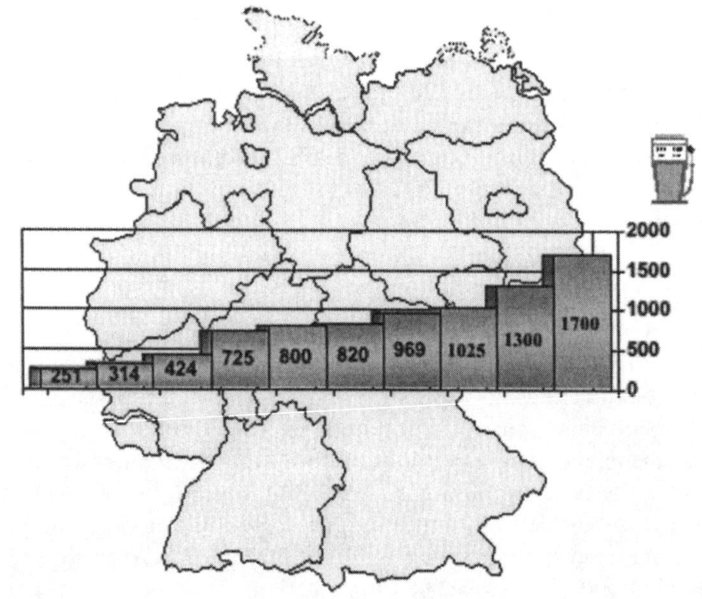

Fig. 4. Biodiesel filling stations in Germany.

i.e., Cold Filter Plugging Point (CFPP), water content, total acid number, and glycerol and glyceride content. The results give an overview of product quality and shed light on problems during storage and transport. In 2002, after implementation of the AGQM licensing system for filling stations, the surveys were limited to stations participating in this system.

Fuel Additives

An import issue for the use of biodiesel is the cold flow performance of the fuel. The parameter defined in the standard is the CFPP, which is improved by the use of middle distillate flow improvers (MDFI).

Products suitable for petrodiesel fuels did not function with biodiesel, so that additive suppliers had to develop new products. Complaints and failures showed that some of these different products were mutually incompatible. Some investigations carried out under control of the AGQM led to recommendations for the use of flow MDFI with the result that significant improvements could be observed during the last winter period.

Since November 2003, the European biodiesel standard includes a requirement for oxidative stability. Although most of the biodiesel producers using native vegetable oils do not have any problem fulfilling this requirement, practical experience showed that stability cannot be guaranteed at the pump. Transportation and storage stress the fuel. In cooperation with suppliers of stabilizing agents, AGQM initiated a test program to examine the effectiveness and performance of these additives, which included laboratory tests as well as long-term storage tests and a fleet test with taxi cabs to prove that the chosen product would not harm injection equipment.

Research and Development

The German government has never given significant support to biodiesel research. Most of the activities were initiated by UFOP and individual companies, without any active coordination. The task of AGQM is not really research and development, but problems previously not observed related to the use of biodiesel have occurred. These problems had to be investigated. For example, owners of filling stations observed destruction of concrete surfaces by biodiesel, the effectiveness of the waste water separation system had to be proven, and blocking of fuel filters had to be examined.

A large amount of work has also been conducted on exhaust emissions. Together with an automobile manufacturer and the German Government Agency for Renewable Resources, AGQM initiated the development of a fuel sensor which enables the engine management system to detect the fuel and optimize emission controls. Other projects funded by AGQM included emissions studies on biodiesel/petrodiesel blends and a comparison between biodiesel and Swedish MKI diesel fuel.

Summary

AGQM has proven to be an important tool of the German biodiesel industry, improving internal cooperation as well as contacts with authorities and important industrial partners. It has been able to coordinate the activities of individual companies and to evaluate synergies among them. It could also serve as a model for biodiesel industries in other countries.

8.3

Status of Biodiesel in Asia, the Americas, Australia, and South Africa

Werner Körbitz

Introduction

This section briefly discusses the status of biodiesel outside the United States and Europe. Countries with biodiesel activity that are discussed include Argentina, Brazil, Canada and Nicaragua in the Americas and China, India, Japan, Malaysia, the Philippines, South Korea and Thailand in Asia as well as Australia and South Africa.

The Americas

Argentina. Argentina is the world's largest exporter of oilseed meals and the third largest exporter of oilseeds as well as edible oils, mainly soybean and sunflower. Furthermore, it is ranked as the fourth largest oilseed producer. Consequently, there is an enormous potential for biodiesel production. Unfortunately, the country's current socioeconomic crisis is hindering investment decisions, which is the main barrier for any significant biodiesel development. There are seven existing biodiesel production units with capacities ranging from 10 to 50 t/d, and at least 11 projects, ranging from small-scale farmers' cooperatives to large-scale production with U.S. $30 million investment, are pending. At present, however, only one small-scale home-brewer unit is effectively producing.

Brazil. In May 2002, the PROBIODIESEL (Programa Brasileiro de Desenvolvimento Tecnológico de Biodiesel) program was announced, with the goal of setting up the regulatory framework for biodiesel development and production. The program is coordinated by the Ministry of Sciences and Technology, Secretariat of Technology and Enterprise Policy. In addition to the production of soy methyl ester (SME), the development of a soy ethyl ester (SEE) is under consideration because Brazil traditionally has a very large national production of bioethanol from sugar cane. Bioethanol has been used as a liquid biofuel for transportation purposes for many years. Both SME and SEE will be used as B-5 blends in fossil diesel. The vegetable industry (ABIOVE: Associação Brasileira de das Indústrias de Óleos Vegetais) will provide 80,000 L free of charge for performing fuel tests. Currently there are four companies able to start biodiesel production, but because commercialization has not yet been authorized, no dedicated biodiesel production plant is currently in operation.

There is one company producing biodiesel as a fuel additive for its product called "AEP 102," which is fossil diesel blended with bioethanol and 2% SME. A provisional biodiesel standard was developed through the Agência Nacional do Petróleo (ANP, National Petroleum Agency); see Appendix B. There is also some interest in using castor oil as biodiesel feedstock because castor grows in the semi-arid northeastern part of the country and may provide a source of income. Neat biodiesel from this feedstock would have high viscosity.

Canada. Canadian scientists at the University of Saskatoon (the home of "canola," the basic 00-rapeseed variety for the last 30 yr) were the organizers of the first Canadian biodiesel conference in March 1994. Biodiesel is not yet a commercial fuel product in Canada, but the recent foundation of the "Biodiesel Canada Association" indicates that commercial activities are in the planning phase and will be accelerated.

At present, biodiesel is fully taxed at the same level as fossil diesel fuel in Canada, with the exception of the province of Ontario in which the provincial road tax does not apply to biodiesel. Due to low taxation of petrodiesel, an exemption from the Canadian federal tax rate on biodiesel of CDN $0.04/L would not present a sufficient incentive, even when combined with a provincial incentive.

Biodiesel is registered as a fuel and fuel additive with the Canadian Environmental Protection Agency and meets the clean diesel standards as established by Environment Canada. Neat (100% or B-100) biodiesel has been designated as an alternative fuel by the required federal and provincial bodies.

Institutions supporting the development of the biodiesel industry include the following: (i) Biodiesel Association of Canada, which was founded in June 2003 by COPA (Canadian Oilseed Processors Association); (ii) the Canadian Renewable Fuels Association, which was founded in 1994, to promote renewable biofuels (bioethanol, biodiesel) for automotive transportation (http://www.greenfuels.org); and (iii) Natural Resources Canada, Office of Energy Efficiency, Transportation Energy Use Division (http://www.oee.nrcan.gc.ca).

Canada is known for its large rapeseed production ("canola"), but sunflower seed is also grown there. Today Canada is the world's fourth largest oilseed exporter. Currently, vegetable oils, but also recycled frying oils and animal fats, are used as feedstock sources.

For the past several years, there have been efforts in Canada to develop a Canadian General Standards Board specification for biodiesel fuels. These efforts have not yet resulted in a published specification; the European standard EN 14214 (the German standard DIN 51606 preceding that) and the US ASTM D 6751-02 FAME remain the orientation standards for quality management. The introduction of a quality assurance system seems to be indispensable for the market introduction of biodiesel in Canada as well.

In April 2001, a first large-scale demonstration plant went into operation. The technology was provided by the University of Toronto. In a quite challenging approach, it is planned to raise capacity in the following steps: 80 t/yr in 2001 to

880 t/yr in 2002 increasing to 158,000 t/yr in 2003 and finally to 480,000 t/yr in 2004.

For 1 y, 155 urban buses will run on biodiesel in downtown Montréal to gain practical experience in the use of biodiesel under real-life conditions, particularly in cold weather, and to demonstrate the feasibility of supplying biodiesel to a mass transit company (the STM). The project will also assess the economic and environmental effects of using this fuel, which is made from recycled sub-food-grade vegetable oil and animal fats (http://www.stcum.qc.ca). Further on-road tests are in progress in Saskatoon with two buses of the Saskatoon Transit Services running on B-5 (http://www.city.saskatoon.sk.ca/org/transit/biobus.asp). In September 2001, the fleet services of Toronto Hydro-Electric systems began a large-scale pilot project using biodiesel in ~80 fleet vehicles. By July 2002, the project was extended to the entire fleet of 400 vehicles (http://www.torontohydro.com/corporate/initiatives/green_fleet/index.cfm#biod).

With its large production of rapeseed but also sunflower, Canada has plenty of virgin vegetable oil available. The one potential key weakness that may impede its development in Canada is biodiesel's winter operability. Additives or special diesel blending fuels may be one solution but they will add cost to the product. Nevertheless, with the tax incentive introduced by the Ontario government, it is likely that one or two biodiesel plants could be commercially viable.

Nicaragua. In the early 1990s, a biodiesel plant with a production capacity of 3000 t/yr was established with the support of an Austrian development program. It is unique because the feedstock is produced from the locally grown bush *Jatropha curcas* or physic nut, which produces an oil highly suitable for biodiesel production according to the European standard EN 14214 (formerly the Austrian fuel standard ON C 1191 for FAME).

South Africa

A study was conducted with the objective of determining what influence biodiesel would have on the economy without affecting food production. It was concluded that biodiesel could replace 20% of imported diesel. Therefore, the government decided to grant a 30% fuel tax reduction for biodiesel. The South African Bureau of Standards drafted a biodiesel standard (see Appendix B) based on the European standard, with some changes such as iodine value and allowing other esters to be used. Feedstocks of interest include soybean oil and the oil of the physic nut (*J. curcas*).

Australia

Today, Australian biodiesel production is still in its infancy, but the production of liquid biofuels is receiving increased attention. The two main facts that can be held responsible for this trend are that petroleum imports account for more than half of total usage, making up the single largest component of the trade deficit, and that motorized transport turned out to be the most significant contributor to urban air pollution in Australia. Therefore a reduction in exhaust emissions from road transport is a key element of the air quality management strategies established by Commonwealth, State, and Territory governments. The federal government recently commissioned a "barriers to entry study" for biodiesel and bioethanol. The interim results appeared at the end of November 2002 with the study and its recommendations released at the end of 2003.

As announced on 30 May 2003, the new biodiesel taxation arrangements include the government applying excise on biodiesel, whether pure or blended, at the same rate as diesel fuel after 18 September 2003; providing domestic biodiesel producers and importers with a subsidy of AUS$ 0.38143/L until June 30, 2008, with a net effective excise rate for biodiesel of zero over this period; adjusting the net effective excise rate for biodiesel in a series of five even annual steps, commencing on July 1, 2008, and ending on July 1, 2012; setting a fuel standard for biodiesel, after which biodiesel will be listed as an eligible alternative fuel for on-road grants under the Energy Grants Credit Scheme; and setting a new excise rate that will apply to biodiesel beginning July 1, 2012. Under the existing excise and subsidy arrangements, the effective zero excise tax rate for already domestically produced bioethanol is extended until June 30, 2008.

In other regulatory measures, the Federal Environment Minister and the Agriculture Minister announced a two-year study with an AUS $5 million budget to address market barriers to the increased use of biofuels (mainly bioethanol and biodiesel) in transport. The study will develop a broad strategy to increase biofuel production to 350 million L/yr by 2010. The study will examine options for addressing market access difficulties including an assessment of the respective merits of nationally mandated minimum biofuel standards for transport fuels and voluntary arrangements. Existing biofuel manufacturers report that, even with an excise exemption, they are having difficulty accessing markets. Additionally there have been other regulatory disputes: Trade Measurement Regulation bodies at the State and Federal level have been refusing to approve the use of "diesel" dispenser pumps (bowsers) for biodiesel.

Current production originates from recycled frying oil (which would otherwise be exported to Asia for soap production) and animal fats (e.g., beef tallow). Various oilseeds are grown in Australia, and probably a multifeedstock blend of oils from rapeseed, sunflower, and soybean will be used for biodiesel production.

Until early 2003, no biodiesel standard was in place in Australia. Each new batch was tested at the small-scale production plants against variables, and monitored for storage stability (as part of a State Government-funded and operated Biodiesel Verification Trial). The next step was to develop a national fuel standard for biodiesel and, following Ministerial approval, the "National Standard for Biodiesel–Discussion Paper 6," which integrated the latest information and data material from the European EN 14214 and the US ASTM 6751-02, was released for public comment by the Department for the Environment and Heritage. The comment period closed on Friday, May 23, 2003, received inputs were integrated and in September 2003, the final Australian biodiesel standard was published (see Appendix B); some parameters are slightly less challenging than the European standard. Earlier, the Australian Greenhouse Organization had commissioned CSIRO with a study of life-cycle emissions and environmental benefits of biofuels (published in March 2000) to obtain information to assess their eligibility under the Diesel and Alternative Fuels Grants Scheme. Both studies provided the foundation for the development of a quality standard.

Reportedly, there are a few small-scale "backyard producers" and three larger production plants, two of which are commercial plants. Total capacity amounts to 48,000 t/yr. Six major potential producers, who plan to establish production capacities >40,000 t/yr, were identified. Additionally, there are another 10 potential smaller/niche market producers. The Biodiesel Association of Australia was founded in late 2000 to build biodiesel from a backyard industry to a standardized, viable addition to Australia's energy resources. A regular BAA newsletter is published.

Biodiesel is not yet available for general supply throughout Australia. Very limited trials are being executed, and there are three service stations that have been selling biodiesel. Currently, biodiesel has a pump price approximately the same as that of fossil diesel (AUS $0.90/L). One company is retailing biodiesel B100 at its Head Office site in Pooraka, South Australia; they will also retail B20 blend after the necessary excise lift has been implemented.

Estimations for 2003 biodiesel production capacity are estimated at ~40,000 t. The further development of Australia's biodiesel industry is highly dependent on the outcomes of the federal "barriers to entry" study. The federal government has set a target of 350 million L of biofuels (bioethanol and biodiesel) by 2012, and if the current six major potential producers implement their plans, total production will reach 350 million L as early as 2006. With an Australian biodiesel fuel standard in place, the engine manufacturers have recently provided statements of support and are starting to look at involvement in trials.

Asia

China. The transportation sector was left out of China's economic plans for many years, and the resulting lack of infrastructure is a major obstacle for the country's energy sector and overall economy. Nevertheless, China is one of the largest consumers of fossil diesel oil worldwide. About 60–70 million tons of fossil diesel oil are used every year, with approximately one third of it being imported to balance the market. The Chinese government emphasized its support for biofuels some time ago, but it seems that with the construction of the world's largest Bioethanol-production plant with ~600,000 t/yr in Changchun, Jilin Province, the development of this biofuel has a higher priority than biodiesel for the time being. In 1998, the Austrian Biofuels Institute completed a study together with the Centre for Renewable

Energy Development (CRED) in Beijing and the Scottish Agricultural College (SAC) within the INCO-programme of the European Union. This study evaluated the feedstock availability for biodiesel production from a variety of potential sources. There are no special regulations or tax exemptions for biodiesel in mainland China and Hong Kong, but following the example set for bioethanol, this may change in the near future.

Currently, rapeseed oil, cottonseed oil, and recycled frying oils are being investigated and used for trial biodiesel production. All existing biodiesel producers are producing according to the European standard EN 14214 (formerly the German DIN E 51 606 FAME standard) according to obtained reports. A vegetable fat chemistry plant that was started in June 2001 in Gushan has established a capacity or 10,000 t/yr as the first commercial biodiesel producer and plans to extend the capacity to 100,000 t/yr by the end of 2004. A petroleum company in Beijing plans to establish a capacity of 50,000 t/yr using the process technology of the Gushan company.

The Hong Kong Government has commissioned the University of Hong Kong to conduct a feasibility study of biodiesel as an automobile fuel in Hong Kong. The report was recently submitted to the government for consideration, and it is expected that price competitiveness will be the most decisive factor concerning the decision for promoting this fuel in Hong Kong and China.

The potential markets are expected to grow rapidly: the vehicle ownership rate in China is ~8.5 vehicles/1000 persons (equal to the level in the United States in 1912), and the number is projected to grow sixfold by 2020 (52 vehicles/1000 persons). Corresponding to these figures, transportation energy demand in China is projected to grow by 6.4%/year from 1999 to 2020, increasing its share of world energy use for transportation from 4.1% in 1999 to 9.1% in 2020. This indicates that China is expected to overtake Japan by 2005 and become the world's second largest consumer of transportation fuels.

India. At present, biodiesel initiatives are focused mainly on research, development, and demonstration projects. On September 12, 2002, the first biodiesel/bioethanol conference was held in New Delhi, and was sponsored by the Ministry of Rural Development and the Petroleum Conservation Research Association. Because it is a tropical country, India has a wide variety of domestic plants that produce oil-bearing seeds of sufficient volume potential, e.g., Sal (*Shorea robusta*), Neem (*Azadirachta indica*) and, specifically, the physic nut (*J. curcas*), to be considered as feedstock for biodiesel production. It is reported that in the near future, a 100 ton/d unit is expected to begin operation near Hyderabad in the state of Andhra Pradesh.

Japan. The city of Kyoto introduced biodiesel made from recycled frying oil into 220 garbage collection trucks in 1997, and has used the B-20 mixture for 81 city buses since 2000. If all of the edible-oil wastes are recycled and reused as biodiesel fuel (BDF), a market of ~30 billion yen will probably be created. To realize such a market, however, it is necessary to establish an integrated recycling system involving citizens, companies, and local administrations.

Malaysia. As reported at the International Biofuel Conference in 1998, which was organized by PORIM (Palm Oil Research Institute of Malaysia), initial trials in production at PORIM's pilot plant and in utilization as a fuel in diesel engines had promising results. This included very detailed tests in bus fleets, which were started by Daimler-Benz as early as 1987. Malaysia's mineral oil company, PETRONAS, is carefully watching and studying further developments in Europe, but has not yet acted publicly. With feedstock limitations in Europe, the export of palm oil may become an additional venue for Malaysia's palm oil industry.

Philippines. In 2001, the Philippine Coconut Authority announced the launch of a nation-wide program to develop the use of coconut oil biodiesel as an alternative fuel. Coconut oil contains 45–53% lauric acid, which is a saturated short-chain fatty acid (12:0) with a rather high level of oxygen of 14.9%. Although higher oxygen levels cause lower energy contents and therefore lower engine performance on the one hand, on the other hand, they give better combustion and therefore lower emission levels.

South Korea. Biodiesel is approved as an alternative fuel. It is expected that a tax exemption will be given in two to three years. Investigation showed that there are two small-scale biodiesel production plants with a total capacity of 8000 t/yr; one company has one large-scale plant (100,000 t/yr) under construction and intends to market a soy oil-based nonbranded biodiesel as a B-20 blend. Biodiesel was provided to vehicle fleets operated by several municipal governments for the test operation. The tests were completed in July 2004.

Thailand. Various mixes of unesterified vegetable (coconut and palm) oil blended with diesel oil or kerosene were introduced in the past year under the name of "biodiesel"; most of them did not meet official standards for commercial use. Other tests are in progress with "real" biodiesel from recycled cooking oils (called "super-biodiesel"), but the quality levels achieved cannot be reported as yet.

Suggested Readings

ABI (Austrian Biofuels Institute): Review on Commercial Biodiesel Production World-Wide, Study for the IEA-Bioenergy, Vienna, Austria, 1997.

ABI (Austrian Biofuels Institute): World-Wide Trends in Production and Marketing of Biodiesel, ALTENER: Seminar on New Markets for Biodiesel in Modern Common Rail Diesel Engines, University for Technology in Graz, Austria, 22 May 2000.

ADM (Archer Daniels Midland): Blending Agriculture into Energy-Economic Opportunity, presented at Saskatoon Inn, Saskatoon, Canada, January 2002.

Energy Information Administration: International Energy Outlook 2002. http://www.eia.doe.gov/oiaf/ieo.

Hopkinson, L., and S. Skinner, Civic Exchange, the Asia Foundation: Cleaner Vehicles and Fuels, the Way Forward, Hong Kong, August 2001.

Körbitz, W., New Trends in Developing Biodiesel, presented at the Asia Bio-Fuels Conference, Singapore, April 2002.

Levelton Engineering, $(S\&T)^2$ Consulting, Inc., Assessment of Biodiesel and Ethanol Diesel Blends, Greenhouse Gas Emissions, Exhaust Emissions and Policy Issues, Ottawa, Canada, 2002.

8.4

Environmental Implications of Biodiesel (Life-Cycle Assessment)

Sven O. Gärtner and Guido A. Reinhardt

Introduction

Biodiesel is generally considered to be environmentally friendly. At first glance, it is CO_2-neutral and biodegradable, preserves fossil fuels, and does not cause significant sulfur-containing emissions upon combustion. In some areas, such a characterization may well be valid, e.g., in the case of direct combustion, which yields exactly the amount of CO_2 that was removed from the atmosphere when cultivating the energy-yielding plants.

When the whole life cycle of biodiesel from production of the biomass *via* conversion to use as an energy source is analyzed, these are not necessarily natural advantages. For example, in the agricultural production of rape and sunflower seeds, two important feedstocks for biodiesel, fertilizers and biocides as well as tractor fuels, are consumed. The production of those resources, in turn, consumes significant amounts of fossil fuels, an observation that also holds in a restricted fashion for the production of soy, another important biodiesel feedstock. The use of fossil fuels has a connection with climate-affecting emissions, so that when considering the entire life cycle, the CO_2 balance is not neutral initially. CO_2 is included because of its effect on climate. However, because CO_2 is only one of several gases that affect climate, the question arises whether, through the presence of other climate-affecting gases, a positive CO_2 balance is diminished, neutralized, or even overcompensated. This is especially the case for dinitrogen oxide (N_2O; nitrous oxide; laughing gas), which arises from fertilizer production and agricultural ecosystems and which is not liberated in significant amounts in the production chain of fossil fuels.

Furthermore, in connection with the production of agricultural raw materials, environmental impacts such as contamination of ground and surface water with biocides (pesticides and herbicides) and their degradation products as well as nitrates and phosphates must be discussed. These environmental impacts do not occur in the case of fossil fuels. The use of areas otherwise left to nature but now used for the production of biomass must also be mentioned. Palm and coconut oils play a special role in this respect when deforestation of tropical rain forests occurs. Thus, fossil fuels can even have some positive environmental impacts compared with biodiesel.

These examples show that the ecological (dis)advantages of biodiesel cannot be evaluated *ad hoc* but rather must be determined very carefully and with consideration of the complete system, not simply certain segments. How to accomplish this and the results thereof will be detailed below.

How Are the Environmental Implications of Biodiesel Assessed?

Generally, numerous tools for evaluating environmental impacts exist, and these should provide answers to different questions. For describing the ecological (dis)advantages of biodiesel compared with conventional diesel fuel, life-cycle assessment (LCA), which was developed for comparing products and systems, is especially suitable. This evaluation tool now exists in the form of a standard [ISO 14040-43; (1)] and the first LCA for biodiesel using this standard have been carried out. Basically, LCA are conducted in four steps (see Fig. 1).

For the goal and scope definition, the objectives are precisely defined as are the frame of the investigation, the time, the spatial and physical boundaries, selected procedures, and many other facets. One of the most important elements is to define precisely the life-cycle boundary. Figure 2 depicts the system boundaries for rapeseed-based biodiesel (rapeseed oil methyl ester; RME) compared with conventional diesel fuel. All processes of the complete life cycle that have significance for the production of RME are considered ("cradle to grave" assessment). This begins with the manufacture of fertilizer and diesel fuel for the growing of rape and, *via* production, leads to the use of biodiesel and its effects. On a detailed level, these are hundreds of individual processes whose inputs from and outputs to the environment must be analyzed.

Thus, all co-products have to be taken into consideration. In LCA, they are considered in the form of so-called allocation procedures or through credits. In the allocation procedures, the environmental implications between a product and its co-products are split, whereas in the credit procedure the desired product receives a credit. The credit corresponds to the environmental implications of an equivalent product that can be used instead of the considered co-products, all having the same utility.

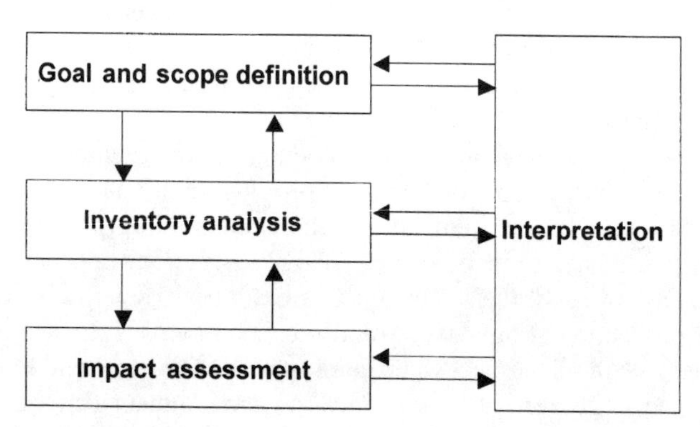

Fig. 1. Life-cycle assessment according to ISO 14040-43. *Source:* Reference 1.

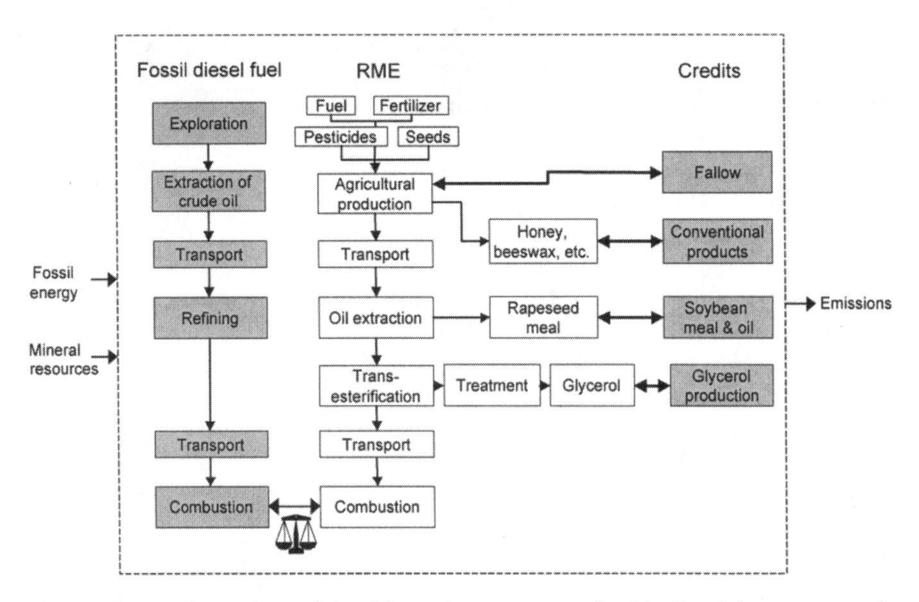

Fig. 2. System boundary of the life-cycle assessment for biodiesel from rapeseed. *Source:* Reference 3. RME, rapeseed oil methyl ester.

Therefore, the credit procedures describe reality much more closely than the distribution process. For the life cycle of rape (see Fig. 2) the following credits result:

1. Agrarian reference system: This system answers the question "What would happen to the rape-producing land, if there were no rape to cultivate?" Under European conditions, rape is replacing the rotating set-aside of agriculturally used land. The expenditures for its maintenance and emissions can be credited to rape.
2. Honey and other bee products: Rape, a blooming cultivar, is a good source of honey. Varying with conditions, the honey can replace different products such as foreign honey, preserves, or other spreads. Other bee products such as wax and *Gelee royale*, whose production can be credited to rape, replace materials produced elsewhere.
3. Animal feed: Rape meal, derived from the extraction of oil from the seed, is generally used as a high-value protein feed and can then replace soy meal. The expenditures and emissions from soy production are credited to rape when comparing these in terms of biodiesel production in Europe.
4. Glycerol: Glycerol is produced during the transesterification of rapeseed oil and other fats and oils. This can be used for numerous purposes and replaces synthetically produced glycerol. For more details on the uses of glycerol, see Chapter 11.

In the second step of the LCA, for each individual process, the expenditures (use of energy, equipment, and raw materials) and all of the environmental effects (e.g., emissions) are determined and compiled for the complete life-cycle comparison. In the third step of the LCA, the individual inventory parameters are merged into impact categories to obtain environmental impacts from the individual entities. For example, all climate-affecting gases are calculated in relation to CO_2 regarding their environmental impacts. Table 1 lists the corresponding equivalence factors for some selected impacts categories.

The detailed inventory of the LCA of biodiesel fuel, with RME serving as an example, is illustrated in Table 2 for finite energy sources, CO_2 equivalents, NO_x emissions, and SO_2 equivalents. All other LCA are calculated basically in the same fashion. First, the individual results for RME and diesel fuel are calculated, and then the ecological effect is assessed on balance. Negative values signalize results in favor of RME; positive values indicate results in favor of conventional diesel fuel. The energy balance indicates the potential to reduce consumption of finite energy sources by substituting RME for diesel fuel. Note that the energy consumed is balanced with respect to savings in finite resources. Therefore, in this assessment, the energy bound as biomass is not listed. The same applies to CO_2 fixed in the plant because only carbon that has been stored during the plant's growth is released. The comparison is based on the same amount of useful energy in both life cycles. The results for all precisely determinable parameters are given in Table 3. The values are given for the standard assessment "Rapeseed Biodiesel versus Ordinary Diesel Oil" for a modern diesel car (EURO 4 standard) in Europe.

In addition to quantitative parameters, however, numerous other parameters can be determined only qualitatively. In particular, it must be decided whether accident risks or environmental harm by mishandling should also be included in the assessment. For a few cases, there is no clear distinction possible even if the criteria and the system borders are precisely defined. For example, it is clearly a mishandling of a biocide if it is disposed of into surface water. What, however, if the weather unforeseeably changes after application of the biocide and it rains heavily? It is justified to some extent to consider such risks in an LCA. A list of all quantitative and qualitative environmental impacts of the substitution of diesel by RME based on topical knowledge will be given in the next segment.

Interpretation is the fourth and final step of the LCA (see Fig. 1). How can the environmentally relevant parameters be evaluated once they are determined? The ecological advantages and disadvantages of biodiesel compared with diesel fuel (see Table 3) show the difficulties in assessing the two types of fuel. RME has significantly better energy efficiency and less emission of SO_2, whereas conventional diesel fuel has a better NO_x and N_2O balance. Which is preferred? There have been several suggestions for handling this problem; in the interim, national and international committees for the standardization of LCA prefer models that integrate diverse quantifiable ecological parameters and a verbal, nonquantitative discussion of advantages and disadvantages in their final assessment. Such an evaluation method is discussed below.

TABLE 1
Selected Equivalence Factors for Some Impact Categories[a,b]

Effect category	Parameter	Substance	Formula	Factor
Resource demand	Cumulated finite energy demand	Crude oil, natural gas, coal, uranium ore...		
		Ores, limestone, clay...		
Greenhouse effect	CO_2 equivalents	Carbon dioxide	CO_2	1
		Dinitrogen oxide	N_2O	296
		Methane	CH_4	23
Ozone depletion	CFC 11 equivalents	CFC 11	CCl_3F	1
		CFC 12	CCl_2F_2	0.9
		CFC 22	$CClF_2H$	0.05
		Dinitrogen oxide	N_2O	NA
Acidification	SO_2 equivalents	Sulfur dioxide	SO_2	1
		Nitrogen oxides	NO_x	0.7
		Ammonia	NH_3	1.88
		Hydrochloric acid	HCl	0.88
Eutrophication	PO_4 equivalents	Nitrogen oxide	NO_x	0.13
		Ammonia	NH_3	0.346
Smog	C_2H_4 equivalents	Non-methane hydrocarbons	NMHC	0.5
		Methane	CH_4	0.007
Human and ecotoxicity		Diesel particulate		
		Carbon monoxide	CO	
		Dioxins (TCDD)		
		Dust		
		Benzene	C_6H_6	
		Benzo-a-pyrene		
		Hydrochloric acid	HCl	
		Ammonia	NH_3	

[a]*Sources:* References 3, 12, and 13
[b]Abbreviations: CFC, chlorofluorocarbon; TCDD, 2,3,7,8-tetrachlorodibenzo-*p*-dioxin.

Results

LCA of biofuels were carried out concurrently with the development of LCA methodology. In the late 1990s, the first more or less complete biodiesel LCA appeared in Europe on RME and in the United States on soybean methyl ester (SME). Soon LCA for biodiesel from sunflower and coconut oils followed. The results for RME are presented and discussed below as an example, and the various biodiesel fuels are then compared.

Table 4 lists all environmental (dis)advantages of RME vs. conventional diesel fuel. There are several advantages and disadvantages, so that it is not immediately clear which fuel is better when considering environmental aspects. Also, several of the listed aspects can be quantitated, whereas others can be described only qualitatively. When dealing with the quantifiable aspects, as described above, the final assessment is made on a verbal, nonquantitative level. The most commonly used

TABLE 2
Energetic Expenditures (Finite Energy) and Selected Emissions (CO_2 Equivalents, NO_x, SO_2 Equivalents) for Biodiesel from Rapeseed and Diesel Fuel[a,b]

Life-cycle step	Cumulated finite energy demand (MJ/kg)	CO_2 equivalents (g/kg)	NO_x (g/kg)	SO_2 equivalents (g/kg)
Plant production				
Harrowing	0.86	66	0.638	0.488
Plowing	0.66	50	0.486	0.372
Sowing	0.33	25	0.263	0.200
Harvest	0.67	51	0.482	0.369
Seeds	0.01	2	0.004	0.017
N fertilizer	7.19	1,124	2.303	4.216
P fertilizer	0.95	64	0.235	0.598
K fertilizer	0.31	20	0.034	0.032
Ca fertilizer	0.04	6	0.010	0.009
Biocides	0.33	15	0.019	0.059
Field emissions	0.00	619	0.000	11.079
Subtotal	11.36	2,042	4.474	17.441
Provision				
Bee keeping	0.32	29	0.064	0.121
Storage	1.36	98	0.066	0.186
Transport	0.42	32	0.417	0.313
Oil extraction	3.05	181	0.261	0.410
Hexane	0.16	2	0.003	0.007
Refinement	0.54	31	0.043	0.064
Fuller's earth	0.02	1	0.008	0.011
Phosphoric acid	0.01	1	0.003	0.013
Esterification	2.44	143	0.191	0.303
Methanol	4.81	352	0.136	0.347
Caustic soda	0.12	8	0.009	0.027
Glycerol treatment	0.24	14	0.019	0.026
Subtotal	13.49	893	1.219	1.830
Energetic use				
Transport	0.22	17	0.158	0.12
Use	0.00	216	10.190	7.316
Subtotal	0.22	233	10.348	7.437

(*Continued*) →

procedure is based on the evaluation of the so-called "specific contributions" and the "ecological relevance" of the selected environmental impact categories. This procedure was originally proposed in 1995 by the German Federal Environmental Agency and is applied widely at present.

The determination of the specific contributions is a way of measuring the importance of the individual ecological advantages and disadvantages relative to

TABLE 2
(Continued)

Life-cycle step	Cumulated finite energy demand (MJ/kg)	CO_2 equivalents (g/kg)	NO_x (g/kg)	SO_2 equivalents (g/kg)
Credits RME				
Reference system	–0.83	–67	–0.616	–0.485
Overseas honey	–0.24	–17	–0.059	–0.090
Honey by-products	–0.03	–2	–0.003	–0.006
Soybean meal (agric.)	–4.46	–318	–1.305	–1.485
Soybean meal (transp.)	–2.03	–162	–1.263	–1.697
Glycerol energy	–10.30	–758	–1.015	–4.421
Chlorine	–4.01	–275	–0.293	–0.918
Caustic soda	–2.68	–184	–0.197	–0.614
Propylene	–7.03	–188	–0.247	–0.751
Subtotal	–31.61	–1,971	–4.998	–10.467
Diesel fuel				
Provision	4.82	374	0.649	1.825
Use	42.96	3,392	10.190	8.101
Subtotal	47.78	3,766	10.839	9.925
RME minus diesel fuel	–54.32	–2,569	0.204	6.316

[a]*Source:* Reference 3.
[b]All values refer to 1 kg diesel fuel, i.e., 1 kg diesel fuel equivalents of RME, referred to the same amount of useful energy. Negative values are advantageous for biodiesel.

the overall situation, for example, in Europe or the United States. Here, the method is applied directly to the values of the quantified parameters in Table 3 (see Fig. 3). For a better graphical presentation, the specific contributions in Figure 3 refer to the so-called equivalent value per capita and to the average mileage of 1000 passenger cars. In this case, data for Germany were used. However, this does not mask the fact that the specific contributions listed here are *simply normative means*. The figures are rounded to avoid the pretense of extreme accuracy.

The results can be summarized as follows:

- RME displays a positive energy balance and climate gas balance, i.e., RME preserves fossil energy sources and helps avoid greenhouse gases.
- On the other hand, RME causes more emissions in the impact categories of acidification and eutrophication compared with conventional diesel fuel.
- No clear result is obtained concerning smog and ozone depletion. There is no global aggregation model yet for describing smog. Different models give different results. In the case of ozone depletion, no method for calculating chlorofluorocarbon (CFC)-equivalents that would unambiguously reflect the significance of N_2O for ozone depletion is known (see Table 1).

TABLE 3
Results for the Life-Cycle Assessment of Biodiesel from Rapeseed vs. Diesel Fuel for the Quantifiable Inventory Parameters and Impact Categories, Standard Utilization Options[a,b]

Inventory parameter	Units/ (ha·y)[c]	Rapeseed	Effect category	Units/ (ha·y)	Rapeseed
Crude oil	GJ	–53.9			
Natural gas	GJ	5.0	Cumulated energy demand[d]	GJ	–54.0
Mineral coal	GJ	–1.2			
Lignite	GJ	–1.8	Greenhouse effect		
Uranium ore	GJ	–2.2	(CO_2 equivalents)	t	–3.1
Limestone	kg	114			
Phosphate ore	kg	202	Acidification		
Sulfur	kg	14	(SO_2 equivalents)	kg	9.9
Potassium ore	kg	213			
Rock salt	kg	–297	Eutrophication		
Clay minerals	kg	9	(PO_4 equivalents)	kg	2.3
CO_2 (fossil)	t	–3.8			
CH_4	g	–255	Smog		
N_2O	kg	2.1	(C_2H_4 equivalents)	g	–37
SO_2	kg	–2.6			
CO	g	–185			
NO_x	g	–154			
NMHC	g	–85			
Diesel particulate	g	–25			
Dust	g	275			
HCl	g	–14			
NH_3	kg	6.71			
Formaldehyde	g	–1.62			
Benzene	g	–1.82			
Benzo(a)pyrene	µg	–241			
TCDD-eq.	ng	–29			

[a]*Source:* Reference 3.
[b]Abbreviations: NMHC, nonmethane hydrocarbon; TCDD, 2,3,7,8-tetrachlorodibenzo-*p*-dioxin.
[c]The unit (ha·y) indicates the saved energy and emissions or the additional amounts used or emitted when the amount of biodiesel produced per hectare and year replaces the corresponding amount of fuel in a vehicle engine. Positive numbers indicate a favorable result for the fossil fuel and negative values indicate a favorable result for biodiesel.
[d]Crude petroleum, natural gas, uranium ore, anthracite (hard coal) and lignite.

The results thus obtained are blended with "ecological relevance" and discussed. In other words, scientific methods cannot lead to a final result because values influence the recognition of ecological significance. These values can be of a personal or social nature. For example, the preservation of fossil energy sources and the greenhouse effect currently have the greatest political significance in Europe, so that these values justify a final assessment favoring biodiesel in this case. Such values are not scientifically irrefutable, i.e., a different set of values could lead to a different result. In this case, it is important that the pros and cons of these values be discussed clearly and that the decision-making process be transparent.

TABLE 4
Advantages and Disadvantages of Biodiesel from Rapeseed Compared with Conventional Diesel Fuel[a]

	Advantages for biodiesel from rapeseed	Disadvantages for biodiesel from rapeseed
Resource demand	Savings of finite energy resources	Consumption of mineral resources
Greenhouse effect	Lower emissions of greenhouse gases	
Ozone depletion		More N_2O emissions
Acidification		Higher acidification
Eutrophication		Higher NO_x emissions Risk: eutrophication of surface waters
Human- and eco-toxicity	Lower SO_2 emissions Lower diesel particle emissions in urban areas Less pollution of oceans due to extraction and transport of crude oil Risk: less pollution by oil spillage after accidents Risk: less toxicity/better biodegradability	Risk: pollution of surface waters by pesticides Risk: pollution of ground water by nitrate

[a]*Source:* Reference 3.

We now compare the biodiesel fuels derived from different vegetable oils. Figure 4 depicts the results for the energy and greenhouse gas balances from rape and canola, soy, sunflower, and coconut. The ranges for the two environmental impacts according to Quirin *et al.* (2) are shown; they arise from a comparison of the two corresponding LCA. For the determination of the range for RME, 10 international LCA from Europe (3–5) and Australia (6) were used: for SME, LCA from North America (7) and Australia (6); for biodiesel from sunflower oil, LCA from Europe (4,8,9); for canola, LCA from the United States (10) and Australia (6); and for biodiesel from coconut oil, an LCA from the Philippines (11).

The numbers in the figure reflect the complete results of the life-cycle comparisons, i.e., biodiesel minus diesel. Area was selected as a reference because it became obvious that the global area for planting biomass is one of the greatest bottlenecks. The ranges reflect different yields in different areas, different uses of co-products (for example, rape meal from rapeseed oil production), or different uncertainties in the base data.

The comparison of all biodiesel fuels showed the following:

- All biodiesel fuels possess positive energy and climate gas balances, i.e., all biodiesel fuels preserve fossil energy sources compared with conventional diesel regardless of the source of the vegetable oils and help avoid greenhouse gases.

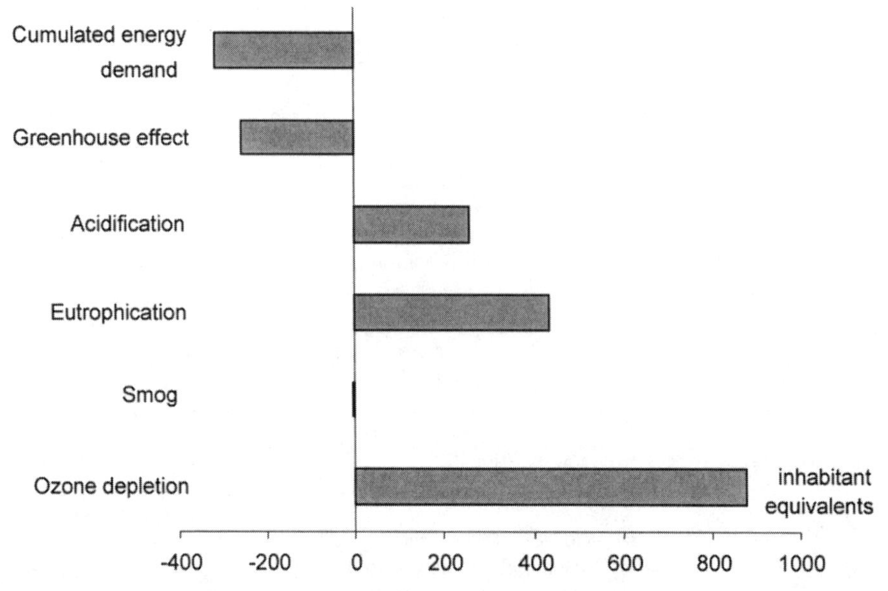

Fig. 3. Environmental impacts for the life-cycle assessment of biodiesel from rapeseed vs. diesel fuel, standard utilization options. Negative values are advantages for biodiesel. *Source:* Reference 3.

- The savings effect is greatest for rape and sunflower, followed by canola, whereas SME is at the lower end. However, not all oil crops can be planted in all countries. In those areas in which climate conditions permit the planting of several oil crops, the more efficient oil crops could be selected.

The conclusions reached here from energy and climate gas balances can, in our view, generally be applied to other impact categories, and therefore the whole LCA, if the preservation of fossil resources and the greenhouse effect are seen as having great environmental significance. If acidification or eutrophication is assigned more importance, the result may well be the opposite. It also should be noted that in some individual cases, as a result of certain local or regional conditions (especially fertile soil, climate, infrastructure), the result of an LCA may well be very different. Averaged results that may be helpful for the political decision-maker or the consumer are discussed here.

Conclusions

Biodiesel has advantages and disadvantages compared with fossil diesel fuel. The advantages that can be described quantitatively are the preservation of nonrenewable fossil resources and the diminished greenhouse effect; the ecological disadvantages are eutrophication and acidification. For ozone depletion, smog, and

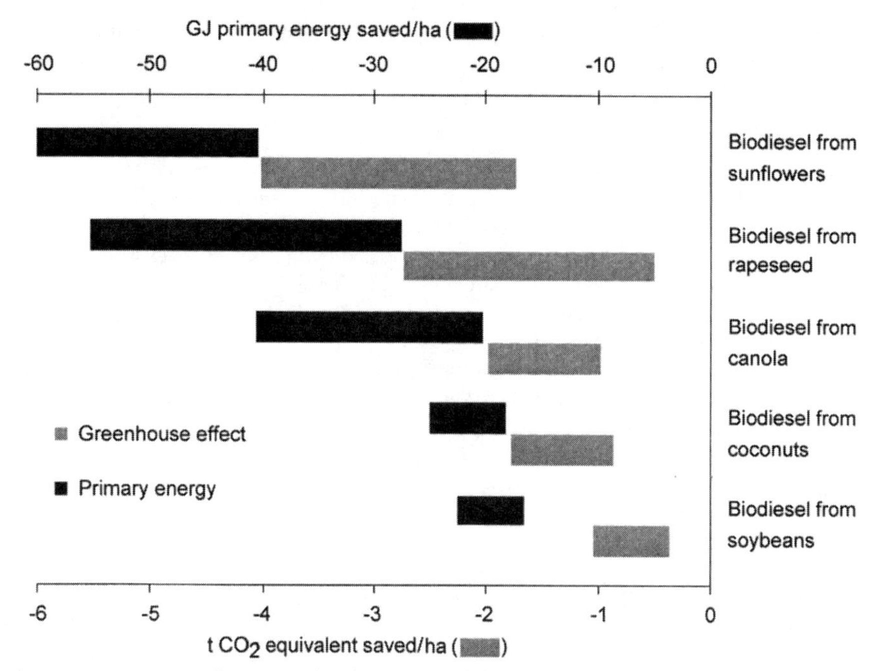

Fig. 4. Environmental impacts for the biodiesel from different plant oil vs. conventional diesel fuel. The negative values are advantages for biodiesel. *Source:* Reference 2.

human toxicity, no scientifically clear result can be obtained. Beyond these aspects, several other ecological (dis)advantages of a qualitative nature exist (see Table 4). Thus, an objective decision for or against one of the fuels is not possible; however, a decision can be made using a subjective set of values.

If preservation of finite energy sources and the reduction of climate-affecting gases has political priority, then biodiesel is superior to its fossil fuel counterpart. Other value systems can lead to other results, so that it should be possible to document reproducibly the set of values and the verbal arguments.

The results from the comparison of biodiesel fuels from different sources are similar. All biodiesel fuels, whether they are derived from rape, soy, sunflower, or tropical palm, preserve fossil energy sources, reduce greenhouse gases, and are disadvantageous with respect to acidification, albeit in different amounts. For a direct comparison, again a set of values must be used. If preservation of finite energy sources and reduction of climate-affecting gases enjoy political priority, then rape and sunflower methyl esters clearly have higher potential than, for example, soy. It must be noted that not all oil crops can be produced in all countries or regions due to climatic reasons. Therefore, a political set of values as well as the predominant climate and infrastructure will determine the source of biodiesel.

In summary, LCA is an important tool for describing environmental (dis)advantages and therefore serves in decision-making. Usually, it would have to be complemented by a socioeconomic assessment to ensure complete and sustainable long-term production of biodiesel.

References

1. International Standardization Organization, ISO 14040-43. Environmental Management—Life Cycle Assessment—Principles and Framework. German and English version by Deutsches Institut für Normung (DIN, ed.), Beuth Verlag, Berlin, 1997–2000.
2. Quirin, M., S.O. Gärtner, U. Höpfner, M. Pehnt, and G.A. Reinhardt, CO_2-neutrale Wege zukünftiger Mobilität durch Biokraftstoffe: eine Bestandsaufnahme [CO_2 Mitigation Through Biofuels in the Transport Sector: Status and Perspectives], Study commissioned by the FVV, Frankfurt, Germany, 2004.
3. Gärtner, S.O., and G.A. Reinhardt, Erweiterung der Ökobilanz von RME [Life Cycle Assessment of Biodiesel: Update and New Aspects], Project commissioned by the UFOP (Union for the Promotion of Oil and Protein Plants), 2003.
4. Final report for the European Commission: IFEU Heidelberg (ed.), Bioenergy for Europe: Which Ones Fit Best? A Comparative Analysis for the Community. Under support of DG XII, in collaboration with BLT, CLM, CRES, CTI, FAT, INRA, and TUD, Heidelberg, 2000.
5. Ceuterick, D., and C. Spirinckx, Comparative LCA of Biodiesel and Fossil Diesel Fuel, VITO (Vlaamse Instelling voor Technologisch Onderzoek), Mol, Belgium, 1999.
6. Beer, T., G. Morgan, J. Lepszewicz, P. Anyon, J. Edwards, P. Nelson, H. Watson, and D. Williams, Comparison of Transport Fuels. Life-Cycle Emission Analysis of Alternative Fuels for Heavy Vehicles, CSIRO (Commonwealth Scientific and Industrial Research Organization), Australia, 2001.
7. Sheehan, J., V. Camobreco, J. Duffield, M. Graboski, and H. Shapouri, Life Cycle Inventory of Biodiesel and Petroleum Diesel for Use in an Urban Bus, NREL/SR-580-24089, National Renewable Energy Laboratory, Golden, CO, 1998.
8. EUCAR, CONCAWE, and JRC, Well-to-Wheels Analysis of Future Automotive Fuels and Powertrains in the European Context, 2003. Downloadable at http://ies.jrc.cec.eu.int/Download/eh/31.
9. Direction of Agriculture and Bioenergies of the French Environmental and Energy Management Agency (ADEME) and French Direction of the Energy and Mineral Resources (DIREM), Bilans énergétiques et gaz à effet de serre des filières de production de biocarburants en France [Energy Balance and Greenhouse Gases of the Life Cycle of Biofuel Production in France], 2002.
10. Levelton Engineering Ltd. (S&T)^2 Consulting Inc., Assessment of Biodiesel and Ethanol Diesel Blends, Greenhouse Gas Emissions, Exhaust Emissions, and Policy Issues, Ottawa, Canada, 2002.
11. Tan, R.R., A.B. Culaba, and M.R.I. Purvis, Carbon Balance Implications of Coconut Biodiesel Utilization in the Philippine Automotive Transport Sector, *Biomass Bioenergy 26:* 579–585 (2002).
12. Heijungs, R., J.B. Guinée, G. Huppes, R.M. Lamkreijer, H.A. Udo de Haes, A. Wegener Sleeswijk, A.M.M. Ansems, P.G. Eggels, R. van Duin, and H.P. de Goede, Environmental Life Cycle Assessment of Products. Guide (Part 1) and Backgrounds (Part 2), prepared by CML, TNO and B&G, Leiden, NL.
13. International Panel on Climate Change, Climate Change 2001—Third Assessment Report. Cambridge, UK, 2001.

8.5

Potential Production of Biodiesel

Charles L. Peterson

Introduction

At the present time, the United States is almost totally dependent upon petroleum for liquid energy. One report stated that 71.5% of U.S. energy comes from oil and natural gas, whereas only 2% comes from biomass. Currently, the United States uses ~19.76 million bbl (1 bbl = 42 gal; 42 gal = 159 L) of petroleum/d (0.83 billion gal/d), 25.5% of total world consumption. In 2002, 58% of the petroleum was imported. For agriculturally produced renewable fuels, such as biodiesel, to make a significant contribution to this mammoth energy consumption will require every foreseeable agriculturally produced energy source that can be developed. This chapter reviews the status and contribution of only one of these fuels, i.e., biodiesel from vegetable oils and animal fats. Others, such as ethanol, are important components of the biomass complex.

Biodiesel

Biodiesel can be thought of as a solar collector that operates on CO_2 and water through the process of photosynthesis. The photosynthesis process captures the energy from sunlight to produce a hydrocarbon, i.e., vegetable oil. CO_2 is used by the plant in the creation of the organic material, and then the CO_2 is released in the combustion process when the fuel is used by a diesel engine. Photosynthesis is carried out by many different organisms, ranging from plants to bacteria. Energy for the process is provided by light, which is absorbed by pigments such as chlorophylls and carotenoids. Thus, through the process of photosynthesis, the energy of sunlight is converted into a liquid fuel that with some additional processing can be used to power a diesel engine. The photosynthesis process requires one major element—land. The crop must be planted over a wide area; to be economically feasible, it must compete advantageously with other crops that the landowner might choose to plant.

Vegetable oils have the potential to serve as a substitute for petroleum diesel fuel. Of the >350 known oil-bearing crops, those with the greatest production potential are sunflower, safflower, soybean, cottonseed, rapeseed, canola, corn, and peanut oil. Modifying these oils to produce the methyl or ethyl esters was shown to be essential for successful engine operation over the long term. Development of vegetable oil as an alternative fuel would make it possible to provide energy for agriculture from renewable sources located in the area close to where it could be used.

According to the 2002 Census of Agriculture (1), the harvested cropland in the United States consisted of 363.3 million acres (includes cropland used as pasture; ~147.1 million hectares) with some additional land idling, summer fallowed, or having crop failures (Table 1). If one crop, rapeseed, was considered on every acre of the available harvested land at a production rate of 1 ton/acre equivalent to 100 gal/acre of oil and ~1200 lb/acre (1345 kg/ha; 1 lb/acre = 1.121 kg/ha) of meal, 36.3 billion gal (137.4 billion L) of oil would be produced. Another 3.7 billion gal (14 billion L) could be produced on the idle land. In 2002, 72.4 million acres (29.3 million ha) of U.S. cropland, 16.7% of all cropland, was planted in soybeans. The average yield for soybeans is 38 bushels per acre with an oil content of 1.4 gal/bushel or 53.2 gal (201.4 L) per acre (0.405 ha). Thus, if all land were planted in soybeans, the United States could produce 23 billion gal (87 billion L; ~68.9 million t) of soybean oil. These calculations do not take into account the methyl or ethyl alcohol required in the transesterification process (~10% on a volume basis of the vegetable oil produced). This estimate of the maximum vegetable oil production is equivalent to 1.21 times the current annual consumption of petrodiesel used for on-highway transportation for rapeseed and 0.70 times for soybean oil.

Computations of the land that could realistically be used for vegetable oil production are complicated. Certainly land must be available for domestic food production. It is logical to assume that some production of food for export will continue to be required. It is also reasonable that crop rotations will mean that only a portion of the land would be in vegetable oil production in any one year. The idle cropland reported in 2002 could produce 3.7 billion gal of vegetable oil per year or 11% of the diesel used in transportation. In an earlier report (2), the author made an estimate of additional cropland potentially available for vegetable oil production by comparing crop production for several of the major crops with domestic use. Any production over domestic use was termed excess and, using the national average production for that crop, an estimate of excess crop production land of 62 million acres was calculated. This land could produce an additional 6.2 billion gal of vegetable oil or yield an additional 18.7% of on-highway diesel fuel for consumption at the expense of foreign exports of the commodities currently produced on that land.

A discussion of the potential production of biodiesel will consider four questions: how much petrodiesel is used, the quantity of fats and oils currently produced, how these fats and oils are currently used, and how much used oil is available for biodiesel production.

Use of Petrodiesel. As shown in Table 2 for the year 2000, total use of oil and kerosene in the United States amounted to 57.1 billion gal (216.1 billion L).

Current production of Edible Fats and Oils in the United States. In 1997, the United States produced 29,985 million pounds (~3.945 billion gal = 14.93 billion L; ~13.4 million t) of edible fats and oils. Of this, 70% was from soybeans, 10% from corn, 10% from lard and tallow, 3% from cottonseed, 1% from peanuts, 2%

TABLE 1
Major Crops and Harvested Acres for 2002 in the United States[a]

Crop	Acres	Ha	Crop	Acres	Ha
Barley	4,015,654	1,625,340	Tobacco	428,631	173,596
Corn	74,914,518	30,340,380	Field and grass seed	1,422,133	575,964
Oats	1,996,916	808,751	Forage crops	64,041,337	25,936,741
Popcorn	309,879	125,501	Dry edible beans	1,691,775	685,169
Millet	282,664	114,479	Dry edible peas	281,871	88,642
Rice	3,197,641	1,295,045	Lentils	198,997	80,594
Rye	285,366	115,573	Potatoes	1,266,087	512,765
Sorghum	7,161,357	2,900,350	Sweet potatoes	92,310	37,386
Wheat	45,519,976	18,435,590	Hops	29,309	11,870
Canola	1,208,251	489,342	Mint	108,798	44,063
Flaxseed	641,288	259,721	Pineapple	10,211	4,135
Peanuts	1,223,093	495,352	Berries	206,034	83,444
Safflower	182,292	73,828	Vegetables	3,433,269	1,390,474
Soybeans	72,399,844	29,321,937	Orchards	5,330,439	2,158,828
Sunflower for oil	1,500,828	607,835	Pasture	60,557,805	24,525,911
Sunflower nonoil	332,607	134,705	Failed crops	17,069,564	6,913,173
Cotton	12,456,162	5,044,746	Summer fallow	16,559,229	6,706,487
Sugarbeets	1,365,769	553,136	Idle land	37,281,096	15,098,843
Sugarcane	978,393	396,249			

[a] *Source:* Reference 1. The total harvested cropland comprises 363.3 million acres (147.1 million ha).

TABLE 2
Annual Sales of Diesel Fuel in the Year 2000 in the United States Only

Use	(gal × 10^9)	(L × 10^9)
On-highway diesel	33.13	125.4
Off-highway	2.8	10.6
Farm	3.1	11.7
Electric power	1.13	4.3
Military	0.23	0.87
Railroad	3.0	11.4
Heating (residential, commercial, industrial)	11.5	43.5
Total fuel oil and kerosene	57.1	216.1

from canola, 0.3% from safflower, and 2% from sunflower. The United States imported 3630 million pounds (0.48 billion gal = 1.82 billion L) and exported 6040 million pounds (0.79 billion gal = 2.99 billion L) of edible fats and oils. Current vegetable oil production is equivalent to 1.26 times the on-farm use of petrodiesel, ~12% of on-highway diesel use, or ~7% of total fuel oil and kerosene. Table 3 contains an estimate of the annual production of fats and oils.

Current Use of Edible Fats and Oils. For the 1999–2000 recordkeeping year, 6450 million pounds (2925 million kg) were used for baking or frying fats, 1727 million pounds (783 million kg) for margarine, 8939 million pounds (4055 million kg) for salad or cooking oil, and 436 million pounds (198 million kg) for other edible uses totaling 17,551 million pounds (7960 million kg; 2.3 billion gal) for edible food products. In the year 2000, the United States also used 1896 million pounds (860 million kg) for industrial fatty acid production, 3253 million pounds (1475 million kg) for animal feeds, 366 million pounds (166 million kg) for soap production, 100 million pounds for paint and varnish (45 million kg), 138 million pounds (63 million kg) for resin and plastics, 120 million pounds (54 million kg) for lubricants and similar oils, and 471 million pounds (214 million kg) for other industrial

TABLE 3
Total Annual Production of U.S. Fats and Oils[a]

	Vegetable oil production			Animal fats	
	(gal × 10^9)	(L × 10^9)		(gal × 10^9)	(L × 10^9)
Soybean	2.44	9.24	Inedible tallow	0.51	1.93
Peanuts	0.29	1.10	Lard and grease	0.17	0.64
Sunflower	0.13	0.49	Yellow grease	0.35	1.32
Cottonseed	0.13	0.49	Poultry fat	0.30	1.14
Corn	0.32	1.21	Edible tallow	0.21	0.79
Others	0.09	0.34			
Total vegetable oil	3.15	11.92	Total animal fat	1.55	5.87

[a]*Source:* Reference 3.

uses. This industrial use totaled 6344 million pounds (2877 million kg; 0.834 billion gal). Total edible and industrial use is 3.134 billion gal or 79% of total production. Exports account for another 8%, which leaves 0.5 billion gal unaccounted for.

A recent article from the USDA suggests that the soybean oil surplus is ~1 billion lb (454 million kg; 0.133 billion gal). This might be considered an approximation of the amount of feedstock readily available for biodiesel use.

The Potential of "Used Oil" for Biodiesel Production. Estimates of potential production often ignore "double counting." It should be recognized that all used oils started out as new oil so that the production figures must be reduced by the amount of used oil applied. For example, yellow grease was originally produced as vegetable oil or animal fat. We should not include both in our estimates. It was estimated (3) that 0.35 billion gal of yellow grease and 1.2 billion gal/yr of other animal fats are collected in the United States each year. Yellow grease is used oil from fast-food restaurants, delicatessens, and similar sources.

Estimates from the National Renewable Energy Laboratory (NREL) (4) are that ~9 lb/person/yr (1.16 gal/person/yr = 4.4 L/person/yr) of used oil and 13 lb/person/yr (1.69 gal/person/yr = 6.4 L/person/yr) of trap greases and similar oils are produced each year. The U.S. population estimated by the Census Bureau is 293,444,408 on April 8, 2004. Using the 1.16 gal/person/yr estimate of NREL this would suggest 0.34 billion gal (1.29 billion L) of used oil per year and 0.5 billion gal (1.89 billion L) of trap greases. The Jacobsen Fats and Oils Bulletin lists 40 companies dealing in yellow grease. The collecting and recycling of used oils is a highly competitive business. An analysis of one of these companies reported that the most competitive part of their business was obtaining the product. Yellow grease is used in the manufacture of products such as soap, textiles, cleansing creams, inks, glues, solvents, clothing, paint thinner, rubber, lubricants, and detergents. Its principal use is as a livestock feed additive. It makes the feed less dusty and adds lubrication to the feed, thereby reducing wear on milling machinery. It is a dense source of energy, which is important for animals such as cattle and horses that have a hard time eating more than they already do.

It is likely that many of the current applications of used oil will continue to take precedence for a major portion of these oils. Mad cow disease has reduced the use of some used oil products for feed, opening an opportunity for more to be used as biodiesel feedstock. If 25–30% of used oil were available for biodiesel, this would be ~100 million gal (~380 million L)/yr. The use of trap greases for biodiesel may add to that total, but they will require further development before generally use because of the potential for contamination with chemicals, pesticides, sewage components, water, and their high free fatty acid content.

Improving Production Potential

Additional acreage, improved varieties, and the use of idle cropland could all increase total vegetable oil production. Each of these methods for improving production has specific challenges to overcome. To have additional acreage for fats and oils, the price must be competitive with the displaced crop. Improved varieties require time and money for research. Idle cropland must be made available for crop production. This idle land is generally low in productive capacity and erodible, i.e., land that is potentially lower yielding.

Plant scientists at the University of Idaho have developed yellow mustard varieties that have the potential to significantly reduce the cost of the oil used in biodiesel production. This reduced cost of the oil is made possible by producing cultivars with specific properties remaining in the meal after the oil is removed. One of the potential uses for the meal is as a soil fumigant to replace chemicals currently in use today such as methyl bromide, which will soon be removed from the market due to its toxicity.

Biodiesel has the potential to be a very large agriculturally produced commodity. However, biodiesel can never displace a significant portion of petrodiesel because of the limited capacity for producing vegetable oil and because there are more important food uses for the major portion of our edible fats and oils.

World Vegetable Oil Production. The production of vegetable oil in the entire world is estimated at 26.9 billion gal (80.64 million t; Table 4). The world production of vegetable oil is equivalent to 81% of U.S. on-highway petrodiesel use or 47% of the total fuel oil and kerosene use of the United States in a year. It would require more than the entire world production of these vegetable oils to replace the U.S. on-highway petrodiesel use.

Potential Production of Biodiesel. It would be very ambitious to produce the amount of petrodiesel used on the farm. That would require all of the vegetable oil currently produced in the United States and ~15% of total production land area. It would be very ambitious to have a 0.5 billion gal (1.9 billion L)/yr biodiesel industry. This would be only 1.5% of U.S. on-highway petrodiesel or <1% of total fuel

TABLE 4
World Vegetable Oil Production: 2002/2003

	(gal × 10^9)	(L × 10^9)	(t × 10^6)
Soybean	8.8	33.31	6.38
Palm	7.4	28.01	25.21
Sunflower	2.4	9.08	8.17
Rapeseed	3.3	12.49	11.24
Cottonseed	1.0	3.79	3.41
Peanut	1.3	4.92	4.43
Coconut	0.95	3.60	3.24
Olive	0.69	2.61	2.35
Palm kernel	0.93	3.52	3.17
Total	26.9	101.82	80.64

oil and kerosene use. A 0.5 billion gal/yr industry would require all of the surplus vegetable oil (0.13 billion gal), half of the used oil (0.17 billion gal), and all of the oil that could be produced on the 37 million acres (13.77 million ha) of idle cropland (~0.3 billion gal) or the equivalent by displacing current crops. It is apparent that a challenge for biodiesel production will occur at ~0.2–0.3 billion gal (0.76–1.14 billion L) when the acquisition of additional feedstocks will become very difficult.

The other side of this argument is that a 0.2–0.5 billion gal (0.76–1.9 billion L) biodiesel industry would have a very significant beneficial effect on agriculture and rural communities. The industry would have to grow by 10–25 times its current size. It would provide an outlet for surplus vegetable oil crops, and land currently being used to produce surplus crops could be switched to vegetable oil to provide additional feedstock for biodiesel.

The challenge for this sustained growth is economics. Most of the biodiesel produced in 2003 was subsidized by the Commodity Credit Corporation to 11 companies producing a total of 18.5 million gal (1.9 billion L) at the rate of ~$1.03/gal. To subsidize a 0.5 billion gal industry at the same level would require $500 million or 3.5 times the current legislated limit and which is also shared with ethanol production.

Conclusions

Vegetable oil has potential as an alternative energy source. However, vegetable oil alone will not solve the dependence on foreign oil. Use of this and other alternative energy sources could contribute to a more stable supply of energy. Major production centers have not been developed; however, the number of plants is expanding and many additional ones are under study.

Economically, these fuels compare marginally with traditional petroleum resources; public policy must be revised to encourage development. The state and federal governments have made strides in that direction but much more will be required if vegetable oils are to achieve their potential. Increased vegetable oil production would require a significant commitment of resources. Land for production would have to be contracted, crushing and esterification plants would be required, distribution and storage facilities constructed, and monitoring of users for detection of problems in large-scale use are all needed to encourage development of the industry.

In addition to the oil produced, a vegetable oil crop such as winter rape also produces considerable biomass. It was estimated that a 2000 lb/acre (2242 kg/ha) crop of winter rape produces 100 gal/acre (153 L/ha) of oil, 1250 lb/acre of meal and 5000 lb/acre of biomass normally left on the field at harvest. It was estimated that the energy equivalent of these by-products was 350 gal/acre of petrodiesel fuel, equivalent to 8.33 bbl/acre (1 bbl = 42 gal). The meal can also be used as a high-protein livestock feed. However, if there were a major shift of land into production of vegetable oil crops for energy, these by-products would likely be used for direct combustion or for production of ethanol. Utilization of the entire crop leads to the concept of a complete "energy" crop. Agricultural policy makers must seriously consider ways to encourage the development of these energy crops.

The magnitude of our energy needs provides an inexhaustible market for our total agricultural production capacity at the highest possible level. We could put the farm back to work providing for our food needs and also growing crops and livestock for energy. Energy is the only crop that could never grow in surplus.

References

1. U.S. Department of Agriculture, National Agricultural Statistics Service, 2002 Census of Agriculture; http://www.nass.usda.gov/census/census02/volume1/USVolume104.pdf.
2. Peterson, C.L., M.E. Casada, L.M. Safley Jr., and J. D. Broder, Potential Production of Agricultural Produced Fuels, *App. Eng. Agric. 11:* 767-772.
3. Pearl, G.G., Animal Fat Potential for Bioenergy Use, *Proceedings of Bioenergy 2002*, The 10th Biennial Bioenergy Conference, Boise, ID, Sept. 22–26, 2002.
4. Wiltsee, G., Urban Waste Grease Resource Assessment. NREL/SR-570-26141. National Renewable Energy Laboratory, Golden, CO, 1998.

9

Other Uses of Biodiesel

Gerhard Knothe

Although this book generally deals with the use of biodiesel in surface transportation, esters of vegetable oils possess other applications that are outlined briefly here. Biodiesel has other fuel or fuel-related applications that are noted briefly here.

Cetane improvers based on fatty compounds were reported. The use of nitrate esters of fatty acids in diesel fuel was reported in a patent (1). Multifunctional additives consisting of nitrated fatty esters for improving combustion and lubricity were reported (2,3). Glycol nitrates of acids of chain lengths C_6, C_8, C_{14}, C_{16}, and C_{18} (oleic acid) were also prepared and tested as cetane improvers (4) with C_6–C_{14} glycol nitrates demonstrating better cetane-improving performance due to a balance of carbon numbers and nitrate groups. These compounds (2–4) are more stable and less volatile than ethylhexyl nitrate (EHN), the most common commercial cetane improver, and their cetane-enhancing capability is up to 60% of that of EHN.

Biodiesel can be used as a heating oil (5). In Italy, the esters of vegetable oils serve as heating oil instead of diesel fuel (6). A European standard, EN 14213, was established for this purpose. The specifications of EN 14213 are given in Table B-2 in Appendix B. A salient project in this regard has been the use of biodiesel as the heating oil for the Reichstag building in Berlin, Germany (7).

Another suggested use of biodiesel as fuel has been in aviation (8,9). A major problem associated with this use concerns the low-temperature properties of biodiesel, which make it feasible for use only in lower-flying aircraft (8).

In addition to serving as fuels, the esters of vegetable oils and animal fats can be utilized for numerous other purposes. The methyl esters can serve as intermediates in the production of fatty alcohols from vegetable oils (10). Fatty alcohols are used in surfactants and cleaning supplies. Branched esters of fatty acids are used as lubricants, with their improved biodegradability making them attractive, given environmental considerations (11). Vegetable oil esters also possess good solvent properties. This is shown by their use as a medium for cleaning beaches contaminated with crude oil (petroleum) (12–14). The high flash point, low volatile organic compounds, and benign environmental properties of compounds of methyl soyate make it attractive as a cleaning agent (15). The solvent strength of methyl soyate is also demonstrated by its high Kauri-Butanol value (relating to the solvent power of hydrocarbon solvents), which makes it similar or superior to many conventional organic solvents. Methyl esters of rapeseed oil were suggested as plasticizers in the production of plastics (16) and as high-boiling absorbents for the cleaning of gaseous industrial emissions

(17). Hydrogenated fatty alkyl esters mixed with paraffin-based or carboxylic acid-based candle fuels provide improved combustion performance (18).

References

1. Poirier, M.-A., D.E. Steere, and J.A. Krogh, Cetane Improver Compositions Comprising Nitrated Fatty Acid Derivatives, U.S. Patent 5,454,842 (1995).
2. Suppes, G.J., M. Goff, M.L. Burkhart, K. Bockwinkel, M.H. Mason, J.B. Botts, and J.A. Heppert, Multifunctional Diesel Fuel Additives from Triglycerides, *Energy Fuels 15:* 151–157 (2001).
3. Suppes, G.J., and M.A. Dasari, Synthesis and Evaluation of Alkyl Nitrates from Triglycerides as Cetane Improvers, *Ind. Eng. Chem. 42:* 5042–5053 (2003).
4. Suppes, G.J., Z. Chen, Y. Rui, M. Mason, and J.A. Heppert, Synthesis and Cetane Improver Performance of Fatty Acid Glycol Nitrates, *Fuel 78:* 73–81 (1999).
5. Mushrush, G., E.J. Beal, G. Spencer, J.H. Wynne, C.L. Lloyd, J.M. Hughes, C.L. Walls, and D.R. Hardy, An Environmentally Benign Soybean Derived Fuel as a Blending Stock or Replacement for Home Heating Oil, *J. Environ. Sci. Health A36:* 613–622 (2001).
6. Staat, F., and E. Vallet, Vegetable Oil Methyl Ester as a Diesel Substitute, *Chem. Ind.:* 863–865 (1994).
7. Anonymous, Vegetable Oil to Heat New Reichstag Building, *inform 10*: 886 (1999).
8. Dunn, R.O., Alternative Jet Fuels from Vegetable Oils, *Trans. ASAE 44:* 1751–1757 (2001).
9. Wardle, D.A., Global Sale of Green Air Travel Supported Using Biodiesel, in *Renewable and Sustainable Energy Reviews*, Vol. 7, 2003, pp. 1–64.
10. Peters, R.A., Fatty Alcohol Production and Use, *inform 7:* 502–504 (1996).
11. Willing, A., Oleochemical Esters—Environmentally Compatible Raw Materials for Oils and Lubricants from Renewable Resources, *Fett/Lipid 101:* 192–198 (1999).
12. Miller, N.J., and S.M. Mudge, The Effect of Biodiesel on the Rate of Removal and Weathering Characteristics of Crude Oil Within Artificial Sand Columns, *Spill Sci. Technol. Bull. 4:* 17–33 (1997).
13. Mudge, S.M., and G. Pereira, Stimulating the Biodegradation of Crude Oil with Biodiesel. Preliminary Results, *Spill Sci. Technol. Bull. 5:* 353–355 (1999).
14. Pereira, M.G., and S.M. Mudge, Cleaning Oiled Shores: Laboratory Experiments Testing the Potential Use of Vegetable Oil Biodiesels, *Chemosphere 54:* 297–304 (2004).
15. Wildes, S., Methyl Soyate: A New Green Alternative Solvent, *Chem. Health Safety 9:* 24–26 (2002).
16. Wehlmann, J., Use of Esterified Rapeseed Oil as Plasticizer in Plastics Processing, *Fett/Lipid 101:* 249–256 (1999).
17. Bay, K., H. Wanko, and J. Ulrich, Biodiesel: High-Boiling Absorbent for Gas Purification (Biodiesel: Hoch siedendes Absorbens für die Gasreinigung), *Chem. Ing. Technik. 76:* 328–333 (2004).
18. Schroeder, J., I. Shapiro, and J. Nelson, Candle Mixtures Comprising Naturally Derived Alkyl Esters, PCT Int. Appl. WO 46,286 (2004), *Chem. Abstr. 141:* 8856j.

10

Other Alternative Diesel Fuels from Vegetable Oils

Robert O. Dunn

Introduction

In the past, engine performance and durability studies showed that combustion of neat (100%) vegetable oil and vegetable oil/No. 2 petrodiesel (DF2) blends led to incomplete combustion, nozzle coking, engine deposits, ring sticking, and contamination of crankcase lubricant (1–5). Contamination and polymerization of lubricating oil by unsaturated triacylglycerol (TAG) led to an increase in lubricant viscosity. Many durability problems were traced to poor fuel atomization aggravated by the relatively high kinematic viscosity (ν) (6,7). As shown in the tables in Appendix A, ν data for most TAG are typically an order of magnitude greater than that of DF2. Injection systems in modern diesel engines are very sensitive to ν. Although durability problems were less severe for indirect injection engines (1,8,9), the majority of engines deployed in North America are the direct injection type in which heating and mixing with air occur in the main combustion chamber and the design requires fuels with tighter control of properties such as ν and cetane number (CN). In addition, recent studies with preheated palm oil (10,11) suggested that although problems associated with high ν and clogged fuel filters were mitigated by preheating to 100°C, performance was unaffected and emissions did not improve.

One practical solution for most performance-related problems with TAG is the reduction of ν by physical or chemical modification. The following four approaches were examined for reducing ν (6,12): (i) dilution in petrodiesel; (ii) conversion to biodiesel; (iii) pyrolysis; and (iv) formulation of microemulsions or co-solvent blends. Biodiesel has made much progress toward commercialization and is reviewed extensively elsewhere in this handbook. This chapter examines the remaining three approaches with an emphasis on microemulsions and co-solvent blends.

Definitions

The American Society for Testing and Materials (ASTM) standard fuel specification for petrodiesel, D 975 (13,14), may be used to evaluate the suitability of alternative fuels for compression-ignition engines. Test methods for selected fuel properties of diesel fuels are listed in Table 1. Also listed in Table 1 are limits that must be met to keep the engine and fuel system equipment within the manufacturer's warranty. Important fuel properties include ν at 40°C, distillation temperature, cloud point

TABLE 1
ASTM Test Methods and Limits for Selected Fuel Properties[a]

Property	Unit	ASTM method	Limits
Kinematic viscosity (ν) at 40°C	mm^2/s	D 445	1.9–4.1
Distillation temperature at 90 vol% recovered	°C	D 86	282–338
Cloud point (CP)	°C	D 2500	—[b]
Pour point (PP)	°C	D 97	—[c]
Flash point (FP)	°C	D 93	≥52
Water and sediment	vol%	D 2709	≤0.05
Carbon residue at 10% residue	wt%	D 524	≤0.35
Ash	wt%	D 482	≤0.01
Sulfur	wt%	D 2622	≤0.05
Copper strip corrosion rating, 3 h at 50°C	—	D 130	No. 3 (max)
Cetane number	—	D 613	≥40

[a]*Source:* References 13 and 14 for information on low-sulfur (500 ppm) No. 2 diesel fuel (DF2). ASTM, American Society for Testing and Materials.
[b]CP not specified by ASTM D 975. In general, CP may be used to estimate low operating temperature for fuels not treated with cold flow improver additives. When additives are present, alternative methods D 4539 or D 6731 should be employed.
[c]PP not specified by ASTM D 975. PP generally occurs 4–6°C below CP according to Liljedahl *et al.* (105).

(CP), pour point (PP), flash point (FP), water and sediment, carbon residue, ash, sulfur, copper strip corrosion residue, and CN.

Unless otherwise noted, default terms are defined as follows: (i) *alcohol* is simple alcohol with one or two carbon atoms (C_1–C_2); (ii) *neat* is 100% (unblended) TAG or petrodiesel; (iii) *oil* is TAG, petrodiesel, or combinations thereof; (iv) *X/Y/Z* is a mixture of components *X*, *Y*, and *Z* in which each component comprises one or more compounds; and (v) [*A/B*] is a mixture of compounds within a single component. An *amphiphile* is a compound whose molecular structure contains a *hydrophilic* headgroup attached to a *lipophilic* hydrocarbon chain tailgroup. Some, but not all amphiphiles are referred to as surface active agents (surfactants) or detergents. For purposes of this review, the term *amphiphile* is used generically in reference to any agent added to promote the formation of stable single-phase translucent (isotropic) mixtures.

Dilution with Petrodiesel

Results for this approach have been mixed, yielding engine problems similar to those found for combustion of neat vegetable oils. Most studies concluded that vegetable oil/petrodiesel blends are not suitable for long-term fueling of direct injection diesel engines. One experimental model showed that blends should contain a maximum of 34% TAG in DF2 to achieve proper fuel atomization (15). A later study (16) reported an optimum mixing ratio of 30 vol% for blends of rapeseed oil in petrodiesel to ensure adequate thermal efficiency in combination with preventing sedimentation during storage.

A 25/75 (vol) high-oleic safflower oil/DF2 mixture was reported to have ν = 4.92 mm^2/s at 40°C in excess of the maximum limit specified by D 975 (4.1 mm^2/s) (3,4). The blend passed the 200-h Engine Manufacturers Association (EMA) engine durability test with no measurable increases in deposits or contamination of the lubrication oil. A 25/75 (vol) sunflower oil/DF2 mixture with ν = 4.88 mm^2/s (40°C) also passed short-term engine durability tests. However, engine tear-downs gave the impression of injector coking, and carbon and lacquer buildup in the combustion chamber were significant enough to predict a rapid failure if testing had continued in excess of 200 h. Thus, this blend was not recommended for long-term use in a direct injection diesel engine (3,4,17,18). The contrasting results between similar types of mixtures were attributed to the degree of unsaturation of the corresponding TAG (2). The more unsaturated oil (sunflower) is highly reactive and tends to oxidize and polymerize as uncombusted fuel accumulates in the crankcase and hot engine parts. Accumulation in the lube oil could lead to lubricant thickening. A lube oil change is called for by the EMA test after 100 h and, at that time, the viscosity of lube oils had not varied greatly in either test (3,4,17,18).

A 20/80 (vol) peanut oil/DF2 mixture failed the 200-h EMA test primarily due to injector coking (19). A 70/30 winter (high docosanoic acid) rapeseed oil/No. 1 petrodiesel blend passed 200-h (non-EMA) durability and 800-h endurance tests showing no adverse effects on wear or lube oil viscosity (20,21). Running a 1/2 (vol) soybean oil (SBO)/DF2 mixture for 600 h resulted in no significant contamination of crankcase lubricant, although increasing SBO to 50 vol% significantly increased lube oil viscosity (22). A 50/50 (vol) SBO/Stoddard solvent (48% paraffins, 52% naphthenes) blend barely passed the 200-h EMA test (23).

A 20/80 sunflower oil/petrodiesel blend ran for a long period of time before exhaust smoke increased due to carbon buildup or power loss ensued (24). Another engine, due to inadequate atomization, showed more of the engine problems associated with neat TAG. A 25/75 sunflower oil/petrodiesel blend yielded satisfactory performance compared with neat petrodiesel (25). Smoke and hydrocarbon emissions in the exhaust decreased, whereas carbon monoxide (CO) and nitrogen oxides (NO_x) were essentially unchanged. Similar results were reported for blends of up to 15% rapeseed oil in DF2 (26). A 50/50 (vol) sunflower oil/DF2 blend reduced NO_x by 20%, hydrocarbons by 5%, and smoke by 10% compared with neat DF2 (27). Slight increases (2%) in CO emissions were also noted. A 20/80 safflower oil/DF2 blend yielded satisfactory performance with reduced CO and hydrocarbon emissions (28). Studies on blends of SBO, sunflower, and rapeseed oils blended with DF2 also showed decreases in polycyclic aromatic hydrocarbon (PAH) emissions (29,30). Semirefined rapeseed oil (acidified in hot water combined with filtration to 5 μm) was studied in mixtures with petrodiesel (31). The maximum inclusion rate was 25 vol%, based on fuel viscosity. A 15% blend had no measurable effect on lubricating oil with respect to viscosity and wear metals analyses, although some injector fouling was noted. For every 1% increase in inclusion rate, power decreased by 0.06% and fuel consumption increased by 0.14%. Smoke opacity and CO emissions also increased for the blends. A 10/90 (vol) waste vegetable oil/petrodiesel blend was subjected to 500-h performance and durability testing in a direct injection diesel engine (32). Results showed a 12% loss in power, slight increase in fuel consumption, slight decrease in combustion efficiency, and no measurable increase in viscosity of the lubricating oil. Smoke levels and carbon deposits were normal compared with petrodiesel. *Jatropha curcas* oil/petrodiesel blends increased fuel consumption and decreased thermal efficiency and exhaust temperature relative to neat petrodiesel (33). Although blends with 40–50 vol% *J. curcas* oil were acceptable without any engine modification and preheating of the fuel, a maximum rate of 20–30% *J. curcas* oil was recommended based on fuel viscosity. A 30/70 (vol) coconut oil/petrodiesel mixture was shown to increase fuel consumption, brake power, and net heat release rate and decrease NO_x, CO, hydrocarbon, smoke, and PAH emissions when run in an indirect injection diesel engine (34). Blends with >30% coconut oil developed lower brake power and net heat release rate due to lower calorific values, although reduced emissions were still noted. Increases in carbon dioxide (CO_2) emissions were observed for blend rates of 10–50%.

Mixtures of DF2, degummed dewaxed SBO, and ethanol (EtOH) passed short-term (25-h) engine performance tests, although a [30/40]/30 (vol) [DF2/SBO]/EtOH blend separated at temperatures near 20°C (35). Results at full load showed only 1.5% loss in power with an increase in thermal efficiency. No excess carbon buildup occurred on cylinder walls or injector tips, although some residues were observed between the exhaust manifold and muffler. Although pressure data indicated slight ignition delay at full governor, no audible knocks were observed during engine operation.

Cold flow properties of various 50/50 vegetable oil/DF2 mixtures were reported (36). Blends with either high-oleic or high-linoleic safflower oil yielded CP = –13°C and PP = 15°C. Similar blends with winter rapeseed oil had CP = –11°C and PP = –18°C. Baranescu and Lusco (18) showed that CP = –15, –13, and –10°C for 25/75, 50/50, and 75/25 (vol) sunflower oil/DF2 blends, respectively.

Pyrolysis

Pyrolysis or cracking involves the cleavage of chemical bonds to form smaller molecules (37). SBO was distilled with a nitrogen sparge and collected 77% of starting material as a mixture of distilled and cracked products (38,39). Similar results were obtained from distillation of high-oleic safflower oil. Pyrolyzed SBO had a C/H mass ratio = 79/12, indicating the presence of oxygenated compounds. In a comparison with SBO, pyrolysis did not significantly affect gross heat of combustion (ΔHg), increased CN to 43, decreased ν to 10.2 mm^2/s at 37.8°C but increased PP to 7°C (see Table A-3 in Appendix A for properties of SBO). Pyrolysis increased ash content and slightly increased carbon residue. Analysis of pyrolyzed distillate showed 31.3 wt% alkanes, 28.3% alkenes, 9.4% diolefins, 2.4% aromatics, 12.2% medium- and long-chain carboxylic acids, 5.5% unresolved unsaturates, plus 10.9% unidentified compounds. The appearance of aromatics was explained by Diels-Alder addition of ethylene to conjugated dienes after the formation of these components by cracking.

Catalytic conversion of TAG using a medium-severity refinery hydro-process yielded a product in the diesel boiling range with CN = 75–100 (40). The main liquid product was a straight-chain alkane. Other products included propane, water, and CO_2. Decomposition of used cottonseed oil from frying processes with Na_2CO_3 catalyst at 450°C yielded a pyrolyzate composed mainly of C_8–C_{20} alkanes plus alkenes and aromatics (41). The product oil had lower ν, FP, PP, and CN and nearly equivalent ΔH_g compared with petrodiesel.

SBO and babassu oil were processed by hydrocracking them with a Ni-Mo/γ-Al_2O_3 catalyst sulfided *in situ* with elemental sulfur under hydrogen pressure (42). Various alkanes, alkyl-cycloalkanes, and alkylbenzenes were observed. Decarboxylation was indicated by the presence of CO_2 and water, and the formation of C_1–C_4 compounds indicated the decomposition of acrolein. Differences were observed between oils with increasing degrees of unsaturation. Sulfided Ni-Mo/γ-Al_2O_3 and Ni/SiO_2 catalysts were studied for hydrocracking vegetable oils at 10–200 bar (9.9–197 atm) and 623–673 K (43). The products were mainly alkanes and other hydrocarbons commonly found in petrodiesel. Hydrogenolysis of palm oil over Ni/SiO_2 or cobalt at 300°C and 50 bar (49.3 atm) gave a nearly colorless oil of mainly C_{15}–C_{17} alkanes (44). The same process applied to rapeseed oil yielded soft solids with 80.4% C_{17} alkanes. Hydrocracking of SBO over Rh-Al_2O_3 catalyst at 693 K and 40 bar (39.5 atm) gave products that were distilled into gasoline and gas-oil fractions (45). Decarboxylation and decarbonylation were also noted in the products. In another study, crude and partially hydrogenated SBO was decomposed by passage over solid acidic Al_2O_3 and basic MgO catalysts (46). The type of catalyst and degree of unsaturation appeared to influence the composition of the products.

Microemulsions and Co-Solvent Blends

Formulating *hybrid* diesel fuels by mixing with low-molecular-weight alcohols is another approach for reducing ν of vegetable oils. Alcohols such as methanol (MeOH) or EtOH have limited solubility in nonpolar vegetable oils; therefore, amphiphilic compound(s) are added to increase solubility, dilute the oil, and reduce ν (2). *Solubilization* is defined as the dispersal of a normally insoluble substance in a solvent by forming a thermodynamically stable isotropic solution when an amphiphilic compound is added (47). The formulation of hybrid diesel fuels by solubilization of vegetable oil/alcohol mixtures through the addition of amphiphiles was initially referred to as microemulsification (2). A *microemulsion* is an equilibrium dispersion of optically isotropic fluid microstructures with an average diameter less than one quarter the wavelength of visible light that spontaneously form upon the addition of amphiphiles to a mixture of otherwise nearly immiscible liquids (48). Unlike (macro)emulsions, microemulsions are thermodynamically stable and do not require agitation to remain in single-phase, translucent solution at constant temperature and pressure (49).

Hybrid fuels may also be formulated by solubilization of alcohol molecules within *micelles* formed in solution (50,51). Micelles are aggregates of amphiphilic molecules formed when the amphiphile concentration exceeds its critical micelle concentration. In oil-external solutions, aggregates form with hydrophilic headgroups oriented toward the micellar interior and hydrocarbon tailgroups extended toward the bulk oil-phase. These micelles are often referred to as reverse or inverse micelles because micelles were initially discovered in water-external/oil-internal type mixtures. Solubilization occurs when solute molecules are absorbed into the micellar structure. Although mechanisms may be similar, some controversy exists concerning whether microemulsions and micelles should be treated as separate or related phenomena.

Finally, hybrid diesel fuels may be formulated by employing a *co-solvent* to solubilize TAG/alcohol mixtures. Co-solvency generally occurs when a mixture of indifferently effective solvents has dissolution properties far exceeding those of either solvent by itself (51). Co-solvency can result from mechanisms similar to those arising from the formation of micelles or microemulsions (47,50–52), although it is generally associated with large co-solvent concentrations, whereas solubilization in micelles or microemulsions occurs at relatively dilute amphiphile concentrations. Earlier work (53–55) demonstrated that solubilization of SBO/ MeOH mixtures by the addition of medium-chain (C_4–C_{12}) *n*-alcohol/long-chain unsaturated fatty alcohols occurs preferentially by co-solvency under most conditions.

Many variations of this approach were applied in the formulation of hybrid diesel fuels. One is to blend microemulsions or co-solvent blends with petrodiesel. Another is to employ ethers such as methyl or ethyl *tert*-butyl ether, tetrahydrofuran, tetrahydropyran, or 1,4-dioxane to solubilize mixtures of SBO or rapeseed oil and EtOH as well as possibly improve exhaust emissions (56,57). Finally, some reports claim that emulsification of 10 vol% water in SBO or rapeseed oil reduces NO_x, CO, and smoke emissions (58,59).

Selection of Components

As defined above, microemulsion and co-solvent blend fuel formulations have three types of components, i.e., TAG, simple alcohol, and amphiphile(s). Selection of the TAG is based primarily on fuel properties such as those listed in Table 1 (see Table A-3 for CN, ΔH_g, ν, CP, PP and FP of several common fats and oils). Also, MeOH or EtOH are generally desirable because they are relatively inexpensive, readily available, and can be obtained from renewable resources. Aqueous EtOH [e.g., "E95" = 95/5 (vol) EtOH/water] has also been employed in many hybrid fuel formulations. However, more expensive, water-in-petrodiesel micro-emulsions made with *n*-butanol were reported to be substantially more stable and lower in viscosity than those made with MeOH or EtOH (60).

Although amphiphiles are selected primarily for their ability to solubilize TAG/alcohol mixtures, the effects of their molecular structure on fuel properties of the final formulation also play a role. Increasing hydrocarbon chain length of the tailgroup can increase CN or ΔH_g (61–63). In contrast to DF2 (CN = 47; ΔH_g = 45.3 MJ/kg), SBO has CN = 37.9 and ΔH_g = 39.6 MJ/kg (64). Adding an amphiphile such as *n*-dodecanol (CN = 63.6; ΔHg = 46.2 MJ/kg) to SBO will enhance both parameters, whereas formulations with *n*-pentanol (CN = 18.2; ΔH_g = 37.8 MJ/kg) (61) may require compensating fuel additives. Increasing amphiphile tailgroup chain length also increases the ν of formulations. This was reported for SBO/MeOH mixtures stabilized by *n*-alcohols (65) and mixed amphiphiles consisting of *n*-alcohol and oleyl alcohol (53), with respect to increasing *n*-alcohol chain length. Similar results were reported for palm oil mixtures with MeOH and EtOH stabilized by *n*-alcohols (66). Other structural factors that influence fuel properties include degrees of branching and unsaturation in the amphiphilic tailgroup, factors that may influence the CN and ν of long-chain hydrocarbons (53,55,67). Other physical properties that should be taken into account include effects on density, water tolerance, and the "critical solution temperature" (CST), defined as the temperature of phase separation into two or more liquid layers.

Two or more mixed amphiphiles may be employed, with each affecting fuel quality and other aspects. For example, long-chain (fatty) alcohols such as those obtained from hydrolysis of vegetable oil fatty acid esters may increase CN and ΔH_g (68,69). Shorter-chain alcohols such as 2-octanol more efficiently enhance water tolerance (65). Other factors include cost and availability. The effects of mixed amphiphiles are discussed below.

Formulating Microemulsions or Co-Solvent Blends from TAG, Amphiphile and Alcohol

The miscibility of three-component hybrid diesel fuel formulations depends on the nature and concentration of the components. An important step in the formulation process is to construct a *ternary* phase diagram to identify regions of isotropic equilibrium behavior. Figure 1 is a ternary diagram for SBO/[9/1 (mol) *n*-octanol/Unadol 40 (alcohol derived from SBO-fatty acids)]/MeOH at 25°C (53). The three sides of the triangle represent binary mixtures SBO/MeOH (bottom), amphiphile/SBO (left side) and MeOH/amphiphile (right side). The curve separates regions of isotropic and *anisotropic* behavior with immiscible compositions located within the gray region. These results show how much amphiphile is necessary to solubilize mixtures with a known SBO/MeOH ratio.

It was demonstrated that medium-chain (C_4–C_{14}) *n*-alcohols were effective in solubilizing MeOH in solution with triolein and SBO (65). Adding a small quantity (<2 vol%) of water resulted in formation of a *detergentless* microemulsion. Higher water concentrations yielded turbid solutions resembling a macroemulsion (particle sizes >150 nm). Similar results were observed in mixtures of aqueous EtOH in hexadecane, triolein, trilinolein, and sunflower oil stabilized by *n*-butanol in which microemulsions

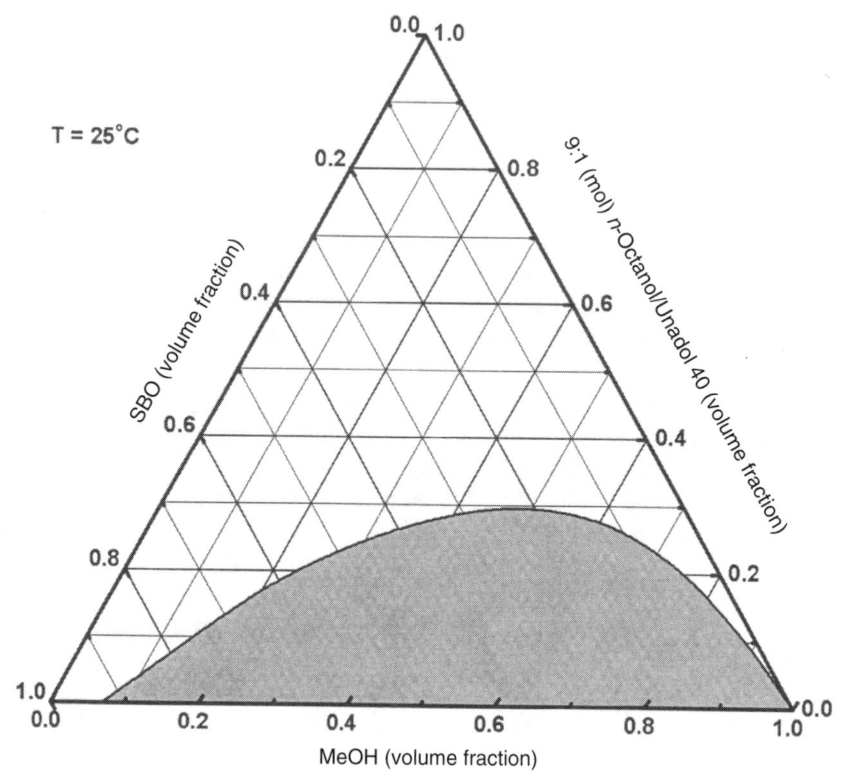

Fig. 1. Ternary phase equilibria at 25°C for soybean oil (SBO)/[9/1 (mol) *n*-octanol/-Unadol 40]/methanol (MeOH). Unadol 40 is a long-chain (mostly C_{18}) unsaturated fatty alcohol derived from the reduction of SBO fatty acids. Mixtures at all compositions inside the gray region are immiscible (anisotropic).

were promoted by solubilization of water (48,70). Similar effects of medium-chain (C_4–C_{12}) *n*-alcohols on the solubilization of MeOH and EtOH with crude and refined palm oil and palm olein were studied (66). Results showed that MeOH or EtOH solubilized in unsaturated palm olein to a much greater extent than saturated crude or refined palm oils. Results also showed that increasing the chain length reduces the quantity of *n*-alcohol necessary to achieve miscibility.

Another study (71) reported that micellar and "premicellar" solutions could be formed from mixtures of triolein and MeOH stabilized by 2-octanol in combination with bis(2-ethylhexyl) sodium sulfosuccinate, triethylammonium linoleate, or tetradecyldimethylammonium linoleate. An earlier work (72) showed that unsaturated C_{18} fatty alcohols form large, polydisperse micelles in MeOH. Critical micelle concentration was affected by the degree of unsaturation and *cis-/trans*-configuration. Viscosity results of mixtures solubilized with SBO were consistent with the formation of nonaqueous detergentless microemulsions. A recent study demonstrated that microemulsions form when water is solubilized in mixtures of petrodiesel with SBO, palm, and ricin oils stabilized by a combination of soap of the soy oil with medium-chain (C_2–C_5) alcohol (73).

Once components are chosen and regions of isotropic solution behavior are identified, appropriate amphiphile/oil (A/Oil) ratios are selected. This ratio may depend on the aforementioned factors governing the selection of TAG and amphiphile components including fuel properties, cost, renewability, and availability.

After selection of the appropriate A/Oil ratio, the effects of dilution with alcohol on ν and other relevant physical properties are determined. The final composition depends primarily on the quantity of alcohol necessary to reduce the ν of a solution of known A/Oil ratio to meet the maximum limit specified in fuel specifications such as ASTM D 975 (4.1 mm^2/s at 40°C).

Figure 2 is a graph showing the effect of increased MeOH volume fraction on ν at 37.8°C of mixtures of 2-octanol, oleyl alcohol, and 4/1 (mol) 2-octanol/oleyl alcohol with SBO and A/Oil = 1/1 (vol). The results show that more MeOH is required to sufficiently reduce ν for oleyl alcohol (0.426 volume fraction) than 2-octanol (0.260). Thus, increasing amphiphile tailgroup chain length increases the volume fraction of alcohol necessary to reduce ν. However, decreasing chain length also raises the temperature of phase separation (T_ϕ), resulting in anisotropic phase behavior at higher temperatures (53,65). Similar results were reported for the effects of *n*-alcohols on the ν of palm olein mixtures with MeOH or EtOH (66).

The mixed-amphiphile formulation shown in Figure 2 required only 0.325 volume fraction MeOH to reduce ν to 4.1 mm^2/s. Thus, substituting with 2-octanol allowed reduction in quantity of MeOH necessary to meet the ν specification. Earlier studies (53,55) reported analogous results for mixed amphiphiles of (C_4–C_{12}) *n*-alcohol and oleyl alcohol. Those studies also showed that decreasing the *n*-alcohol tailgroup chain length reduces the minimum volume fraction of MeOH necessary to sufficiently reduce ν. A comparison of ternary phase diagrams in these two studies showed that adding C_8–C_{12} *n*-alcohol to the amphiphilic component reduced the anisotropic region and total amphiphile necessary to formulate isotropic solutions. Keeping a small portion of oleyl alcohol in the amphiphile component helps maintain isotropic behavior when more MeOH was added to reduce ν. As noted above, fatty alcohol amphiphiles may positively influence CN of the formulation (68).

From the standpoint of reducing ν of hybrid fuel formulations, mechanisms promoting solubilization play an important role. It was noted above that detergentless microemulsions were observed when TAG/*n*-alcohol/MeOH mixtures were doped with small quantities of water and in formulations containing aqueous EtOH (48,65,70). Results also indicated that increasing the water or aqueous EtOH phase volume increases the viscosity of the microemulsions relative to nonaqueous mixtures. Other studies (53–55) reported similar behavior in nonaqueous triolein/- and SBO/[*n*-alcohol/long chain (C_{18}) unsaturated fatty alcohol]/MeOH formulations under certain conditions. Microemulsions or micellar solutions increase in relative viscosity with

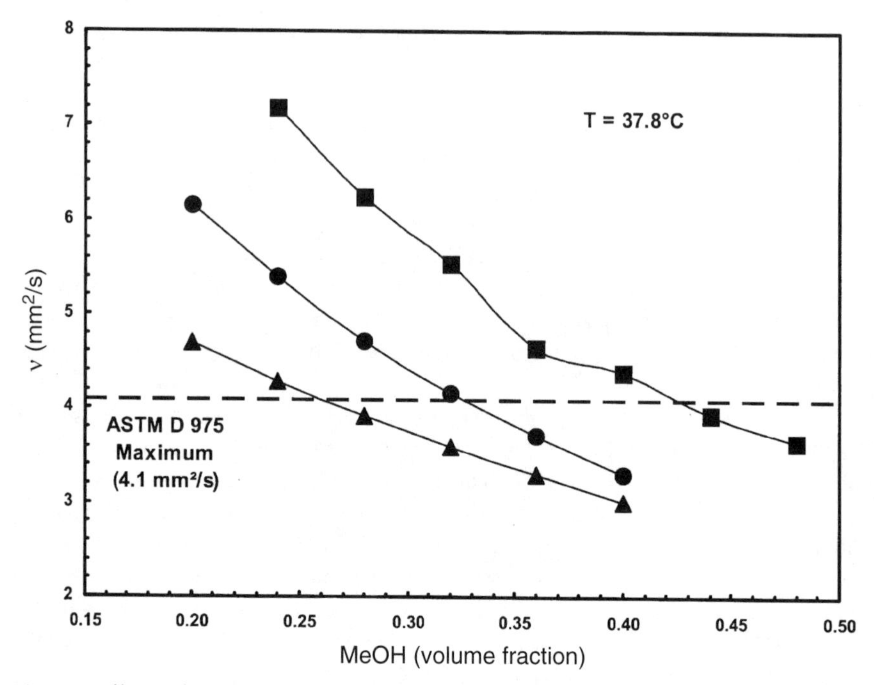

Fig. 2. Effect of MeOH volume fraction on kinematic viscosity (ν) of 1/1 (vol) amphiphile/SBO (A/Oil) mixtures. Legend: ▲ = 2-octanol; ■ = oleyl alcohol; ● = 4/1 (mol) 2-octanol/oleyl alcohol. See Figure 1 for abbreviations.

increasing phase volumes of dispersed or solubilized components (74–76). In contrast, increasing the concentration of solubilized component in a co-solvent blend decreases the relative viscosity (51). Given the objective of decreasing by solubilizing TAG/alcohol mixtures through the addition of amphiphile(s), these effects suggest that vegetable oil-based co-solvent blends are preferable over microemulsions for the formulation of hybrid diesel fuels. Most experimental evidence suggests that nonaqueous TAG/amphiphile/alcohol formulations do not readily form microemulsions under most conditions (53,54).

Ambient temperatures may influence the phase stability of TAG/amphiphile/alcohol formulations. Figure 3 is a graph showing the effects of A/Oil ratio on MeOH solubility in formulations stabilized by 4/1 (mol) 2-octanol/Unadol 40 (77,78). Data curves at 0 and 30°C are the MeOH solubilities where phase separation occurs at MeOH volume fractions above the curves. These results show that increasing the A/Oil ratio increases MeOH solubility at a constant temperature. Decreasing temperature requires an increase in the A/Oil ratio to solubilize MeOH into isotropic solution. Similar results were also reported for MeOH solubility in SBO stabilized by 8/1 (mol) *n*-butanol/oleyl alcohol and 6/1 (mol) 2-octanol/triethylammonium linoleate and E95 in 2/1 (vol) DF2/SBO stabilized by *n*-butanol.

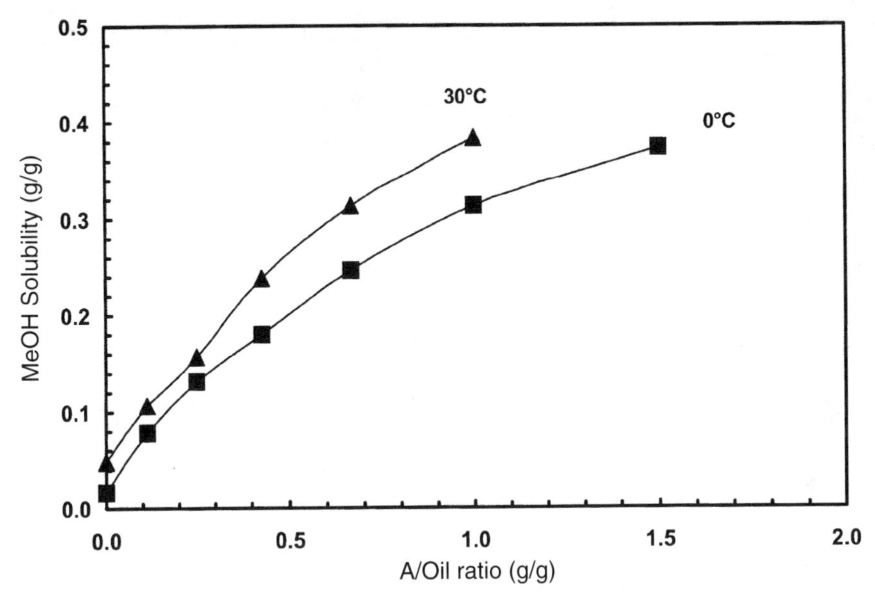

Fig. 3. Effect of A/Oil mass ratio on MeOH solubility (g/g solution) in mixtures of SBO and 4/1 (mol) 2octanol/Unadol 40. Unadol 40 is a long-chain unsaturated fatty alcohol derived from the reduction of SBO fatty acids. See Figures 1 and 2 for abbreviations. Mixtures at compositions exceeding those on the respective curves are immiscible (anisotropic).

The results in Figures 2 and 3 demonstrate that a trade-off exists in deciding how much alcohol should be added to a TAG/amphiphile mixture. Depending on the nature of the amphiphiles and selected A/Oil ratio, increasing alcohol content decreases ν and increases T_ϕ. Therefore, caution should be employed to avoid causing a phase separation to occur before ν has been reduced below the maximum limit specified in ASTM D 975. Similar results were reported for studies on the effects of *n*-alcohols in mixtures of palm olein and MeOH or EtOH (66). The effects of alcohol on other fuel properties such as CN and ΔH_g should also be kept in mind.

Two types of anisotropic phase behavior in microemulsions or co-solvent blends are possible at low temperatures. The first is a CP-type phase separation in which high melting point compounds cause the formation of a cloudy suspension similar to petrodiesel, biodiesel, and other petroleum products. The second is a CST-type separation into two or more translucent liquid layers. Earlier studies (77,79) determined that higher A/Oil ratios favor CP-type separations, whereas lower ratios favor CST-type behavior. Diluting co-solvent blends in DF2 favors CP-type separations and significantly reduces CP (77).

Figure 4 is a graph showing the effect of the A/Oil ratio on T_ϕ of three formulations similar to those discussed above (77). These curves show that increasing the A/Oil ratio decreases T_ϕ with respect to constant MeOH concentration. Mixtures with

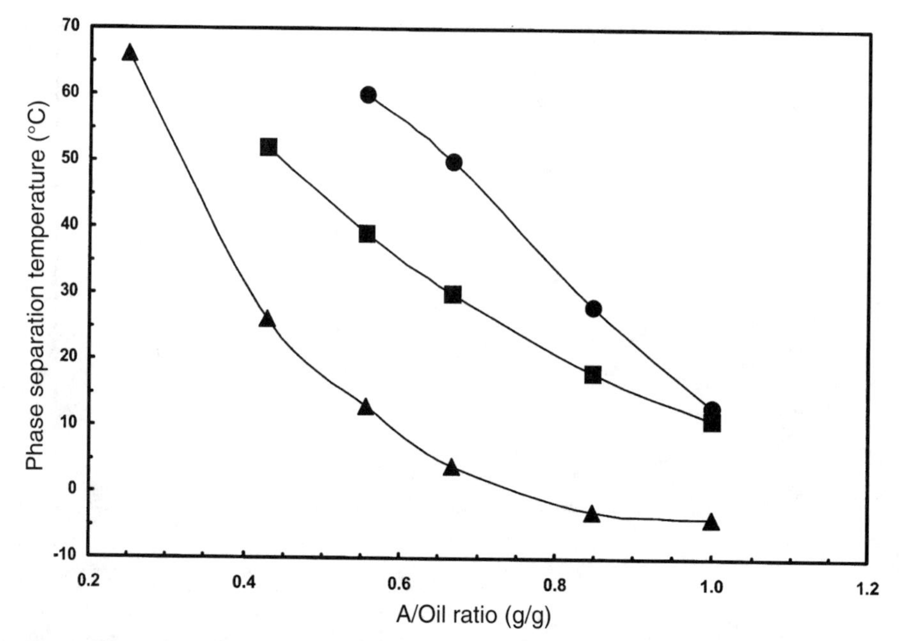

Fig. 4. Effect of A/Oil mass ratio on phase separation temperature (T_ϕ) of SBO/MeOH mixtures stabilized by mixed amphiphiles. Legend: ▲ = 4/1 (mol) 2-octanol/-Unadol 40, SBO/MeOH = 3/1 (mass); ● = 6/1 (mol) 2-octanol/-triethylammonium linoleate, SBO/MeOH = 7/3; ■ = 8/1 (mol) *n*-butanol/oleyl alcohol, SBO/MeOH = 3/1. Unadol 40 is a long-chain unsaturated fatty alcohol derived from the reduction of SBO fatty acids. See Figures 1 and 2 for abbreviations. Mixtures at temperatures exceeding those on the respective curves are miscible (isotropic).

T_ϕ below –5°C underwent CP-type transitions, whereas those at higher MeOH concentrations underwent CST-type transitions. Furthermore, T_ϕ was nearly independent of the A/Oil ratio at higher ratios (e.g., 2-octanol/Unadol 40 at A/Oil > 0.80 g/g).

Most amphiphiles possess a capability for increasing water tolerance (solubility) in hybrid diesel fuel formulations. As discussed above, water promotes formation of microemulsions (and macroemulsions). Water may also influence storage stability due to the effects of hydrolysis or microbial contamination. One study (65) reported the effects of medium-chain alcohols on water tolerance of 6/3/1 (vol) triolein/alcohol/MeOH solutions. Tolerance increased for C_4–C_8 *n*-alcohols and then decreased for C_{10+} *n*-alcohols. A comparison of four branched-chain octanols showed that the 1- and 4-isomers were superior to the 2- and 3-isomers. This work demonstrated the following sequence for water tolerances in triolein/MeOH solutions stabilized by 2-octanol plus: tetradecyldimethylammonium linoleate > bis(2-ethylhexyl) sodium sulfosuccinate > triethylammoniumlinoleate. Studies on the physical properties of microemulsions support the notion that those stabilized by long-chain amphiphiles demonstrate higher degrees of water tolerance than those stabilized by *n*-butanol (80).

Hybrid fuels formulated with TAG are susceptible to oxidative degradation. An enormous volume of research has been devoted to assessing and improving oxidative stability of TAG and their derivatives. The oxidative stability of fats and oils is generally determined in terms of an *induction period* necessary for significant changes to occur in one or more physical properties such as acid value, peroxide value, or viscosity. Oxidation induction periods for sunflower oil were reported to be nearly three times longer than those of its fatty acid mono-alkyl esters (biodiesel) (81). An earlier work (82) reported that storage stability decreases with increasing refinement grades and that the addition of DF2 to crude or degummed vegetable oils did not improve the stability of 1/1 (vol) mixtures.

Although TAG generally have very high FP, mixing with MeOH or EtOH and some amphiphiles such as *n*-butanol or 2-octanol may result in flammable hybrid fuel formulations. A 53.3/33.4/13.3 (vol) SBO/2-octanol/MeOH mixture yielded FlP= 12°C (23), a value well below the minimum limit in specification ASTM D 975 (see Table 1). Hybrid diesel fuels formulated with short- or medium-chain alcohols will require special handling and storage considerations applicable to flammable fluids.

Precombustion and Combustion Chemistry

One conceptual view of TAG-based microemulsion or co-solvent blend hybrid fuels is to consider them as mixtures with two functional parts. The first part is the "fuel," consisting of TAG plus amphiphiles, DF2, and other additives that improve performance. The second part is a "vehicle" that ensures proper delivery of the fuel to the combustion chamber. This part consists of MeOH or EtOH (anhydrous or aqueous) plus other compounds that may have poor fuel properties. Compatibility between the two parts depends on phase equilibria and ν. This concept allows the supposition that improving combustion of the combined parts depends upon improving combustion of the fuel while maintaining compatibility with the vehicle. This approach allows interpretation of mechanisms associated with combustion based on studying the chemistry of TAG and their derivatives.

Review articles (78,83–85) provide detailed descriptions of the combustion of petrodiesel and neat TAG in compression-ignition engines. The combustion of diesel fuels encompasses two general phases. In the first phase, *precombustion*, the fluid is subjected to intense thermodynamic conditions. Products from chemical reactions during this phase may lead to incomplete combustion and influence exhaust emissions from the second phase, *diffusion burning* (86,87). It was shown how the ν of the fuel affects physical characteristics of the spray as fuel is injected into the combustion chamber (6,86,88). These characteristics influence how compounds on the fringes of the spray region degrade and affect atomization and precombustion of the fuel. Finally, apparent discrepancies between CN and observed ignition quality of TAG were attributed to chemical reactions taking place during the precombustion phase.

Reaction temperatures during the precombustion phase also influenced the composition of degradation products present in the combustion chamber at the onset of diffusion burning. For pure unsaturated TAG under 4.2 MPa pressure, reaction temperatures up to 400°C yielded degradation products that were predominantly the corresponding fatty acids and aliphatic hydrocarbons plus smaller concentrations of aldehydes, alcohols, glycerol, succinic acid, and benzoic acid (87,89). In contrast, precombustion at 450°C resulted in more extensive degradation and more diverse classes of compounds including large quantities of volatile and semivolatile compounds characterized as aliphatic hydrocarbons, aromatics with side chains, PAH, and unsaturated aldehydes. No detectable quantities of fatty acids, glycerol, or acrolein were found. Another study (86) examined degradation products of several vegetable oils injected for 400 μs into a nitrogen environment (to prevent autoignition) at 480°C and 4.1 MPa. Product compositions were consistent with those noted above for pure TAG at higher temperatures. Overall, these studies showed that the reaction temperature in the precombustion phase may determine the composition of compounds present at the onset of diffusion burn, influencing subsequent reactions and affecting final exhaust emissions.

Engine Performance, Durability, and Exhaust Emissions

Table 2 is a summary of TAG-based hybrid formulations that were tested for engine performance, durability, and exhaust emissions. Volumetric ratios of components and physical properties are noted for each formulation. Physical properties of neat SBO and DF2 are listed for comparison. DF2 was employed as a reference fuel for each study.

Short-term (3.5 h) engine tests were run on Hybrids A and B, and it was reported that the hybrid agrodiesel formulations produced nearly as much power as the reference fuel despite having 18% lower ΔHg (90). Fuel consumption increased by 16% at maximum power. The presence of oxygenated compounds allowed leaner combustion and yielded a 6% gain in thermal efficiency at full power. Low CN had no adverse effects on short-term performance in terms of an increase in audible engine knocking after warmup for either microemulsion. A relatively large quantity (10 wt%) of primary alkyl nitrate was required to raise the CN of Hybrid A to meet the D 975 minimum limit for petrodiesel.

Hybrids C and D both passed the 200-h EMA durability screening test (23,80). The low-CN Hybrid C formulation required an ether assist for startup during cold weather (≤12°C). In comparison with the reference fuel, combustion of Hybrid C increased carbon deposits on exhaust valves and around the orifices of injector nozzles. Heavy carbon and varnish deposits were also noted inside injectors and on intake valve stems, pistons, and exhaust valve stems. Engine teardowns revealed carbon, lacquer, and varnish deposits similar to but less severe than those noted for Hybrid C, a result attributed to blending with DF2. Although main and rod bearing wear improved for Hybrid D, top ring wear was nearly equivalent to wear caused by the reference fuel.

A hybrid fuel consisting of [33/33]/33/1 (vol) [SBO/ DF2]/*n*-butanol/cetane improver also passed the 200-h EMA test (23). This fuel had properties $\nu = 4.9\ mm^2/s$ (37.8°C), CN = 38.8, ΔH_g = 40.2 MJ/kg, and FP = 38.9°C. Although average net

TABLE 2
Fuel Properties of Hybrid Soybean Oil-Based Microemulsion and Co-Solvent Blend Hybrid Diesel Fuels[a]

Hybrid fuel	ν (37.8°C) (mm^2/s)	CN	ΔHg (MJ/kg)	CP °C	PP °C	FP[b] °C	Carbon residue wt%	Reference
Hybrid A (microemulsion)								
53.3 vol% SBO	6.8	25.1	37.0	0[c]	–65	27.8	0.18	90,97
33.4% *n*-Butanol								
13.3% E95								
Hybrid B (microemulsion)								
52.3 vol% SBO	8.8	29.8	36.7	0[c]	—	22.2	0.21	90
20.5% *n*-Butanol								
6.5% Linoleic acid								
3.3% Triethylamine								
17.4% E95								
Hybrid C (co-solvent blend)								
52.8 vol% SBO	8.3	33.1	37.8	–11	–23	12.2	0.42	23,98
33.0% 2-Octanol								
13.2% MeOH								
1.0% Cetane improver								
Hybrid D (DF2 + co-solvent blend)								
50 vol% DF2	4.0	34.7	41.3	–15	—	28.3	0.14	23,80
25% SBO								
20% *n*-Butanol								
5% E95								
SBO	32.6	37.9	39.6	–4	–12	254	0.27	64
DF2	2.7	47	45.4	–15	–33	52	<0.35	64

[a]Abbreviations: CN, cetane number; E95, 95/5 (vol) EtOH/water; ΔH_g, gross heat of combustion; SBO, soybean oil. See Table 1 for other abbreviations.
[b]Closed-cup (Penske-Martins).
[c]Critical solution temperature (CST); separation into two distinct liquid layers.

consumption of crankcase lubricant decreased, little change in lube oil viscosity was noted during the test. Engine teardowns revealed deposits and buildup similar to those noted for Hybrids C and D. Similar to Hybrid D, main and rod bearing wear decreased, although top ring wear increased relative to the reference fuel. Comparison of this hybrid fuel with those for Hybrids C and D suggested that the extent of durability problems associated with deposits and top ring wear increases with increasing SBO content in the formulation. Results also suggested that durability problems attributed to incomplete combustion may be reduced but not eliminated by dilution of the TAG with simple alcohol and amphiphiles.

Emulsions containing SBO, petrodiesel, and EtOH were tested (91). The hybrid fuel burned faster with higher levels of premixed burning due to longer ignition delays and lower levels of diffusion flame burning than the reference fuel. This resulted in higher brake thermal efficiencies, cylinder pressures, and rates of pressure increase. Although smoke and unburned hydrocarbon emissions decreased, NO_x and CO emissions increased for the hybrid fuels.

A hybrid fuel consisting of 53.3/33.4/13.3 (vol) winterized sunflower oil/*n*-butanol/E95 was reported to pass the 200-h EMA test (70). This fuel had properties $\nu = 6.3$ mm^2/s (40°C), CN = 25.0, $\Delta H_g = 36.4$ MJ/kg, CP = 15°C (CST), and FP = 27°C. Fuel consumption increased by 10%, power decreased by 8%, and specific energy consumption increased by 4%. Performance was not affected by 250+ h of total testing, although incomplete combustion at low-load engine operation was observed. Smoke number and exhaust temperature decreased. Although the injector nozzles showed no measurable reductions in performance, carbon and lacquer buildup caused needle sticking and deposits to form on the cylinder heads, pistons, and rings. Heavier residues on the piston lands, in the piston ring grooves, and in the intake ports were also noted. Contamination of crankcase lubricant was observed based on a 50% increase in lube viscosity over a 60-h period. This hybrid fuel formulation was not recommended for long-term use in a direct-injection engine.

A hybrid fuel consisting of 66.7/[12.5/4.1]/16.7 (vol) DF2/[SBO fatty acids/*N,N*-dimethylethanolamine]/EtOH was studied (92). The fatty acids reacted with dimethylethanolamine to form an amphiphilic agent that promoted formation of microemulsions with EtOH as the interior phase and DF2 as the exterior phase. Fuel properties were $\nu = 3.7$ mm^2/s (37.8°C), CN = 32.9, $\Delta H_g = 40.8$ MJ/kg, CP ≤ -18°C and FP = 15.6°C. After washing with heptane, existent gum decreased to 4.6 mg/100 mL. This fuel gave acceptable performance increasing thermal efficiency by 4–5%. Exhaust temperatures decreased and emissions showed decreases in smoke and CO and increases in unburned hydrocarbons.

Emulsions composed of palm oil and petrodiesel mixed with 5–10% water were tested in an indirect-injection diesel engine for 20 h under steady-state conditions (93,94). Engine performance and fuel consumption were comparable to those for reference fuel. Wear metal particle accumulations in the crankcase lubricant decreased relative to neat petrodiesel.

Mixtures of DF2 and bio-oil produced from light fractions from fast pyrolysis of hardwood stabilized by 1.5 wt% amphiphile were examined (95). Fuel properties were $\nu = 6.5$ mm^2/s (40°C), CN = 33.5, $\Delta H_g = 36.6$ MJ/kg and FP = 62°C. This formulation was very stable at cold temperatures with a CST below 0°C and PP = –41°C. No engine tests were reported.

Patents

Several hybrid fuel formulations containing TAG in the absence of petrodiesel have been patented. A microemulsion consisting of vegetable oil, a lower (C_1–C_3) alcohol, water, and amphiphile comprising the reaction product of a trialkylamine with a long-chain fatty acid was patented (96). Addition of *n*-butanol to the amphiphilic component was optional. Another patent was obtained for a microemulsion consisting of vegetable oil, a C_1–C_3 alcohol, water, and *n*-butanol (97). These formulations had acceptable ν and compared favorably in engine performance with respect to neat DF2. A hybrid fuel consisting of vegetable oil, MeOH or EtOH (anhydrous or aqueous) and a straight-chain octanol isomer had high water tolerance, acceptable ν, and performance characteristics compared with neat DF2 (98). A microemulsion fuel consisting of degummed rapeseed oil, water, and an alkaline soap or a potassium salt of fatty acids was also patented (99).

Inventions taking advantage of microemulsion or co-solvent blend technology in which the fatty ingredient is present as part of the amphiphilic component are also reported in the patent literature. These formulations are typically microemulsions of petrodiesel, water, alcohol (anhydrous or aqueous), and one or more amphiphiles. Several microemulsions with mixed amphiphiles comprising *N,N*-dimethylethanolamine plus a long-chain fatty substance (C_9–C_{22}) were patented (100). These microemulsions contained the fatty compound in small quantities and showed good low-temperature stability and a high tolerance for water, enabling blending with petrodiesel with relatively high concentrations of aqueous alcohol. Microemulsions stabilized by a C_1–C_3 alcohol in combination with an ethoxylated C_{12}–C_{18} alkyl ammonium salt of C_9–C_{24} alkylcarboxylic or alkylarylsulfonic acids were shown to improve CN and combustion properties and reduced smoke, particulate matter, and CO and NO_x emissions (101). Similar microemulsions consisting of DF2 and water stabilized by a combination of MeOH and fatty acid partially neutralized by ammonia, (mono) ethanol amine, dimethylethanolamine, or isopropanol amine were reported to have good phase stability over wide variations in temperature and reduce combustion emissions (102). Improved microemulsion fuels consisting of alcohol-fatty acid esters (i.e., biodiesel), alcohol, and <1% alkali metal soap were patented (103).

Outlook

Aside from biodiesel, the development of alternative fuels from TAG has seen very little activity in the past 20 years. Many formulations passed the 200-h EMA durability test, although durability problems associated with the combustion of neat TAG were diminished rather than eliminated. Because of the durability issues, most of the studies reviewed above recommended against long-term use of vegetable oil-based hybrid diesel fuels.

On the other hand, hybrid fuel formulations such as microemulsions or co-solvent blends are generally less expensive to prepare than biodiesel because no complicated technology other than simply mixing the components at ambient temperature is involved. Thus, a hybrid fuel composed of SBO, 2-octanol, MeOH, and a cetane improver was the least expensive TAG-derived alternative diesel fuel ever to pass the 200-h EMA test (2,104). Other advantages to preparing hybrid fuel formulations include ease of adaptability to large scale in continuous flow processing equipment, no catalysts to recover or regenerate, no excess alcohol(s) to recover, no elevated temperatures required to mix components, and no need for specially designed equipment for separation of co-products. Components including amphiphiles used in hybrid formulations may be 100% renewable and derived from agricultural feedstocks. For the example noted above, 2-octanol is a co-product with 10-undecenoic acid of the high-temperature oxidation of castor oil. Fatty alcohols used in formulations may be derived either from hydrolysis of mono-alkyl esters from TAG or from reduction of fatty acids from alkali refining of vegetable oils.

Despite favorable aspects for ease of conversion and economics, development of microemulsion or co-solvent blend type hybrid fuel formulations will require research to improve fuel properties and resolve engine performance, durability, and exhaust emissions issues related to the inclusion of TAG in such formulations.

References

1. Goering, C.E., M.D. Schrock, K.R. Kaufman, M.A. Hanna, F.D. Harris, and S.J. Marley, Evaluation of Vegetable Oil Fuels in Engines, *Proceedings of the International Winter Meeting of the ASAE*, 1987, Paper No. 87-1586.
2. Schwab, A.W., M.O. Bagby, and B. Freedman, Preparation and Properties of Diesel Fuels from Vegetable Oils, *Fuel 66:* 1372–1378 (1987).
3. Ziejewski, M., H. Goettler, and G.L. Pratt, Comparative Analysis of the Long-Term Performance of a Diesel Engine on Vegetable Oil Based Alternate Fuels, in SAE Technical Paper Series, No. 860301, 1986.
4. Ziejewski, M., H. Goettler, and G.L. Pratt, Influence of Vegetable Oil Based Alternate Fuels on Residue Deposits and Components Wear in a Diesel Engine, in SAE Technical Paper Series, No. 860302, 1986.
5. Engler, C.R., and L.A. Johnson, Effects of Processing and Chemical Characteristics of Plant Oils on Performance of an Indirect-Injection Diesel Engine, *J. Am. Oil Chem. Soc. 60:* 1592–1596 (1983).
6. Ryan, III, T.W., L.G. Dodge, and T.J. Callahan, The Effects of Vegetable Oil Properties on Injection and Combustion in Two Different Diesel Engines, *J. Am. Oil Chem. Soc. 61:* 1610–1619 (1984).
7. Pryde, E.H., and A.W. Schwab, Cooperative Work on Engine Evaluation of Hybrid Fuels, in *Vegetable Oil as Diesel Fuel Seminar III* (ARM-NC-28), U.S.D.A., Peoria, IL, 1983, pp. 90–95.

8. Lilly, L.R.C., ed., *Diesel Engine Reference Book*, Butterworths, London, 1984, pp. 22/9–22/12.
9. Borgelt, S.C., and F.D. Harris, Endurance Tests Using Soybean Oil-Diesel Fuel Mixture to Fuel Small Pre-Combustion Chamber Engines, in *Proceedings of the International Conference on Plant and Vegetable Oils as Fuels*, ASAE Special Publication No. SP 4-82, 1982, pp. 364–373.
10. Bari, S., T.H. Lim, and C.W. Yu, Effects of Preheating of Crude Palm Oil on Injection System, Performance and Emission of a Diesel Engine, *Renewable Energy 27:* 339–351 (2002).
11. de Almeida, S.C.A., C.R. Belchior, M.V.G. Nascimento, L.S.R. Vieira, and G. Fleury, Performance of a Diesel Generator Fueled with Palm Oil, *Fuel 81:* 2097–2102 (2002).
12. Sinha, S., and N.C. Misra, Diesel Fuel Alternative from Vegetable Oils, *Chem. Eng. World 32:* 77–80 (1997).
13. Anonymous, *Annual Book of ASTM Standards*, Vol. 05.01, ASTM International, West Conshohocken, PA, 2003, D975.
14. Westbrook, S.R., in *Significance of Tests for Petroleum Products*, 7th edn., edited by S.J. Rand, ASTM International, West Conshohocken, PA, 2003, pp. 63–81.
15. Msipa, C.K.M., C.E. Goering, and T.D. Karcher, Vegetable Oil Atomization in a DI Diesel Engine, *Trans. ASAE 26:* 1669–1672 (1983).
16. He, Y., and Y.D. Bao, Study on Rapeseed Oil as Alternative Fuel for a Single-Cylinder Diesel Engine, *Renewable Energy 28:* 1447–1453 (2003).
17. Ziejewski, M., and K.R. Kaufman, Laboratory Endurance Test of a Sunflower Oil Blend in a Diesel Engine, *J. Am. Oil Chem. Soc. 60:* 1567–1573 (1983).
18. Baranescu, R.A., and J.J. Lusco, Performance, Durability, and Low Temperature Evaluation of Sunflower Oil as a Diesel Fuel Extender, in *Proceedings of the International Conference on Plant and Vegetable Oils as Fuels*, ASAE Special Publication No. SP 4-82, 1982, pp. 312-328.
19. Goodrum, J.W., Peanut Oil as an Emergency Farm Diesel Fuel, in *Vegetable Oil as Diesel Fuel Seminar III* (ARM-NC-28), U.S.D.A., Peoria, IL, 1983, pp. 112–118.
20. Peterson, C.L., D.L. Auld, and R.A. Korus, Winter Rape Oil Fuel for Diesel Engines: Recovery and Utilization, *J. Am. Oil Chem. Soc. 60:* 1579–1587 (1983).
21. Bettis, B.L., C.L. Peterson, D.L. Auld, D.J. Driscoll, and E.D. Peterson, Fuel Characteristics of Vegetable Oil from Oilseed Crops in the Pacific Northwest, *Agron. J. 74:* 334–339 (1982).
22. Adams, C., J.F. Peters, M.C. Rand, B.J. Schroeder, and M.C. Ziemke, Investigation of Soybean Oil as a Diesel Fuel Extender: Endurance Tests, *J. Am. Oil Chem. Soc. 60:* 1574–1579 (1983).
23. Goering, C.E., in *Effect of Nonpetroleum Fuels on Durability of Direct-Injection Diesel Engines* (Contract 59-2171-1-6-057-0), U.S.D.A., Peoria, IL, 1984.
24. Bruwer, J.J., B.D. Boshoff, F.J.C. Hugo, J. Fuls, C. Hawkins, A.N. Walt, A. Engelbrecht, and L.M. du Plessis, The Utilization of Sunflower Seed Oil as a Renewable Fuel for Diesel Engines, in *Proceedings of the National Energy Symposium of the ASAE*, 1980.
25. Zubik, J., S.C. Sorenson, and C.E. Goering, Diesel Engine Combustion of Sunflower Oil Fuels, *Trans. ASAE 27:* 1251–1256 (1984).
26. Schmidt, A., W. Staetter, A. Marhold, W. Zeiner, and G. Joos, Rape Seed Oil as a Source of Energy (2): Rape Seed Oil as a Diesel Oil Extender—Results of Laboratory and Test Stand Experiments, Erdoel, *Erdgas, Kohle 108:* 415–418 (1992).

27. Ziejewski, M., and H.J. Goettler, Comparative Analysis of the Exhaust Emissions for Vegetable Oil Based Alternative Fuels, in Alternative Fuels for CI and SI Engines, SAE Special Publication SP-900 (Paper No. 920195), 1992, pp. 65–73.
28. Isióigür, A., F. Karaomanoólu, H.A. Aksoy, F. Hamdallahpur, and Ö.L. Gülder, Safflower Seed Oil of Turkish Origin as a Diesel Fuel Alternative, *Appl. Biochem. Biotechnol. 39–40:* 89–105 (1993).
29. Ziejewski, M., H.J. Goettler, L.W. Cook, and J. Flicker, Polycyclic Aromatic Hydrocarbons Emissions from Plant Oil Based Alternative Fuels, in SAE Technical Paper Series, Paper No. 911765, 1991.
30. Mills, G.A., and A.G. Howard, A Preliminary Investigation of Polynuclear Aromatic Hydrocarbon Emissions from a Diesel Engine Operating on Vegetable Oil-Based Alternative Fuels, *J. Instit. Energy 56:* 131–137 (1983).
31. McDonnell, K.P., S.M. Ward, P.B. McNulty, and R. Howard-Hildge, Results of Engine and Vehicle Testing of Semirefined Rapeseed Oil, *Trans. ASAE 43:* 1309–1316 (2000).
32. Dorado, M.P., J.M. Arnal, J. Gómez, A. Gil, and F.J. López, The Effect of a Waste Vegetable Oil Blend with Diesel Fuel on Engine Performance, *Trans. ASAE 45:* 519–523 (2002).
33. Pramanik, K., Properties and Use of Jatropha Curcas Oil and Diesel Fuel Blends in Compression Ignition Engine, *Renewable Energy 28:* 239–248 (2003).
34. Kalam, M.A., M. Husnawan, and H.H. Masjuki, Exhaust Emission and Combustion Evaluation of Coconut Oil-Powered Indirect Injection Diesel Engine, *Renewable Energy 28:* 2405–2415 (2003).
35. Braun, D.E., and K.Q. Stephenson, Alternative Fuel Blends and Diesel Engine Tests, in *Proceedings of the International Conference on Plant and Vegetable Oils as Fuels*, 1982, pp. 294–302.
36. Peterson, C.L., R.A. Korus, P.G. Mora, and J.P. Madsen, Fumigation with Propane and Transesterification Effects on Injector Coking with Vegetable Oils, *Trans. ASAE 30:* 28–35 (1987).
37. Weisz, P.B., W.O. Haag., and P.G. Rodewald, Catalytic Production of High-Grade Fuel (Gasoline) from Biomass Compounds by Shape-Selective Catalysis, *Science 206:* 57–58 (1979).
38. Schwab, A.W., G.J. Dykstra, E. Selke, S.C. Sorenson, and E.H. Pryde, Diesel Fuel from Thermal Decomposition of Soybean Oil, *J. Am. Oil Chem. Soc. 65:* 1781–1786 (1988).
39. Dykstra, G.J., M.S. Thesis, Diesel Fuel from Thermal Decomposition of Vegetable Oils, University of Illinois, Urbana-Champaign, IL, 1985.
40. Stumborg, M., D. Soveran, W. Craig, W. Robinson, and K. Ha, Catalytic Conversion of Vegetable Oil to Diesel Additives, *Energy Biomass Wastes 16:* 721–738 (1993).
41. Zaher, F.A., and A.R. Taman, Thermally Decomposed Cottonseed Oil as a Diesel Engine Fuel, *Energy Sources 15:* 499–504 (1993).
42. Da Rocha Filho, G.N., D. Brodzki, and G. Djéga-Mariadassou, Formation of Alkanes, Alkylcycloalkanes and Alkylbenzenes During the Catalytic Hydrocracking of Vegetable Oils, *Fuel 72:* 543–549 (1993).
43. Gusmão, J., D. Brodzki, G. Djéga-Mariadassou, and R. Frety, Utilization of Vegetable Oils as an Alternative Source for Diesel-Type Fuel: Hydrocracking on Reduced Ni/SiO_2 and Sulphided $Ni\text{-}Mo/\gamma\text{-}Al_2O_3$, *Catal. Today 5:* 533–544 (1989).
44. Cecchi, G., and A. Bonfand, Conversion of Vegetable Oils into Potential Motor Fuel. Preliminary Studies, *Rev. Fr. Corps Gras 34:* 397–401 (1987).

45. Nunes, P.P., D. Brodzki, G. Bugli, and G. Djéga-Mariadassou, Hydrocracking of Soybean Oil Under Pressure: Research Procedure and General Aspects of the Reaction, *Rev. Inst. Fr. Pet. 41:* 421–431 (1986).
46. Dos Anjos, J.R.S., W.D.A. Gonzalez, Y.L. Lam, and R. Frety, Catalytic Decomposition of Vegetable Oil, *Appl. Catal. 5:* 299–308 (1983).
47. Mackay, R.A., in *Nonionic Surfactants: Physical Chemistry*, Surfactant Science Series Vol. 23, edited by M. J. Schick, Marcel Dekker, 1987, pp. 297–368.
48. Schwab, A.W., H.C. Nielson, D.D. Brooks, and E.H. Pryde, Triglyceride/Aqueous Ethanol/1-Butanol Microemulsions, *J. Dispersion Sci. Technol. 4:* 1–17 (1983).
49. Rosen, M.J., in *Surfactants and Interfacial Phenomena*, 2nd edn., Wiley and Sons, New York, 1989, pp. 322–324.
50. Mukerjee, P., in *Solution Chemistry of Surfactants*, Vol. 1, edited by K. L. Mittal, Plenum, New York, 1979, p. 153.
51. McBain, M.E.L., and E. Hutchinson, in *Solubilization and Related Phenomena*, Academic, New York, 1955, pp. 179–194.
52. Friberg, S.E., Microemulsions, Hydrotropic Solutions and Emulsions, a Question of Phase Equilibria, *J. Am. Oil Chem. Soc. 48:* 578–581 (1971).
53. Dunn, R.O., and M.O. Bagby, Solubilization of Methanol and Triglycerides: Unsaturated Long-Chain Fatty Alcohol/Medium-Chain Alkanol Mixed Amphiphile Systems, *J. Am. Oil Chem. Soc. 71:* 101–108 (1994).
54. Dunn, R.O., A.W. Schwab, and M.O. Bagby, Solubilization and Related Phenomena in Nonaqueous Triolein/Unsaturated Long Chain Fatty Alcohol/Methanol Solutions, *J. Dispersion Sci. Technol. 14:* 1–16 (1993).
55. Dunn, R.O., A.W. Schwab, and M.O. Bagby, Physical Property and Phase Studies of Nonaqueous Triglyceride/Unsaturated Long Chain Fatty Alcohol/Methanol Systems, *J. Dispersion Sci. Technol. 13:* 77–93 (1992).
56. Yaginuma, R., S. Moriya, Y. Sato, T. Sako, D. Kodama, D. Tanaka, and M. Kato, Homogenizing Effect of Addition of Ethers to Immiscible Binary Fuels of Ethanol and Oil, *Sekiyu Gakkaishi 42:* 173–177 (1999).
57. Kato, M., H. Tanaka, H. Ueda, S. Moriya, F. Yaginuma, and N. Isshiki, Effect of Methyl tert-Butyl Ether to Homogenize Immiscible Binary-Component Fuels of Three Types: Ethanol Mixed with Diesel Fuel, Soybean Oil, or Rape Oil, *Sekiyu Gakkaishi 35:* 115–118 (1992).
58. Crookes, R.J., F. Kiannejad, G. Sivalingam, and M.A.A. Nazha, Effects of Using Vegetable Oil Fuels and Their Emulsions on the Performance and Emissions of Single- and Multi-Cylinder Diesel Engines, *Archivum Combustionis 13:* 57–74 (1993).
59. Crookes, R.J., M.A.A. Nazha, and F. Kiannejad, Single and Multi Cylinder Diesel-Engine Tests with Vegetable Oil Emissions, in SAE Technical Paper Series Paper No. 922230, 1992.
60. Googin, J.M., A.L. Compere, and W.L. Griffith, Technical Considerations in Choosing Alcohol Fuels for Less-Developed Countries, *Energy Res. 3:* 173–186 (1983).
61. Freedman, B., and M.O. Bagby, Predicting Cetane Numbers of n-Alcohols and Methyl Esters from Their Physical Properties, *J. Am. Oil Chem. Soc. 67:* 565–571 (1990).
62. Klopfenstein, W.E., Effects of Molecular Weights of Fatty Acid Esters on Cetane Numbers as Diesel Fuels, *J. Am. Oil Chem. Soc. 62:* 1029–1031 (1985).
63. Klopfenstein, W.E., Estimation of Cetane Index for Esters and Fatty Acids, *J. Am. Oil Chem. Soc. 59:* 531–533 (1982).

64. Goering, C.E., A.W. Schwab, M.J. Daugherty, E.H. Pryde, and A.J. Heakin, Fuel Properties of Eleven Vegetable Oils, *Trans. ASAE 25:* 1472–1477 and 1483 (1982).
65. Schwab, A.W., and E.H. Pryde, Triglyceride-Methanol Microemulsions, *J. Dispersion Sci. Technol. 6:* 563–574 (1985).
66. Dzulkefly, K., W.H. Lim, S. Hamdan, and K. Norsilah, Solubilization of Methanol and Ethanol in Palm Oil Stabilized by Medium- and Long-Chain Alkanols, *J. Chem. Technol. Biotechnol. 77:* 627–632 (2002).
67. Harrington, K.J., Chemical and Physical Properties of Vegetable Oil Esters and their Effect on Diesel Fuel Performance, *Biomass 9:* 1–17 (1986).
68. Freedman, B., M.O. Bagby, T.J. Callahan, and T.W. Ryan III, Cetane Numbers of Fatty Esters, Fatty Alcohols and Triglycerides Determined by a Constant-Volume Combustion Bomb, in SAE Technical Paper Series, No. 900343, 1990.
69. Freedman, B., M.O. Bagby, and H. Khoury, Correlation of Heats of Combustion with Empirical Formulas for Fatty Alcohols, *J. Am. Oil Chem. Soc. 66:* 595–596 (1989).
70. Ziejewski, M., K.R. Kaufman, A.W. Schwab, and E.H. Pryde, Diesel Engine Evaluation of a Nonionic Sunflower Oil-Aqueous Ethanol Microemulsion, *J. Am. Oil Chem. Soc. 61:* 1620–1626 (1984).
71. Schwab, A.W., and E.H. Pryde, Micellar Solubilization of Methanol and Triglycerides, in Phenomena in Mixed Surfactant Systems, edited by J. F. Scamehorn, ACS, Washington, 1986, pp. 283–296.
72. Dunn, R.O., and M.O. Bagby, Aggregation of Unsaturated Long-Chain Fatty Alcohols in Nonaqueous Systems, *J. Am. Oil Chem. Soc. 72:* 123–130 (1995).
73. de Castro Dantas, T.N., A.C. da Silva, and A.A.D. Neto, New Microemulsion Systems Using Diesel and Vegetable Oils, *Fuel 80:* 75-81 (2001).
74. Schwab, A.W., R.S. Fattore, and E.H. Pryde, Diesel Fuel-Aqueous Ethanol Microemulsions, *J. Dispersion Sci. Technol. 3:* 45–60 (1982).
75. Ito, K., and Y. Yamashita, Viscosity and Solubilization Studies on Weak Anionic Polysoaps in Nonaqueous Solvents, *J. Colloid Sci. 19:* 152–164 (1964).
76. Roscoe, R., Viscosity of Suspensions of Rigid Spheres, *Brit. J. Appl. Phys. 3:* 267–269 (1952).
77. Dunn, R. O., and M.O. Bagby, Low-Temperature Phase Behavior of Vegetable Oil/Co-solvent Blends as Alternative Diesel Fuel, *J. Am. Oil Chem. Soc. 77:* 1315–1323 (2000).
78. Dunn, R.O., G. Knothe, and M.O. Bagby, Recent Advances in the Development of Alternative Fuels from Vegetable Oils and Animal Fats, *Recent Res. Devel. Oil Chem. 1:* 31–56 (1997).
79. Dunn, R.O., Low-Temperature Flow Properties of Vegetable Oil/Cosolvent Blend Diesel Fuels, *J. Am. Oil Chem. Soc. 79:* 709–715 (2002).
80. Goering, C.E., and B. Fry, Engine Durability Screening Test of a Diesel/Soy Oil/Alcohol Microemulsion Fuel, *J. Am. Oil Chem. Soc. 61:* 1627–1632 (1984).
81. Du Plessis, L.M., J.B.M. De Villiers, and W.H. Van der Walt, Stability Studies on Methyl and Ethyl Fatty Acid Esters of Sunflowerseed Oil, *J. Am. Oil Chem. Soc. 62:* 748–752 (1985).
82. Du Plessis, L. M., Plant Oils as Diesel Fuel Extenders: Stability Tests and Specifications of Different Grades of Sunflower Seed and Soyabean Oils, *CHEMSA 8:* 150–154 (1982).
83. Knothe, G., and R.O. Dunn, Biofuels Derived from Vegetable Oils and Fats, in *Oleochemical Manufacture and Applications*, edited F.D. Gunstone and R.J. Hamilton, Sheffield Academic Press, Sheffield, 2001, pp. 106–163.

84. Knothe, G., R.O. Dunn, and M.O. Bagby, Biodiesel: The Use of Vegetable Oils and Their Derivatives as Alternative Diesel Fuels, in *Fuels and Chemicals from Biomass*, edited by B.C. Saha and J. Woodward, ACS, Washington, 1997, pp. 172–208.
85. Graboski, M.S., and R.L. McCormick, Combustion of Fat and Vegetable Oil Derived Fuels in Diesel Engines, *Prog. Energy Combust. Sci. 24:* 125–164 (1998).
86. Ryan III, T.W., and M.O. Bagby, Identification of Chemical Changes Occurring the Transient Injection of Selected Vegetable Oils, in New Developments in *Alternative Fuels and Gasolines for SI and CI Engines*, SAE Special Publication SP-958 (Paper No. 930933), 1993, pp. 201–210.
87. Knothe, G., M.O. Bagby, T.W. Ryan III, T.J. Callahan, and H.G. Wheeler, Vegetable Oils as Alternative Diesel Fuels: Degradation of Pure Triglycerides During the Precombustion Phase in a Reactor Simulating a Diesel Engine, in *Alternative Fuels for CI and SI Engines*, SAE Special Publication SP-900, (Paper No. 920194), 1992, pp. 37–63.
88. Ryan III, T.W., T.J. Callahan, L.G. Dodge, and C.A. Moses, in *Development of a Preliminary Specification for Vegetable Oil Fuels for Diesel Engines, USDA Grant No. 59-2489-1-6-060-0*, Southwest Research Institute, San Antonio, 1983.
89. Knothe, G., M.O. Bagby, T.W. Ryan III, and T.J. Callahan, Degradation of Unsaturated Triglycerides Injected into a Pressurized Reactor, *J. Am. Oil Chem. Soc. 68:* 259–267 (1991).
90. Goering, C.E., A.W. Schwab, R.M. Campion, and E.H. Pryde, Soyoil-Ethanol Microemulsions as Diesel Fuel, *Trans. ASAE 26:* 1602–1604,1607 (1983).
91. Faletti, J.J., S.C. Sorenson, and C.E. Goering, Energy Release Rates from Hybrid Fuels, *Trans. ASAE 27:* 322–325 (1984).
92. Boruff, P.A., A.W. Schwab, C.E. Goering, and E.H. Pryde, Evaluation of Diesel Fuel/Ethanol Microemulsions, *Trans. ASAE 25:* 25, 47–53 (1982).
93. Masjuki, H., M.Z. Abdulmuin, H.S. Sii, L.H. Chua, and K.S. Seow, Palm Oil Diesel Emulsion as a Fuel for Diesel Engine: Performance and Wear Characteristics, *J. Energy, Heat Mass Transfer 16:* 295–304 (1994).
94. Sii, H.S., H. Masjuki, and A.M. Zaki, Dynamometer Evaluation and Engine Wear Characteristics of Palm Oil Diesel Emulsions, *J. Am. Oil Chem. Soc. 72:* 905–909 (1995).
95. Ikura, M., M. Stanciulescu, and E. Hogan, Emulsification of Pyrolysis Derived Bio-Oil in Diesel Fuel, *Biomass Bioenergy 24:* 221–232 (2003).
96. Schwab, A.W., and E.H. Pryde, U.S. Patent No. 4,451,267 (1984).
97. Schwab, A.W., and E.H. Pryde, U.S. Patent No. 4,526,586 (1985).
98. Schwab, A.W., and E.H. Pryde, U.S. Patent No. 4,557,734 (1985).
99. Martin, J., and J.-L. Vanhemelryck, Eur. Patent Appl. EP 587,551 (1994).
100. Schwab, A.W., U.S. Patent 4,451,265 (1984).
101. Sexton, M.D., A.K. Smith, J. Bock, M.L. Robbins, S.J. Pace, and P.G. Grimes, Eur. Patent Appl. EP 475,620 (1992).
102. Schon, S.G., and E.A. Hazbun, U.S. Patent No. 5,004,479 (1991).
103. Hunter, H.F., U.S. Patent 5,380,343 (1995).
104. Goering, C.E., A.W. Schwab, R.M. Campion, and E.H. Pryde, Evaluation of Soybean Oil-Aqueous Ethanol Microemulsions for Diesel Engines, in *Proceedings of the International Conference on Plant and Vegetable Oils as Fuels*, ASAE Special Publication No. SP 4-82, 1982, pp. 279–286.
105. Liljedahl, J.B., W.M. Carleton, P.K. Turnquist, and D.W. Smith, in *Tractors and Their Power Units*, 3rd ed., John Wiley & Sons, NY, 1979, p. 75.

11

Glycerol

Donald B. Appleby

Introduction

Glycerol [Chemical Abstracts Registry #56-81-5; also propane-1,2,3-triol, glycerin (USP); see Fig. 1 in Chapter 1], a trihydric alcohol, is a clear, water-white, viscous, sweet-tasting hygroscopic liquid at ordinary room temperature above its melting point. Glycerol was first discovered in 1779 by Scheele, who heated a mixture of litharge and olive oil and extracted it with water. Glycerol occurs naturally in combined form as glycerides in all animal and vegetable fats and oils, and is recovered as a by-product when these oils are saponified in the process of manufacturing soap, when the oils or fats are split in the production of fatty acids, or when the oils or fats are esterified with methanol (or another alcohol) in the production of methyl (alkyl) esters. Since 1949, it has also been produced commercially by synthesis from propylene [115-07-1]. The latter currently accounts for ~25% of the U.S. production capacity and ~12.5% of capacity on a global basis.

The uses of glycerol number in the thousands, with large amounts going into the manufacture of drugs, cosmetics, toothpastes, urethane foam, synthetic resins, and ester gums. Tobacco processing and foods also consume large amounts either as glycerol or glycerides.

Glycerol occurs in combined form in all animal and vegetable fats and oils. It is rarely found in the free state in these fats; rather, it is usually present as a triglyceride combined with such fatty acids, and these are generally mixtures or combinations of glycerides of several fatty acids. Coconut and palm kernel oils containing a high percentage (70–80%) of C_6–C_{14} fatty acids yield larger amounts of glycerol than do fats and oils containing mainly C_{16} and C_{18} fatty acids (see Table A-2 in Appendix A for fatty acid profiles). Glycerol also occurs naturally in all animal and vegetable cells in the form of lipids such as lecithin and cephalins. These complex fats differ from simple fats in that they invariably contain a phosphoric acid residue in place of one fatty acid residue.

The term "glycerol" applies only to the pure chemical compound 1,2,3-propanetriol. The term "glycerin" applies to the purified commercial products normally containing >95% of glycerol. Several grades of glycerin are available commercially. They differ somewhat in their glycerol content and in other characteristics such as color, odor, and trace impurities.

Properties

The physical properties of glycerol are listed in Table 1. Glycerol is completely soluble in water and alcohol, slightly soluble in diethyl ether, ethyl acetate, and dioxane, and insoluble in hydrocarbons (1). Glycerol is seldom seen in the crystallized state because of its tendency to supercool and its pronounced freezing point depression when mixed with water. A mixture of 66.7% glycerol and 33.3% water forms a eutectic mixture with a freezing point of –46.5°C.

Glycerol, the simplest trihydric alcohol, forms esters, ethers, halides, amines, aldehydes, and unsaturated compounds such as acrolein. As an alcohol, glycerol also has the ability to form salts such as sodium glyceroxide.

Manufacture

Until 1949, all glycerol was obtained from fats and oils. Currently ~80% of U.S. production and 90% of global production are from natural glycerides. A variety of processes for synthesizing glycerol from propylene [115-07-1] exists today, but only one is still used in commercial production to any significant degree. The first synthetic glycerol process, put onstream in 1948, followed the discovery that propylene could be

TABLE 1
Physical Properties of Glycerol

Property	Value
Melting point (°C)	18.17
Boiling point (°C)	
0.53 kPa	14.9
1.33 kPa	166.1
13.33 kPa	222.4
101.3 kPa	290
Specific gravity, 25/25°C	1.2620
Vapor pressure (Pa)	
50°C	0.33
100°C	526
150°C	573
200°C	6100
Surface tension (20°C, mN/m)	63.4
Viscosity (20°C, mPa·s)	1499
Heat of vaporization (J/mol)	
55°C	88.12
95°C	76.02
Heat of solution to infinite dilution (kJ/mol)	5.778
Heat of formation (kJ/mol)	667.8
Thermal conductivity [W/(m·K)]	0.28
Flash point (°C)	
Cleveland open cup	177
Pensky-Martens closed cup	199
Fire point (°C)	204

chlorinated in high yields to allyl chloride [107-05-1]. Because allyl chloride could be converted to glycerol by several routes, the synthesis of glycerol from propylene [115-07-1] became possible (2). The production of synthetic glycerol peaked in the 1960s and 1970s, when it accounted for 50–60% of the market, but as the availability of natural glycerin increased, most synthetic producers closed their plants leaving only one producer with a worldwide capacity of 134,000 metric tons as the only significant synthetic glycerin producer. Synthetic glycerin now accounts for <10% of global production.

Glycerol from glycerides (natural glycerol) is obtained from three sources: soap manufacture, fatty acid production, and fatty ester production. In soap manufacture, fat is boiled with a caustic soda (sodium hydroxide) solution and salt. Fats react with the caustic to form soap and glycerol. The presence of salt causes a separation into two layers: the upper layer is soap and the lower layer, which is referred to as spent lye (generally, "lye" is a term for a solution of sodium or potassium hydroxide), contains glycerol, water, salt, and excess caustic. Continuous saponification processes for producing soap are now common and produce a spent lye similar to batch or kettle processes.

In producing fatty acids, the most common process is continuous, high-pressure hydrolysis in which a continuous, upward flow of fat in a column streams countercurrently to water at 250–260°C and 5 MPa (720 psi). The fat is split by the water into fatty acids and glycerol. The fatty acids are withdrawn from the top of the column, and the glycerol containing the aqueous phase (called sweet water) falls and is withdrawn from the bottom. The concentration of the sweet water by evaporation results in a product called hydrolyser crude. The fatty acids from splitting are used to make soap, reduced to the corresponding fatty alcohol, or marketed as fatty acids.

A third source of natural glycerol is the transesterification of oils or fats with alcohol to produce fatty esters (see Fig. 1 in Chapter 1). The glycerol is separated from the resulting esters, usually methyl esters, by water washing. Acidification with hydrochloric acid and removal of residual methanol produce a crude glycerol with a salt content of a few percent. The methyl esters have historically been principally reduced to the corresponding fatty alcohols, but with the advent of biodiesel, the fuel industry has become nearly as large a consumer of methyl esters as the detergent industry. Table 2 provides a breakdown of the global production of glycerol by source. In 2001, biodiesel production, nearly all of which occurred in Europe, accounted for 11% of production or nearly 90,000 metric tons. Before 1995, this source of glycerol was inconsequential.

Recovery. The spent lye resulting from current soap-making processes generally contains 8–15% glycerol; sweet waters from the hydrolysis of fats contain as much as 20% glycerol; crude glycerol from esterification contains 80% or more glycerol. The grade of fat used directly affects the treatment required to produce glycerol of an acceptable commercial quality. The chemicals most commonly used to remove impurities from spent lye and sweet water are hydrochloric acid and caustic soda.

TABLE 2
Global Sources of Glycerol Production[a]

Process	Global production (%)
Fatty acids	41
Soap manufacture	21
Methyl esters for detergent alcohols	14
Methyl esters as biodiesel	11
Synthetic	9
Other	4

[a]*Source:* Reference 3.

The treatment of spent lye consists of a series of operations designed to remove nearly all of the organic impurities (4,5). The spent lye commonly is treated with mineral or fatty acids to reduce the content of free caustic and soda ash and to reduce the pH to 4.6–4.8 (6). Sulfates are to be avoided because they are associated with foaming and heat exchanger fouling during subsequent refining. After cooling, the solid soap is skimmed, and an acid and a coagulant are added, followed by filtration. The addition of caustic soda removes the balance of coagulant in solution and adjusts the pH to a point at which the liquor is least corrosive to subsequent process treatment. Spent lye from modern liquid-liquid countercurrent extraction used with continuous saponification systems requires little treatment other than reduction of free alkali by neutralization with hydrochloric acid. The dilute glycerol is now ready for concentration to 80% soap lye crude glycerol.

The sweet water from continuous and batch autoclave processes for splitting fats contains little or no mineral acids and salts and requires very little in the way of purification compared with spent lye from kettle soap making (7). The sweet water should be processed promptly after splitting to avoid degradation and loss of glycerol by fermentation. Any fatty acids that rise to the top of the sweet water are skimmed. A small amount of alkali is added to precipitate the dissolved fatty acids and neutralize the liquor. The alkaline liquor is then filtered and evaporated to an 88% crude glycerol. Sweet water from modern noncatalytic, continuous hydrolysis may be evaporated to ~88% without chemical treatment.

Ester crude glycerol is usually of high quality when good-quality refined oils are used as raw materials; however, recent trends in the processing of low-quality and/or high free fatty acid oils such as yellow grease, tallow, or recycled frying oils for biodiesel production result in a much lower quality crude glycerin containing various impurities, salts, odor, and color bodies that are difficult to remove in the refining process. In either case, salt residue from the esterification catalyst is typically present at a concentration ≥1%. Crude glycerol originating from esterification or splitting of 100% vegetable oils is segregated from other glycerol throughout processing to produce kosher glycerin, which typically commands a higher value in the marketplace.

Concentration. The quality of crude glycerol directly affects the refining operation and glycerin yield. Specifications for crude glycerol usually limit ash content, i.e., a measure of salt and mineral residue; matter organic nonglycerol, which includes fatty acids and esters; trimethylene glycol, i.e., propane-1,3-diol; water; and sugars (5).

Dilute glycerol liquors, after purification, are concentrated to crude glycerol by evaporation. This process is carried out using conventional evaporation under vacuum heated by low-pressure steam. In the case of soap-lye glycerol, means are supplied for recovery of the salt that forms as the spent lye is concentrated. Multiple effort evaporators are typically used to conserve energy while concentrating the liquors to a glycerol content of 85–90%.

Refining. The refining of natural glycerol is generally accomplished by distillation, followed by treatment with active carbon. In some cases, refining is accomplished by ion exchange.

Distillation. In the case of spent-lye crude, the composition is ~80% glycerol, 7% water, 2% organic residue, and <10% ash. Hydrolysis crude is generally of a better quality than soap-lye crude with a composition of ~88% glycerol, <1% ash (little or no salt), and <1.5% organic residue.

Distillation equipment for soap-lye and esterification crude requires salt-resistant metallurgy. The solid salt that results when glycerol is vaporized is removed by filtration or as bottoms from a wiped film evaporator. Scraped-wall evaporators are capable of vaporizing glycerol very rapidly and almost completely, such that a dry, powdery residue is discharged from the base of the unit (5). Distillation of glycerol under atmospheric pressure is not practicable because it polymerizes and decomposes glycerol to some extent at the normal boiling point of 204°C. A combination of vacuum and steam distillation is used in which the vapors are passed from the still through a series of condensers or a packed fractionation section in the upper section of the still. Relatively pure glycerol is condensed. High-vacuum conditions in modern stills minimize glycerol losses due to polymerization and decomposition.

Bleaching and Deodorizing. The extensive use of glycerol and glycerol derivatives in the food industry underscores the importance of the removal of both color and odor (also necessary requirements of USP and extra-quality grades). Activated carbon (1–2%) and a diatomite filter aid are added to the glycerol in a bleach tank at 74–79°C, stirred for 1–2 h, and then filtered at the same temperature, which is high enough to ensure easy filtering and yet not so high as to lead to darkening of the glycerol.

Ion Exchange. Most natural glycerol is refined by the methods described above. However, several refiners employ or have employed ion-exchange systems. When ionized solids are high, as in soap-lye crude, ion exclusion treatment can be used to separate the ionized material from the nonionized (mainly glycerol). A granular resin such as Dowex 50WX8 may be used for ion exclusion. For ion exchange, crude or distilled glycerol may be treated with a resin appropriate for the glycerol content and impurities present. Macroreticular resins such as Amberlite 200, 200C, IRA-93, and IRA-90 may be used with undiluted glycerol. However, steam deodorization is often necessary to remove odors imparted by the resin. Ion exchange and ion exclusion are not widely used alternatives to distillation (5).

Grades. Two grades of crude glycerol are marketed: (i) soap-lye crude glycerol (also sometimes referred to as salt crude) obtained by the concentration of lye from kettle or continuous soap-making processes, which contains ~80% glycerol and (ii) hydrolysis crude glycerol resulting from hydrolysis of fats, which contains ~88–91% glycerol and a small amount of organic salts. Because glycerol from methyl ester production contains salt, it is usually marketed as salt crude. The value one can obtain for crude glycerol in the marketplace will be a direct function of its quality as determined by the ease of its subsequent refining.

Several grades of refined glycerol, such as high gravity and USP, are marketed; specifications vary depending on the consumer and the intended use. However, many industrial uses require specifications equal to or more stringent than those of USP; therefore, the vast majority of commercially marketed glycerin meets USP specifications at a minimum.

USP-grade glycerol is water-white, and meets the requirements of the USP. It is classified as GRAS (generally recognized as safe) by the FDA (Food and Drug Administration), and is suitable for use in foods, pharmaceuticals, and cosmetics, or when the highest quality is demanded or the product is designed for human consumption. It has a minimum specific gravity (25°C/25°C) of 1.249, corresponding to >95% glycerol; however, 99.5% minimum glycerol is the most common purity found in the marketplace. However, simply meeting the specifications as laid out in the USP compendium is not sufficient to be considered USP glycerol. Good Manufacturing Practices for bulk pharmaceuticals must be strictly adhered to including batch traceability. Additionally, a manufacturer, importer, or reseller must comply will all other applicable federal and local regulations including the Federal Bioterrorism Act of 2003.

The European Pharmacopoeia (PH.EUR.) grade is similar to the USP, but the common PH.EUR. grade has a minimum glycerol content of 99.5%. The chemically pure (CP) grade designates a grade of glycerol that is about the same as the USP but with the specifications varying slightly as agreed by buyer and seller. The high-gravity grade is a pale-yellow glycerol for industrial use with a minimum specific gravity (25°C/25°C) of 1.2595. Dynamite grade, which has the same specific gravity as the high-gravity variety but is more yellow, has all but disappeared from the market. All of these grades satisfy the federal specifications for glycerol (0-G491B-2). Virtually all grades of glycerin can be produced in kosher form provided the raw materials are 100% vegetable-based, and the appropriate rabbinical oversight and certification are obtained.

Economic Aspects

Until recently, commercial production and consumption of glycerol were generally considered a fair barometer of industrial activity because it enters into such a large number of industrial processes. In the past, it generally tended to rise in periods of prosperity and fall in recession times. However, the advent of the biodiesel industry has changed this market dynamic because the fuel consumption of methyl esters is driven by different factors such as farm policy, tax credits, environmental and energy security legislation and regulations, as well as petroleum prices, thereby severing the link between glycerol generation and general economic activity.

In any discussion of the commercial aspects of the glycerol market, it is imperative to keep in mind that, unlike most industrial chemicals, the supply of glycerin is not determined by its demand, but rather by the global demand for soaps, detergents, fabric softeners, and more recently biodiesel, as well as all of the other goods in which fatty acids, fatty alcohols, and their derivatives are used.

Glycerol production in the United States (Table 3) rose from 19,800 metric tons in 1920 to a peak of 166,100 t in 1967 (8). During the next 20 yr, North American production held reasonably steady in the 130,000–140,000 metric tons/yr range. During the 1990s however, production expanded again to challenge the 1967 record. At the same time, world production of glycerol continued to expand from 650,000 metric tons/yr in 1995 to ~800,000 t in 2001. Much of this increased production can be accounted for by the development of a European biodiesel industry during that time period.

The by-product nature of glycerol production leads to rather volatile price swings in the marketplace because producers must stimulate or restrict demand to meet the available supply which, as previously noted, is determined by factors outside the glycerol market. Since 1920, the price of refined glycerol in the United States has varied from a low of \$0.22/kg in the early 1930s to a high of nearly \$2.30/kg in 1995. Within 4 yr of this peak, prices as low as \$0.64/kg were reported in the North American market as large imports from Asia and Europe hit U.S. shores.

TABLE 3
North American Glycerin Production (100% glycerol basis)

Year	Production of crude (t)
1920	19,800
1940	71,600
1950	102,300
1960	136,900
1970	153,900
1980	136,860
1990	133,450
2000	156,950

Uses

Glycerol is used in nearly every industry. The largest single use is in drugs and oral-care products including toothpaste, mouthwash, and oral rinses (Table 4). Its use in tobacco processing and urethane foams remains at a fairly even consumption level. Use in foods, drugs, and cosmetics is growing, whereas alkyd resin usage has shrunk considerably.

Foods. Glycerol as a food is easily digested and nontoxic; its metabolism places it with the carbohydrates, although it is present in a combined form in all vegetable and animal fats. In flavoring and coloring products, glycerol acts as a solvent, and its viscosity lends body to the product. Raisins saturated with glycerol remain soft when mixed with cereals. It is used as a solvent, a moistening agent, and an ingredient of syrups as a vehicle. In candies and icings, glycerol retards crystallization of sugar. Glycerol is used as a heat-transfer medium in direct contact with foods in quick freezing, and as a lubricant in machinery used for food processing and packaging. Emulsifiers represent a large volume food use in the form of glycerol esters. Mixed mono- and diglycerides are the most commonly employed, but distilled monoglycerides are sometimes used (see Derivatives/Esters below). The polyglycerols and polyglycerol esters have increasing use in foods, particularly in shortenings and margarines.

Drugs and Cosmetics. In drugs and medicines, glycerol is an ingredient of many tinctures and elixirs, and glycerol of starch is used in jellies and ointments. It is employed in cough medicines and anesthetics, such as glycerol-phenol solutions, for ear treatments, and in bacteriological culture media. Its derivatives are used in tranquilizers (e.g., glyceryl guaiacolate [93-14-1]), and nitroglycerin [55-65-0] is a vasodilator in coronary spasm. In cosmetics, glycerol is used in many creams and

TABLE 4
North American Glycerol Use[a]

Use	1978 (t)	1998 (t)
Alkyd resins	21,510	9400
Cellophane and casings	6380	5365
Tobacco and triacetin	20,300	24,235
Explosives	2890	2220
Oral care and pharmaceuticals	19,760	44,215
Cosmetics	4340	21,645
Foods, including emulsifiers	14,830	42,180
Urethanes	13,810	17,780
Miscellaneous and distributor sales	26,730	18,130
Total	130,550	185,170

[a]*Source:* Reference 9.

lotions to keep the skin soft and replace skin moisture. It is widely used in toothpaste to maintain the desired smoothness and viscosity, and lend a shine to the paste.

Tobacco. In processing tobacco, glycerol is an important part of the casing solution sprayed on tobacco before the leaves are shredded and packed. Along with other flavoring agents, it is applied at a rate of ~2.0 wt% of the tobacco to prevent the leaves from becoming friable and thus crumbling during processing; by remaining in the tobacco, glycerol helps to retain moisture, thus preventing the drying out of the tobacco and influencing its burning rate. It is used also in the processing of chewing tobacco to add sweetness and prevent dehydration, and as a plasticizer in cigarette papers.

Wrapping and Packaging Materials. Meat casings and special types of papers, such as glassine and greaseproof paper, need plasticizers to give them pliability and toughness; as such, glycerol is completely compatible with the base materials used, is absorbed by them, and does not crystallize or volatilize appreciably.

Lubricants. Glycerol can be used as a lubricant in places in which an oil would fail. It is recommended for oxygen compressors because it is more resistant to oxidation than mineral oils. It is also used to lubricate pumps and bearings exposed to fluids such as gasoline and benzene, which would dissolve oil-type lubricants. In food, pharmaceutical, and cosmetic manufacture, in which there is contact with a lubricant, glycerol may be used to replace oils.

Glycerol is often used as a lubricant because its high viscosity and ability to remain fluid at low temperatures make it valuable without modification. To increase its lubricating power, finely divided graphite may be dispersed in it. Its viscosity may be decreased by the addition of water, alcohol, or glycols and increased by polymerization or mixing with starch; pastes of such compositions may be used in packing pipe joints, in gas lines, or in similar applications. For use in high-pressure gauges and valves, soaps are added to glycerol to increase its viscosity and improve its lubricating ability. A mixture of glycerin and glucose is employed as a nondrying lubricant in the die-pressing of metals. In the textile industry, glycerol is frequently used in connection with so-called textile oils, in spinning, knitting, and weaving operations.

Urethane Polymers. An important use for glycerol is as the fundamental building block in polyethers for urethane polymers. In this use, it is the initiator to which propylene oxide, alone or with ethylene oxide, is added to produce trifunctional polymers which, upon reaction with diisocyanates, produce flexible urethane foams. Glycerol-based polyethers have found some use, too, in rigid urethane foams.

Other Uses. In the late 1990s, when glycerol pricing reached low levels due to the large increase in European biodiesel production, at least one consumer products company formulated glycerin into liquid laundry detergents as a partial replacement for propylene glycol. Glycerol is also used in cement compounds, caulking compounds, embalming fluids, masking and shielding compounds, soldering compounds, asphalt, ceramics, photographic products, and adhesives.

Derivatives

Glycerol derivatives include acetals, amines, esters, and ethers. Of these, the esters are the most widely employed. Alkyd resins are esters of glycerol and phthalic anhydride. Glyceryl trinitrate [55-63-0] (nitroglycerin) is used in explosives and as a heart stimulant. Included among the esters also are the ester gums (rosin acid ester of glycerol), mono- and diglycerides (glycerol esterified with fatty acids or glycerol transesterified with oils), used as emulsifiers and in shortenings. The salts of glycerophosphoric acid are used medicinally.

Mixtures of glycerol with other substances are often named as if they were derivatives of glycerol, e.g., boroglycerides (also called glyceryl borates) are mixtures of boric acid and glycerol. Derivatives such as acetals, ketals, chlorohydrins, and ethers can be prepared but are not made commercially, with the exception of polyglycerols.

The polyglycerols, ethers prepared with glycerol itself, have many of the properties of glycerol. Diglycerol, [627-82-7], is a viscous liquid [287 mm^2/s (= cSt) at 65.6°C], ~25 times as viscous as glycerol. The polyglycerols offer greater flexibility and functionality than glycerol. Polyglycerols up to and including triacontaglycerol (30 condensed glycerol molecules) have been prepared commercially; the higher forms are solid. They are soluble in water, alcohol, and other polar solvents. They act as humectants, much like glycerol, but have progressively higher molecular weights and boiling points. Products based on polyglycerols are useful in surface-active agents, emulsifiers, plasticizers, adhesives, lubricants, antimicrobial agents, medical specialties, and dietetic foods.

Esters. The mono- and diesters of glycerol and fatty acids occur naturally in fats that have become partially hydrolyzed. The triacylglycerols are primary components of naturally occurring fats and fatty oils. Mono- and diacylglycerols are made by the reaction of fatty acids or raw or hydrogenated oils, such as cottonseed and coconut, with an excess of glycerol or polyglycerols. Commercial glycerides are mixtures of mono- and diesters, with a small percentage of the triester. They also contain small amounts of free glycerol and free fatty acids. High purity monoacylglycerols are prepared by molecular or short-path distillation of glyceride mixtures.

The higher fatty acid mono- and diesters are oil-soluble and water-insoluble. They are all edible, except the ricinoleate and the erucinate, and find their greatest use as emulsifiers in foods and in the preparation of baked goods (10). A mixture of mono-, di-, and triglycerides is manufactured in large quantities for use in superglycerinated shortenings. Mono- and diglycerides are important modifying agents in the manufacture of alkyd resins, detergents, and other surface-active agents. The monoglycerides are also used in the preparation of cosmetics, pigments, floor waxes, synthetic rubbers, coatings, textiles (11), for example.

Tailored triglycerides with unique nutritional properties have grown in importance in recent years. These compounds are produced from glycerol esterification with specific high-purity fatty acids. A triglyceride consisting primarily of C_8, C_{10}, and C_{22} fatty acid chains designated "caprenin" was marketed as a low-energy substitute for cocoa butter (12). By starting with behenic monoglyceride made from glycerol and behenic acid, the shorter caprylic and capric acids can be attached to the behenic monoglyceride to deliver a triglyceride having only one long fatty acid chain (13,14).

Acetins. The acetins are the mono-, di-, and triacetates of glycerol that form when glycerol is heated with acetic acid. Monoacetin (glycerol monoacetate [26446-35-5]) is a thick hygroscopic liquid, and is sold for use in the manufacture of explosives, in tanning, and as a solvent for dyes. Diacetin (glycerol diacetate [25395-31-7]) is a hygroscopic liquid, and is sold in a technical grade for use as a plasticizer and softening agent and as a solvent. Triacetin, melting point (m.p.) = –78°C, has a very slight odor and a bitter taste. Glycerol triacetate [102-76-1] occurs naturally in small quantities in the seed of *Euonymus europaeus*. Most commercial triacetin is USP grade. Its primary use is as a cellulose plasticizer in the manufacture of cigarette filters; its second largest use is as a component in binders for solid rocket fuels. Smaller amounts are used as a fixative in perfumes, as a plasticizer for cellulose nitrate, in the manufacture of cosmetics, and as a carrier in fungicidal compositions.

Identification and Analysis

The methods of analysis of the American Oil Chemists' Society (AOCS) are the principal procedures followed in the United States and Canada and are official in commercial transactions; in Europe, however, methods published by the European Oleochemical and Allied Products Group of CEFIC (known as APAG) are used. When the material is for human consumption or drug use, it must meet the specifications of the USP (15), European Pharmacopoeia, or Japanese Pharmacopoeia. Commercial distilled grades of glycerol do not require purification before analysis by the usual methods. The determination of glycerol content by the periodate method (16), which replaced the acetin and dichromate methods previously used, is more accurate and more specific as well as simpler and more rapid.

Glycerol is most easily identified by heating a drop of the sample with ~1 g powdered potassium bisulfate and noting the very penetrating and irritating odor of the acrolein that is formed. Due to the toxicity of acrolein, a preferred method is the Cosmetics, Toiletry, and Fragrance Association (CTFA) method GI-1, an infrared spectrophotometric method. Glycerol may be identified by the preparation of crystalline derivatives such as glyceryl tribenzoate, m.p. 71–72°C; glycerol tris(3,5-dinitrobenzoate), m.p. 190–192°C; or glycerol tris(*p*-nitrobenzoate), m.p. 188–189°C (17).

The concentration of distilled glycerol is easily determined from its specific gravity (18) by the pycnometer method (19) with a precision of ±0.02%. Determination of the refractive index also is employed (but not as widely) to measure glycerol concentration to ±0.1% (20). The preferred method of determining water in glycerol is the Karl Fischer volumetric method (21). Water can also be determined by a special quantitative distillation in which the distilled water is absorbed by anhydrous magnesium perchlorate (22). Other tests such as ash, alkalinity or acidity, sodium chloride, and total organic residue are included in AOCS methods (15,18,20).

Handling and Storage

Most crude glycerol is shipped to refiners in standard tank cars or tank wagons. Imported crude arrives in bulk, in vessels equipped with tanks for such shipment, or in drums. Refined glycerol of a CP or USP grade is shipped mainly in bulk in tank cars or tank wagons. These are usually stainless steel, aluminum, or lacquer-lined. However, pure glycerol has little corrosive tendency, and may be shipped in standard, unlined steel tank cars, provided they are kept clean and in a rust-free condition. Some producers offer refined glycerol in 250- to 259-kg (208-L or 55-gal) drums of a nonreturnable type (lCC-17E). These generally have a phenolic resin lining.

Storage. For receiving glycerol from standard 30.3 m^3 (8000-gal) tank cars (36.3 t), a storage tank of 38–45 m^3 [$(10–12) \times 10^3$ gal)] capacity should be employed. Preferably it should be of stainless steel (304 or 316), of stainless- or nickel-clad steel, or of aluminum. Certain resin linings such as Lithcote have also been used. Glycerol does not seriously corrode steel tanks at room temperature, but gradually absorbed moisture may have an effect. Therefore, tanks should be sealed with an air-breather trap.

Handling Temperatures. Optimum temperature for pumping is in the range 37–48°C. Piping should be stainless steel, aluminum, or galvanized iron. Valves and pumps should be bronze, cast-iron with bronze trim, or stainless steel. A pump of 3.15 L/s (50 gal/min) capacity unloads a tank car of warm glycerol in ~4 h.

Health and Safety Factors

Glycerol has had GRAS status since 1959 and is a miscellaneous or general-purpose food additive under the CFR (Code of Federal Regulations) (23), and it is permitted in certain food-packaging materials.

Oral 50% lethal dose levels were determined in mice at 470 mg/kg (24) and guinea pigs at 7750 mg/kg (25). Several other studies (26–28) showed that large quantities of both synthetic and natural glycerol can be administered orally to experimental animals and humans without the appearance of adverse effects. Intravenous administration of solutions containing 5% glycerol to animals and humans was found to cause no toxic or otherwise undesirable effects (29). The aquatic toxicity (TLm96) for glycerol is >1000 mg/L (30), which is defined by the National Institute for Occupational Safety and Health as an insignificant hazard.

References

1. *Physical Properties of Glycerin and Its Solutions*, Glycerin Producers' Association, New York, 1975.
2. *Ullmann's Encyclopedia of Industrial Chemistry*, 5th edn., Vol. A1, John Wiley & Sons, Hoboken, NJ, p. 425.
3. *HBI Report*, Oleoline Glycerine Market Report, March 2002.
4. Patrick, T.M., Jr., E.T. McBee, and H.B. Baas, *J. Am. Chem. Soc. 68:* 1009 (1946).
5. Sanger, W.E., *Chem. Met. Eng. 26:* 1211 (1922).
6. Woollatt, E., *The Manufacture of Soaps, Other Detergents and Glycerin*, John Wiley & Sons, Inc., New York, 1985, pp. 296–357.
7. Trauth, J.L., *Oil Soap 23:* 137 (1946).
8. *SDA Glycerin and Oleochemicals Statistics Report*, The Soap and Detergent Association, New York, 1992.
9. *SDA Glycerine End Use Survey*, The Soap and Detergent Association, Washington, 2002.
10. Nash, N.H., and V.K. Babayan, *Food Process 24:* 2 (1963); *Baker's Dig. 38:* 46 (1963).
11. Parolla, A.E., and C.Z. Draves, *Am. Dyestuff Rep. 46:* 761 (Oct. 21, 1957); *47:* 643 (Sept. 22, 1958).
12. *Caprenin*, U.S. FDA GRAS Petition 1GO373, U.S. Food and Drug Administration, Washington, 1990.
13. Kluesener, B.W., G.K. Stipp, and D.K. Yang, Procter & Gamble, U.S. Patent 5,142,071 (1992).
14. Stipp, G.K., and B.W. Kluesener, Procter & Gamble, U.S. Patent 5,142,072 (1992).
15. *The United States Pharmacopoeia* (USP XX-NF XV), The United States Pharmacopeial Convention, Rockville, MD, 1980.
16. *Official and Tentative Methods*, 3rd ed., American Oil Chemists' Society, Champaign, IL, 1978, Ea6-51.
17. Miner, C.S., and N.N. Dalton, *Glycerol*, ACS Monograph 117, Reinhold Publishing, New York, 1953, pp. 171–175.
18. Bosart, L.W., and A.O. Snoddy, *Ind. Eng. Chem. 19:* 506 (1927).
19. *Official and Tentative Methods*, 3rd ed., American Oil Chemists' Society, Champaign, IL, 1978, Ea7-50.
20. Hoyt, L.T., *Ind. Eng. Chem. 26:* 329 (1934).
21. *Official and Tentative Methods*, 3rd edn., American Oil Chemists' Society, Champaign, IL, 1978, Ea8-58.
22. Spaeth, C.P., and G.F. Hutchinson, *Ind. Eng. Chem. Anal. Ed. 8:* 28 (1936).
23. *Code of Federal Regulations*, Title 21, Sect. 182.1320, Washington, 1993.
24. Smyth, H.F., J. Seaton, and L. Fischer, *J. Ind. Hyg. Toxicol. 23:* 259 (1941).
25. Anderson, R.C., P.N. Harris, and K.K. Chen, *J. Am. Pharm. Assoc. Sci. Ed. 39:* 583 (1950).
26. Johnson, V., A.J. Carlson, and A. Johnson, *Am. J. Physiol. 103:* 517 (1933).
27. Hine, C.H., H.H. Anderson, H.D. Moon, M.K. Dunlap, and M.S. Morse, *Arch. Ind. Hyg. Occup. Med. 7:* 282 (1953).
28. Deichman, W., *Ind. Med. Ind. Hyg. Sec. 9:* 60 (1940).
29. Sloviter, H.A., *J. Clin. Investig. 37:* 619 (1958).
30. Hann, W., and P.A. Jensen, *Water Quality Characteristics of Hazardous Materials*, Texas A&M University, College Station, 1974, p. 4.

Appendix A

This appendix consists of four tables that were referenced in the preceding chapters:

1. Properties of fatty acids and esters of relevance to biodiesel.
2. Major fatty acids (wt%) in some oils and fats used or tested as alternative diesel fuels.
3. Fuel-related properties of various fats and oils.
4. Fuel-related physical properties of esters of oils and fats.

The data in these tables were drawn from the 27 references listed below. All reference numbers correlate with the corresponding numbers in the tables and not with the reference numbers in chapters.

References

1. Gunstone, F.D., J.L. Harwood, and F.B. Padley, *The Lipid Handbook*, 2nd edn., Chapman & Hall, London, 1994.
2. *Handbook of Chemistry and Physics*, 66th edition, CRC Press, Boca Raton, FL, 1985, edited by R.C. Weast, M.J. Astle, and W.H. Beyer.
3. Freedman, B., and M.O. Bagby, Heats of Combustion of Fatty Esters and Triglycerides, *J. Am. Oil Chem. Soc. 66:* 1601–1605 (1989).
4. Klopfenstein, W.E., Effect of Molecular Weights of Fatty Acid Esters on Cetane Numbers as Diesel Fuels, *J. Am. Oil Chem. Soc. 62:* 1029–1031 (1985).
5. Freedman, B., and M.O. Bagby, Predicting Cetane Numbers of n-Alcohols and Methyl Esters from Their Physical Properties, *J. Am. Oil Chem. Soc. 67:* 565–571 (1990).
6. Knothe, G., A.C. Matheaus, and T.W. Ryan, III, Cetane Numbers of Branched and Straight-Chain Fatty Esters Determined in an Ignition Quality Tester, *Fuel 82:* 971–975 (2003).
7. Freedman, B., M.O. Bagby, T.J. Callahan, and T.W. Ryan, III, Cetane Numbers of Fatty Esters, Fatty Alcohols and Triglycerides Determined in a Constant Volume Combustion Bomb, SAE Technical Paper Series No. 900343, 1990.
8. Knothe, G., M.O. Bagby, and T.W. Ryan, III, Cetane Numbers of Fatty Compounds: Influence of Compound Structure and of Various Potential Cetane Improvers, SAE Technical Paper Series No. 971681; published in SP-1274, SAE Spring Fuels and Lubricants Meeting, Dearborn, Michigan, May 1997.
9. Gouw, T.H., J.C. Vlugter, and C.J.A. Roelands, Physical Properties of Fatty Acid Methyl Esters. VI. Viscosity, *J. Am. Oil Chem. Soc. 43:* 433–434 (1966).
10. Allen, C.A.W., K.C. Watts, R.G. Ackman, and M.J. Pegg, Predicting the Viscosity of Biodiesel Fuels from Their Fatty Acid Ester Composition, *Fuel 78:* 1319–1326 (1999).
11. Applewhite, T.H., Fats and Fatty Oils, in *Kirk-Othmer, Encyclopedia of Chemical Technology*, edited by M. Grayson and D. Eckroth, 3rd edn., Vol. 9, John Wiley & Sons, New York, 1980, pp. 795–831.
12. Goering, C.E., A.W. Schwab, M.J. Daugherty, E.H. Pryde, and A.J. Heakin, Fuel Properties of Eleven Vegetable Oils, *Trans. ASAE 25:* 1472–1477 (1982).
13. Ali, Y., M.A. Hanna, and S.L. Cuppett, Fuel Properties of Tallow and Soybean Oil Esters, *J. Am. Oil Chem. Soc. 72:* 1557–1564 (1995).

TABLE A-1
Properties of Fatty Acids and Esters of Relevance to Biodiesel

Trivial (systematic) name; acronym	M.W.	m.p. (°C)	b.p.[a,b] (°C)	Cetane number	Kinematic viscosity[c] (40°C; mm^2/s = cSt)	HG[d] (kg-cal/mol)
Caprylic (Octanoic) acid; 8:0	144.213	16.5	239.3			
Methyl ester	158.240		193	33.6 (98.6)[e]	1.16[j]; 0.99[k];	1313
Ethyl ester	172.268	–43.1	208.5		1.37 (25°C)[k]	1465
Butyl ester	200.322			39.6 (98.7)[e]		
Capric (Decanoic) acid; 10:0	172.268	31.5	270	47.6 (98.0)[e]		1453.07 (25°C)
Methyl ester	186.295		224	47.2 (98.1)[e]; 47.9[f]	1.69[j]; 1.40[k]	1625
Ethyl ester	200.322	–20	243–245	51.2 (99.4)[e]	1.99 (25°C)[j]	1780
Propyl ester	214.349			52.9 (98)[e]		
Iso-propyl ester	214.349			46.6 (97.7)[e]		
Butyl ester	228.376			54.6 (98.6)[e]		
Lauric (Dodecanoic) acid; 12:0	200.322	44	131^{1}			1763.25 (25°C)
Methyl ester	214.349	5	266^{766}	61.4 (99.1)[e]; 60.8[f]	2.38[j]; 1.95[k];	1940
Ethyl ester	228.376	–1.8fr	163^{25} 2098		2.88[k]	2098
Myristic (Tetradecanoic) acid; 14:0	228.376	58	250.5^{100}			2073.91 (25°C)
Methyl ester	242.403	18.5	295^{751}	66.2 (96.5)[e]; 73.5[f]	3.23[j]; 2.69[k]	2254
Ethyl ester	256.430	12.3	295	66.9 (99.3)[e]		2406
Butyl myristate	284.484			69.4 (99.0)[e]		
Palmitic (Hexadecanoic) acid; 16:0	256.430	63	350			2384.76 (25°C)
Methyl ester	270.457	30.5	$415–418^{747}$	74.5 (93.6)[e]; 85.9[g]; 74.3[f]	4.32[j]; 3.60[k]	
Ethyl ester	284.484	19.3/24	191^{10}	93.1[g]		2550
Propyl ester	298.511	20.4	190^{12}	85.0[g]		2717

Iso-propyl ester	298.511	13–14	160^{2}	82.6[g]		
Butyl ester	312.538	16.9		91.9[g]		
2-Butyl ester	312.538			84.8[g]		
Iso-butyl ester	312.538	22.5, 28.9	199^{5}	83.6[g]		
Triacylglycerol	807.339	66.4	310–320	89[h]		7554
Palmitoleic [9(*Z*)-Hexadecenoic] acid; 16:1	254.412					
Methyl ester	268.439			51.0[g]		2521
Stearic (Octadecanoic) acid; 18:0	284.484	71	360[d]	61.7[i]		2696.12 (25°C)
Methyl ester	298.511	39	$442–443^{747}$	86.9 (92.1)[e]; 101[i]; 75.6[f]	56.1[j]; 4.74[k]	
Ethyl ester	312.538	31–33.4	199^{10}	76.8[i]; 97.7[g]		2859
Propyl ester	312.538			69.9[i]; 90.9[g]		3012
Iso-propyl ester	312.538			96.5[g]		
Butyl ester	326.565	27.5	343	80.1[i]; 92.5[g]		
2-Butyl ester	326.565			97.5[g]		
Iso-butyl ester	326.565			99.3[8]		
Triacylglycerol	891.501	73		85[h]		8558
Oleic [9(*Z*)-Octadecenoic] acid; 18:1	282.468	16	286^{100}	46.1[i]		2657.4 (25°C)
Methyl ester	296.495	–20	218.5^{20}	55[i]; 59.3[g]	4.45[j]; 3.73[k]	2828
Ethyl ester	310.522		$216–217^{151}$	53.9[i]; 67.8[g]	5.50(25°C)	
Propyl ester	324.547			55.7[i]; 58.8[g]		
Iso-propyl ester	324.547			86.6[g]		
Butyl ester	338.574			59.8[i]; 61.6[g]		
2-Butyl ester	338.574			71.9[g]		
Iso-butyl ester	338.574			59.6[g]		
Triacylglycerol	885.453	–5.5	$235–240^{18}$	45[h]		8389

(Continued)

TABLE A-1
(Cont.)

Trivial (systematic) name; acronym	M.W.	m.p. (°C)	b.p.[a,b] (°C)	Cetane number	Kinematic viscosity[c] (40°C; mm^2/s = cSt)	HG[d] (kg-cal/mol)
Linoleic (9*Z*,12*Z*-Octadecadienoic) acid; 18:2	280.452	–5	$229–230^{16}$	31.4[i]		
Methyl ester	294.479	–35	215^{20}	42.2[i]; 38.2[g]	3.64[j];2.65[k]	2794
Ethyl ester	308.506		$270–275^{180}$	37.1[i]; 39.6[g]		
Propyl ester	322.533			40.6[i]; 44.0[g]		
Butyl ester	336.560			41.6[i]; 53.5[g]		
Triacylglycerol	879.405			32[h]		
Linolenic (9*Z*,12*Z*,15*Z*-Octadecatrienoic) acid; 18:3	278.436	–11	$230–232^{17}$	20.4[i]		
Methyl ester	292.463	–57/–52	$109^{0.018}$	22.7[i]	3.27[j];2.65[k]	2750
Ethyl ester	306.490		$174^{2.5}$	26.7[i]		
Propyl ester	320.517			26.8[i]		
Butyl ester	324.544			28.6[i]		
Triacylglycerol	873.357			23[i]		
Erucic (13*Z*-Docosenoic) acid; 22:1	338.574	33–4	265^{15}		7.21[j];5.91[k]	3454
Methyl ester	352.601		$221–222^{5}$			
Ethyl ester	366.628		$229–230^{5}$			

[a]*Source:* References 1 and 2 for melting point and boiling point data.
[b]Superscripts denote pressure (mm Hg) at which the boiling point was determined.
[c]Viscosity values determined at 40°C unless indicated otherwise. *Source:* Reference 8 for kinematic viscosities. Some dynamic viscosity values are also given, see footnote j.
[d]Source: References 2 and 3 for heat of combustion values.
[e]Source: Reference 4. Number in parentheses indicates purity (%) of the material used for cetane number determinations as given in Reference 4.
[f]*Source:* Reference 5.
[g]*Source:* Reference 6.
[h]*Source:* Reference 7 for estimated cetane numbers.
[i]*Source:* Reference 8.
[j]*Source:* Reference 9.
[k]*Source:* Reference 10 for dynamic viscosity (mPa·s = cP).

TABLE A-2
Major Fatty Acids in Some Oils and Fats Used or Tested as Alternative Diesel Fuels[a]

Oil or fat	Iodine value	Saponification value	Fatty acid composition (wt%) 8:0	10:0	12:0	14:0	16:0	18:0	18:1	18:2	18:3	22:1
Babassu	10–18	245–256	2.6–7.3	1.2–7.6	40–45	11–27	5.2–11	1.8–7.4	9–20	1.4–6.6		
Canola	110–126	188–193					1.5–6	1–2.5	52–66.9	16.1–31	6.4–14.1	1–2
Coconut	6–12	248–265	4.6–9.5	4.5–9.7	44–51	13–20.6	7.5–10.5	1–3.5	5–8.2	1.0–2.6	0–0.2	
Corn	103–140	187–198				0–0.3	7–16.5	1–3.3	20–43	39–62.5	0.5–1.5	
Cottonseed	90–119	189–198				0.6–1.5	21.4–26.4	2.1–5	14.7–21.7	46.7–58.2		
Linseed	168–204	188–196					6–7	3.2–5	13–37	5–23	26–60	
Olive	75–94	184–196			0–1.3	7–20	0.5–5.0	55–84.5	3.5–21			
Palm	35–61	186–209		0–0.4	0.5–2.4	32–47.5	3.5–6.3	36–53	6–12			
Peanut	80–106	187–196				0–0.5	6–14	1.9–6	36.4–67.1	13–43		0–0.3
Rapeseed	94–120	168–187				0–1.5	1–6	0.5–3.5	8–60	9.5–23	1–13	5–64
Safflower	126–152	175–198					5.3–8.0	1.9–2.9	8.4–23.1	67.8–83.2		
Safflower, high-oleic	90–100	175–195					4–8	2.3–8	73.6–79	11–19		
Sesame	104–120	187–195					7.2–9.2	5.8–7.7	35–46	35–48		
Soybean	117–143	189–195					2.3–13.3	2.4–6	17.7–30.8	49–57.1	2–10.5	0–0.3
Sunflower	110–143	186–194					3.5–7.6	1.3–6.5	14–43	44–74		
Tallow (beef)	35–48	218–235				2.1–6.9	25–37	9.5–34.2	14–50	26–50		

[a]*Source:* Values combined from References 1 and 11.
[b]These oils and fats may contain small amounts of other fatty acids not listed here. For example, peanut oil contains 1.2% 20:0, 2.5% 22:0, and 1.3% 24:0 fatty acids (1).

TABLE A-3
Fuel-Related Properties of Various Fats and Oils[a]

Oil or fat	CN	HG (kJ/kg)	Kinematic viscosity (37.8°C; mm^2/s)	CP (°C)	PP (°C)	FlP (°C)
Babassu	38.0					
Castor		39500	297	—	–31.7	260
Coconut						
Corn	37.6	39500	34.9	–1.1	–40.0	277
Cottonseed	41.8	39468	33.5	1.7	–15.0	234
Crambe	44.6	40482	53.6	10.0	–12.2	274
Linseed	34.6	39307	27.2	1.7	–15.0	241
Palm	42.0					
Peanut	41.8	39782	39.6	12.8	–6.7	271
Rapeseed	37.6	39709	37.0	–3.9	–31.7	246
Safflower	41.3	39519	31.3	18.3	–6.7	260
High-oleic safflower	49.1	39516	41.2	–12.2	–20.6	293
Sesame	40.2	39349	35.5	–3.9	–9.4	260
Soybean	37.9	39623	32.6; 28.05[a]	–3.9; –9[b]	–12.2; –16[b]	254
Sunflower	37.1	39575	37.1	7.2	–15.0	274
Tallow[b]	—	40054	51.15	—	—	201
No. 2 DF	47.0	45343	2.7	–15.0	–33.0	52

[a]*Source:* Reference 12 unless otherwise noted. CN, cetane number; CP, cloud point; DF, diesel fuel; FlP, flash point; HG, gross heat of combustion; PP, pour point.
[b]*Source:* Reference 13.

TABLE A-4
Fuel-Related Physical Properties of Esters of Oils and Fats

Oil or fat; ester	Cetane number	HG (kJ/kg)	Kinematic viscosity (40°C; mm^2/s)	Cloud point (°C)	Pour point (°C)	Flash point[a] (°C)	Reference
Coconut							
Methyl							
Ethyl	67.4	38158	3.08	5	–3	190	14
Corn							
Methyl	65	38480[b]	4.52	–3.4	–3	111	15
Cottonseed							
Methyl	51.2	—	6.8[c] (21°C)	—	–4	110	16
Olive							
Methyl	61	37287[b]	4.70	–2	–3	>110	15
Mustard, yellow (33% C22:1)							
Ethyl	54.9	40679	5.66	1	–15	183	14
Palm							
Ethyl	56.2	39070	4.50 (37.8°C)	8	6	19?	17
Rapeseed (low-erucic; canola)							
Methyl	56	37300[b]	4.53	CFPP: –6		169	18
Methyl	53.7	8850	4.96	CFPP: –6			19
Methyl	47.9	39870	4.76 (37.8°C)	–3	–9	166	17
Ethyl	67.4	40663	6.02	1	–12	170	14

(Continued)

TABLE A-4
Fuel-Related Physical Properties of Esters of Oils and Fats

Oil or fat; ester	Cetane number	HG (kJ/kg)	Kinematic viscosity (40°C; mm²/s)	Cloud point (°C)	Pour point (°C)	Flash point[a] (°C)	Reference
Safflower							
Methyl	49.8	40060			–6	180/149	20
Ethyl	62.2	39872	4.31	–6	–6	178	14
Soybean							
Methyl	49.6	39823/37372[b]	4.18 (40°C)	–1.1	–3.9	190.6	21
Methyl		40080	4.06[c]	3	–7	127	13
Methyl	55.9	39753	3.99	1	0	185	14
Methyl	51.5	39871/37388[b]	4.27				22
Methyl			4.30	0	–2		23
Methyl				–2	–3		24
Methyl	48.7	39720	4.40 (37.8°C)	0	–3	120	17
Ethyl			4.40	–2	–6		24
Ethyl				1	–4		25
Iso-propyl				–9	–12		24
2-Butyl				–12	–15		24
Sunflower							
Methyl	58	38472[b]	4.39	1.5	3	110	15
Methyl	54	38100	4.79 (37.8°C)	0	–3	85?	17

Tallow							
Methyl	61.8	39961/37531[b]	4.99 (40°C)	15.6	12.8	187.8	21
Methyl		39949	4.11[c]	12	9	96	13
Ethyl			5.20	15	3		23
Propyl			7.30	12	9		25
Iso-propyl			6.40	9	3		23
Iso-propyl			7.10	8	0		25
Butyl			6.90	9	6		25
Iso-butyl			7.40	8	3		25
2-Butyl			6.80	9	0		25
Hydrogenated soybean							
Ethyl	65.1	40093	5.54	7	6	174	14
Yellow grease							
Methyl	62.6	39817/37144[b]	5.16				22
Grease							
Ethyl			6.20	5	–1		23
Used frying oil							
Methyl	59	37337[b]	4.50	1	–3	>110	15
Waste olive oil							
Methyl ester	58.7 (CI)		5.29	–2	–6		26
Soybean soapstock	51.3		4.30	6			27

[a]Some flash points are very low. These may be typographical errors in the references or the materials may have contained residual alcohols.
[b]Lower heating value. In some cases, both higher and lower values are given.
[c]Dynamic viscosity.

14. Peterson, C.L., J.S. Taberski, J.C. Thompson, and C.L. Chase, The Effect of Biodiesel Feedstock on Regulated Emissions in Chassis Dynamometer Tests of a Pickup Truck, *Trans. ASAE 43:* 1371–1381 (2000).
15. Serdari, A., K. Fragioudakis, S. Kalligeros, S. Stournas, and E. Lois, Impact of Using Biodiesels of Different Origin and Additives on the Performance of a Stationary Diesel Engine, *J. Eng. Gas Turbines Power (Trans. ASME) 122:* 624–631 (2000).
16. Geyer, S.M., M.J. Jacobus, and S.S. Lestz, Comparison of Diesel Engine Performance and Emissions from Neat and Transesterified Vegetable Oils, *Trans. ASAE 27:* 375–381 (1984).
17. Avella, F., A. Galtieri, and A. Fiumara, Characteristics and Utilization of Vegetable Derivatives as Diesel Fuels, *Riv. Combust. 46:* 181–188 (1992).
18. Bouché, T., M. Hinz, R. Pittermann, and M. Herrmann, Optimizing Tractor CI Engines for Biodiesel Operation, SAE Technical Paper Series 2000-01-1969, 2000.
19. Krahl, J., K. Baum, U. Hackbarth, H.-E. Jeberien, A. Munack, C. Schütt, O. Schröder, N. Walter, J. Bünger, M.M. Müller, and A. Weigel, Gaseous Compounds, Ozone Precursors, Particle Number and Particle Size Distributions, and Mutagenic Effects Due to Biodiesel, *Trans. ASAE 44:* 179–191 (2001).
20. Isigigür, A., F. Karaosmanoglu, H.A. Aksoy, F. Hamdullahpur, and Ö.L. Gülder, Performance and Emission Characteristics of a Diesel Engine Operating on Safflower Seed Oil Methyl Ester, *Appl. Biochem. Biotechnol. 45–46:* 93–102 (1994).
21. Yahya, A., and S.J. Marley, Physical and Chemical Characterization of Methyl Soy Oil and Methyl Tallow Esters as CI Engine Fuels, *Biomass Bioenergy 6:* 321–328 (1994).
22. Canakci, M., and J.H. Van Gerpen, The Performance and Emissions of a Diesel Engine Fueled with Biodiesel from Yellow Grease and Soybean Oil, ASAE Paper No. 01-6050, 2001.
23. Wu, W.-H., T.A. Foglia, W.N. Marmer, R.O. Dunn, C.E. Goering, and T.E. Briggs, Low-Temperature Property and Engine Performance Evaluation of Ethyl and Isopropyl Esters of Tallow and Grease, *J. Am. Oil Chem. Soc. 75:* 1173–1178 (1998).
24. Lee, I., L.A. Johnson, and E.G. Hammond, Use of Branched-Chain Esters to Reduce the Crystallization Temperature of Biodiesel, *J. Am. Oil Chem. Soc. 72:* 1155–1160 (1995).
25. Foglia, T.A., L.A. Nelson, R.O. Dunn, and W.N. Marmer, Low-Temperature Properties of Alkyl Esters of Tallow and Grease, *J. Am. Oil Chem. Soc. 74:* 951–955 (1997).
26. Dorado, M.P., E. Ballesteros, J.M. Arnal, J. Gómez, and F.J. López Giménez, Testing Waste Olive Oil Methyl Ester as a Fuel in a Diesel Engine, *Energy Fuels 17:* 1560–1565 (2003).
27. Haas, M.J., K.M. Scott, T.L. Alleman, and R.L. McCormick, Engine Performance of Biodiesel Fuel Prepared from Soybean Soapstock: A High-Quality Renewable Fuel Produced from a Waste Feedstock, *Energy Fuels 15:* 1207–1212 (2001).

Appendix B: Biodiesel Standards

This appendix contains the specifications of the following biodiesel standards:

Table B-1: ASTM D6751 (United States): Standard Specification for Biodiesel (B100) Blend Stock for Distillate Fuels.

Table B-2: EN 14213 (Europe): Heating Fuels: Fatty Acid Methyl Esters (FAME). Requirements and Test Methods.

Table B-3: EN 14214 (Europe): Automotive Fuels: FAME for Diesel Engines. Requirements and Test Methods.

Table B-4: Provisional Australian biodiesel standard.

Table B-5: Provisional Brazilian biodiesel standard ANP 255.

Table B-6: Provisional South African biodiesel standard.

Table B-7: A list of analytical standards developed usually for the sake of inclusion in full biodiesel standards.

The European standard EN 14214, which went into effect in 2003, supersedes the biodiesel standards in European countries that are members of the European Committee for Standardization (CEN). Therefore, no standards from individual European countries are given. CEN standards apply in the following member countries: Austria, Belgium, Cyprus, Czech Republic, Denmark, Finland, France, Estonia, Germany, Greece, Hungary, Iceland, Ireland, Italy, Latvia, Lithuania, Luxemburg, Malta, the Netherlands, Norway, Poland, Portugal, Slovakia, Slovenia, Spain, Sweden, Switzerland, and the United Kingdom. The European standard EN 590 for conventional diesel fuel contains a provision that conventional diesel fuel can contain up to 5% FAME meeting the standard EN 14214.

In addition to the biodiesel standards, analytical standards have been developed in the United States and Europe for the purpose of including them as prescribed methods in biodiesel standards. Table B-7 lists such relevant analytical standards. The biodiesel standards can vary by properties included, methods, and limits. The biodiesel standards given are elaborated upon in the full standards available from the corresponding standardization organizations.

The European standard EN 14214 contains a separate section (not given in the table) for low-temperature properties. National standardizing committees are given the option of selecting among six CFPP (cold-filter plugging point; method EN 116) classes for moderate climates and five for arctic climates. The total temperature range for these CFPP classes is from +5°C to –44°C. Because the requirements regarding low-temperature properties vary, the standard ASTM D6751 has a report requirement for its cloud point parameter.

A large number of specifications in the standards deal with completion of the transesterification reaction. These are the specifications related to free and total glycerol or content of mono-, di-, and triglycerides. Analytical standards based on gas chromatography are utilized in the standards for these specifications (see also Chapter 5). Standards related to these analyses have been developed and are included in Table B-7.

TABLE B-1
Biodiesel Standard ASTM D6751 (United States)

Property	Test method	Limits	Unit
Flash point (closed cup)	D 93	130.0 min	°C
Water and sediment	D 2709	0.050 max	% volume
Kinematic viscosity, 40°C	D 445	1.9–6.0	mm^2/s
Sulfated ash	D 874	0.020 max	% mass
Sulfur	D 5453	0.0015 max or 0.05 max[a]	% mass
Copper strip corrosion	D 130	No. 3 max	
Cetane number	D 613	47 min	
Cloud point	D 2500	Report	°C
Carbon residue (100% sample)	D 4530	0.050 max	% mass
Acid number	D 664	0.80 max	mg KOH/g
Free glycerin	D 6584	0.020 max	% mass
Total glycerin	D 6584	0.240 max	% mass
Phosphorus content	D 4951	0.001 max	% mass
Distillation temperature, atmospheric equivalent temperature, 90% recovered	D 1160	360 max	C

[a]The limits are for Grade S15 and Grade S500 biodiesel, respectively. S15 and S500 refer to maximum sulfur specifications (ppm).

TABLE B-2
European Standard EN 14213 for Biodiesel as Heating Oil

		Limits		
Property	Test method	min	max	Unit
Ester content	EN 14103	96.5		% (m/m)
Density; 15°C	EN ISO 3675 EN ISO 12185	860	900	kg/m^3
Viscosity; 40°C	EN ISO 3104 ISO 3105	3.5	5.0	mm^2/s
Flash point	EN ISO 3679	120		°C
Sulfur content	EN ISO 20846		10.0	mg/kg
EN ISO 20884				
Carbon residue (10% dist. residue)	EN ISO 10370		0.30	% (m/m)
Sulfated ash	ISO 3987		0.02	% (m/m)
Water content	EN ISO 12937		500	mg/kg
Total contamination	EN 12662		24	mg/kg
Oxidative stability, 110°C	EN 14112	4.0		h
Acid value	EN 14104		0.50	mg KOH/g
Iodine value	EN 14111		130	g iodine/100 g
Content of FAME with ≥4 double bonds			1	% (m/m)
Monoglyceride content	EN 14105		0.80	% (m/m)
Diglyceride content	EN 14105		0.20	% (m/m)
Triglyceride content	EN 14105		0.20	% (m/m)
Free glycerine	EN 14105, EN 14106		0.02	% (m/m)
Cold-filter plugging point	EN 116			°C
Pour point	ISO 3016		0	°C
Heating value	DIN 51900-1 DIN 51900-2 DIN 51900-3	35		MJ/kg

TABLE B-3
Biodiesel Standard EN 14214 (Europe)

Property	Test method	Limits		Unit
		min	max	
Ester content	EN 14103	96.5		% (m/m)
Density; 15°C	EN ISO 3675 EN ISO 12185	860	900	kg/m^3
Viscosity; 40°C	EN ISO 3104 ISO 3105	3.5	5.0	mm^2/s
Flash point	EN ISO 3679	120	°C	
Sulfur content	EN ISO 20846 EN ISO 20884		10.0	mg/kg
Carbon residue (10% dist. residue)	EN ISO 10370		0.30	% (m/m)
Cetane number	EN ISO 5165	51		
Sulfated ash	ISO 3987		0.02	% (m/m)
Water content	EN ISO 12937		500	mg/kg
Total contamination	EN 12662		24	mg/kg
Copper strip corrosion (3 hr, 50°C)	EN ISO 2160		1	
Oxidative stability, 110°C	EN 14112	6.0		hr
Acid value	EN 14104		0.50	mg KOH/g
Iodine value	EN 14111		120	g iodine/100 g
Linolenic acid content	EN 14103		12	% (m/m)
Content of FAME with ≥4 double bonds			1	% (m/m)
Methanol content	EN 14110		0.20	% (m/m)
Monoglyceride content	EN 14105		0.80	% (m/m)
Diglyceride content	EN 14105		0.20	% (m/m)
Triglyceride content	EN 14105		0.20	% (m/m)
Free glycerine	EN 14105, EN 14106		0.02	% (m/m)
Total glycerine	EN 14105		0.25	% (m/m)
Alkali metals (Na + K)	EN 14108, EN 14109		5.0	mg/kg
Earth alkali metals (Ca + Mg)	prEN 14538		5.0	mg/kg
Phosphorus content	EN 14107		10.0	mg/kg

TABLE B-4
Australian Biodiesel Standard [Fuel Standard (Biodiesel) Determination 2003; Approved Under the Fuel Quality Standards Act 2000 by the Australian Minister for the Environment and Heritage]

		Limits			
Property	Test method	min	max	Unit	Effective date
Sulfur	ASTM D5453		50	mg/kg	Sept. 18, 2003
			10	mg/kg	Feb. 1, 2006
Density, 15°C	ASTM D1298	860	890	kg/m^3	Sept. 18, 2003
	EN ISO 3675				
Distillation T90	ASTM D1160		360	°C	Sept. 18, 2003
Sulfated ash	ASTM D 874		0.20	% mass	Sept. 18, 2003
Viscosity; 40°C	ASTM D445	3.5	5.0	mm^2/s	Sept. 18, 2003
Flash point	ASTM D93	120		°C	Sept. 18, 2003
Carbon residue					Sept. 18, 2003
10% dist. residue	EN ISO 10370		0.30	% mass	
100% dist. sample	ASTM D4530		or 0.05	% mass	
Water and sediment	ASTM D2709		0.50	% vol	Sept. 18, 2003
Copper strip corrosion	ASTM D130		No. 3		Sept. 18, 2003
Ester content	EN 14103	96.5		% (m/m)	Sept. 18, 2003
Phosphorus	ASTM D4951		10	mg/kg	Sept. 18, 2003
Acid value	ASTM D664		0.80	mg KOH/g	Sept. 18, 2003
Total contamination	EN 12662		24	mg/kg	Sept. 18, 2004
	ASTM D5452				
Free glycerol	ASTM D6584		0.02	% mass	Sept. 18, 2004
Total glycerol	ASTM D6584		0.25	% mass	Sept. 18, 2004
Cetane number	EN ISO 5165				
ASTM	D 613		51		Sept. 18, 2004
Cold-filter plugging point	TBA				Sept. 18, 2004
Oxidation stability	EN 14112				
	ASTM D2274 (as relevant to biodiesel)	6		°C	Sept. 18, 2004
Metals: Group I (Na, K)	EN 14108		5	mg/kg	Sept. 18, 2004
	EN 14109				
Metals: Group II (Ca, Mg)	prEN 14538		5	mg/kg	Sept. 18, 2004

TABLE B-5
Provisional Brazilian Biodiesel Standard ANP (Agência Nacional do Petróleo) 255 (released September 2003)

Property	Limits	Method
Flash point (°C)	100 min	ISO/CD 3679
Water and sediments	0.02 max	D 2709
Kinematic viscosity,[a] 40°C (mm^2/s)	ANP 310[b]	D 445; EN/ISO 3104
Sulfated ash (%, m/m)[a]	0.02 max	D 874; ISO3 987
Sulfur (%, m/m)	0.001 max	D 5453; EN/ISO 14596
Copper corrosion 3 hr, 50°C[a]	No. 1 max	D 130; EN/ISO 2160
Cetane number	45 min	D 613; EN/ISO 5165
Cloud point[a]	ANP 310[b]	D 6371
Carbon residue	0.05 max	D 4530; EN/ISO 10370
Acid number[a] (mg KOH/g)	0.80 max	D 664; EN 14104
Free glycerin (%, m/m)	0.02 max	D 6854; EN 14105-6
Total glycerin (%, m/m)	0.38 max	D 6854; EN 14105
Distillation recovery 95% (°C)	360 max	D 1160
Phosphorus (mg/kg)	10 max	D 4951; EN 14107
Specific gravity[a]	ANP 310[b]	D 1298/4052
Alcohol (%, m/m)	0.50 max	EN 14110
Iodine number	take note	EN 14111
Monoglycerides (%, m/m)	1.00 max	D 6584; EN 14105
Diglycerides (%, m/m)	0.25 max	D 6584; EN 14105
Triglycerides (%, m/m)	0.25 max	D 6584; EN 14105
Na + K (mg/kg)	10 max	EN 14108-9
Aspect		—
Oxidation stability 110°C (hr)	6 min	EN 14112

[a]Brazilian ABNT NBR methods are also available for this property.
[b]ANP 310 = current specifications for petrodiesel.

TABLE B-6
Provisional South African Biodiesel Standard

Property	Requirements	Test method
Ester content, % mass fraction, min	96.5	EN 14103
Density, 15°C, kg/m^3	860	ISO 3675, ISO 12185
Kinematic viscosity at 40°C, mm^2/s	3.5–5.0	ISO 3104
Flash point, °C, min	120	ISO 3679
Sulfur content, mg/kg, max	10.0	ISO 20846, ISO 20884
Carbon residue (on 10% distillation residue), % mass fraction, max	0.3	ISO 10370
Cetane number, min	51.0	ISO 5165
Sulfated ash content, % mass fraction, max	0.02	ISO 3987
Water content, % mass fraction, max	0.05	ISO 12937
Total contamination, mg/kg, max	24	EN 12662
Copper strip corrosion (3 hr at 50°C), rating, max	1	ISO 2160
Oxidation stability, at 110°C, hr, min	6	EN 14112
Acid value, mg KOH/g, max	0.5	EN 14104
Iodine value, g of iodine/100 g of FAME, max	140	EN 14111
Linolenic acid methyl ester, % mass fraction, max	12	EN 14103
Polyunsaturated (≥4 double bonds) methyl esters, % mass fraction, max	1	
Methanol content, % mass fraction, max	0.2	EN 14110
Monoglyceride content, % mass fraction, max	0.8	EN 14105
Diglyceride content, % mass fraction, max	0.2	EN 14105
Triglyceride content, % mass fraction, max	0.2	EN 14105
Free glycerol, % mass fraction, max	0.02	EN 14105, EN 14106
Total glycerol, % mass fraction, max	0.25	EN 14105
Group I metals (Na + K), mg/kg, max	5.0	EN 14108, EN 14109
Group II metals (Ca + Mg), mg/kg, max	5.0	prEN 14538
Phosphorus content, mg/kg, max	10.0	EN 14107
Cold Filter Plugging Point (CFPP)		EN 116
Winter, °C, max	–4	
Summer, °C, max	+3	

TABLE B-7
Analytical Standards Developed Usually for the Sake of Inclusion in Full Biodiesel Standards

Standard[a]	Title
ASTM D 6584	Determination of Free and Total Glycerine in B-100 Biodiesel Methyl Esters by Gas Chromatography
EN 14078	Liquid petroleum products. Determination of fatty acid methyl esters (FAME) in middle distillates. Infrared spectroscopy method
EN 14103	Fat and oil derivatives. FAME. Determination of ester and linolenic acid methyl ester contents
EN 14104	Fat and oil derivatives. FAME Determination of acid value
EN 14105	Fat and oil derivatives. FAME. Determination of free and total glycerol and mono-, di-, triglyceride contents
EN 14106	Fat and oil derivatives. FAME. Determination of free glycerol content
EN 14107	Fat and oil derivatives. FAME. Determination of phosphorus content by inductively coupled plasma (ICP) emission spectrometry
EN 14108	Fat and oil derivatives. FAME. Determination of sodium content by atomic absorption spectrometry
EN 14109	Fat and oil derivatives. FAME. Determination of potassium content by atomic absorption spectrometry
EN 14110	Fat and oil derivatives. FAME. Determination of methanol content
EN 14111	Fat and oil derivatives. FAME. Determination of iodine value
EN 14112	Fat and oil derivatives. FAME. Determination of oxidation stability (accelerated oxidation test)
EN 14331	Liquid petroleum products. Separation and characterization of fatty acid methyl esters (FAME) by liquid chromatography/gas chromatography (LC/GC)
prEN 14538	Fat and oil derivatives. FAME. Determination of Ca and Mg content by optical emission spectral analysis with inductively coupled plasma (ICP OES)

[a]The letters "pr" before some European standards indicate that these standards are provisional, i.e., under development.

Appendix C: Internet Resources

A wealth of information on biodiesel is available on the internet. However, caution is advised in many cases regarding correctness of information given on websites. The website address may be an indicator of the quality of the information available on that site, although even quality sites may contain incorrect or misleading information. Some web addresses of trade organizations promoting the cause of biodiesel, government agencies or other entities dealing with biodiesel are given here. Again, this does not imply that all information available on these sites (or those linked to them) is correct.

Country/ region	Organization	Website
Australia	Biodiesel Association of Australia	www.biodiesel.org.au
	Biodiesel standard	http://www.deh.gov.au/atmosphere/biodiesel/
Austria	Austrian Biofuels Institute (Österreichisches Biotreibstoff Institut)	www.biodiesel.at
Canada	Biodiesel Association of Canada	www.biodiesel-canada.org
	Canadian Renewable Fuels Association	http://www.greenfuels.org/bioindex.html
Europe	European Biodiesel Board	www.ebb-eu.org
Germany	Association for the Promotion of Oil and Protein Plants (Union zur Förderung von Oel- und Proteinpflanzen e. V.)	www.ufop.de (English: www.ufop.de/hilfe.html)
	Working Group for Quality Management of Biodiesel (Arbeitsgemeinschaft Qualitätsmanagement Biodiesel)	www.agqm-biodiesel.de/
United Kingdom	British Association for Biofuels and Oils	www.biodiesel.co.uk
United States	National Biodiesel Board	www.biodiesel.org
	EPA emissions report	http://www.epa.gov/otaq/models/biodsl.htm
	EPA biodiesel fact sheet	http://www.epa.gov/otaq/consumer/fuels/altfuels/biodiesel.pdf
	Dept. of Energy legislative website	http://www.eere.energy.gov/vehiclesandfuels/epact/
	Iowa State University	http://www.me.iastate.edu/biodiesel

Index